D1187536

DESTRUCTION OF HAZARDOUS CHEMICALS IN THE LABORATORY

DESTRUCTION OF HAZARDOUS CHEMICALS IN THE LABORATORY

Second Edition

George Lunn

Eric B. Sansone

Program Resources, Inc./DynCorp
Environmental Control and Research Program
NCI-Frederick Cancer Research and
Development Center
Frederick, Maryland

A WILEY-INTERSCIENCE PUBLICATION
JOHN WILEY & SONS, INC.
New York • Chichester • Brisbane • Toronto • Singapore

Extreme care has been taken with the preparation of this work. However, neither the publisher nor the authors warrants the procedures against any safety hazards. Neither the publisher nor the authors shall be held responsible or liable for any damages resulting in connection with or arising from the use of any information in this book.

This text is printed on acid-free paper.

Library of Congress Cataloging in Publication Data:

Lunn, George.

Destruction of hazardous chemicals in the laboratory / George Lunn, Eric B. Sansone. — 2nd ed.

p. cm.

"A Wiley-Interscience publication."

Includes index.

ISBN 0-471-57399-X

1. Hazardous wastes—Safety measures. 2. Chemical laboratories-Safety measures. I. Sansone, E. B. (Eric Brandfon), 1939- . II. Title.

TD1050.S24L86 1994

604.7—dc20 93-35634

Printed in the United States of America.

10 9 8 7 6

Preface

This is the second edition of *Destruction of Hazardous Chemicals in the Laboratory*, originally published in 1990. Most of the existing monographs have been modified to a greater or lesser extent to take into account recent developments in the literature. Entirely new monographs have been included on the removal of metal ions (including lead, mercury, and uranium) and biological stains from solution and the degradation of mycotoxins, enzyme inhibitors (including diisopropyl fluorophosphate and phenylmethylsulfonyl fluoride), polycyclic heterocyclic hydrocarbons, and highly reactive reagents, such as butyllithium, chlorosulfonic acid, peracids, and phosgene.

Hazardous materials, such as complex metal hydrides, are frequently used to prepare super-dry solvents. So as to provide alternatives to these materials we have added a review on the use of much less hazardous reagents, such as molecular sieves (Appendix II). In recent years a number of technologies have begun to emerge that show great promise for treating the complex waste streams produced by biomedical research institutions. Dr. Steven W. Rhodes has contributed an overview of developments in this field, which is included as "Emerging Technologies Applicable to the Treatment of Hazardous Waste in Biomedical Research Institutions" (Appendix IV).

The format is essentially the same as that previously used, although CAS Registry Numbers have now been added at appropriate places in the Monographs.

As before, this book is a collection of detailed procedures that can be used to degrade and dispose of a wide variety of hazardous chemicals. The procedures are applicable to the amounts of material typically found in the chemical laboratory. Exotic reagents and special apparatus are not required: The procedures can readily be carried out, often by technicians, in the laboratory where the hazardous materials are used.

Bulk quantities of hazardous materials and solutions in various solvents can also be degraded using the procedures described in this book. Methods for cleaning up spills are frequently indicated, as are solvents for wipe tests to ensure complete surface decontamination. A listing of hazardous compounds, indexed by name, molecular formula, and CAS Registry Number provides ready access to the information.

The safe handling and disposal of hazardous chemicals is an essential requirement for working with these substances. We hope that this book will contribute to encouraging the use of tested and sound practices.

George Lunn

Eric B. Sansone

Frederick, Maryland
March, 1994

Acknowledgments

The primary impetus for our undertaking the kind of work that led to this book was provided by the Division of Safety, National Institutes of Health (NIH), under the leadership first of Dr. W.E. Barkley and later of Dr. R.W. McKinney. Dr. M. Castegnaro, of the International Agency for Research on Cancer (IARC), has organized collaborative studies with the support of the Division of Safety, NIH in which we have taken part. These studies have contributed materially to this work.

This research was sponsored by the National Cancer Institute under contract No. NO1-CO-74102 with Program Resources, Inc./DynCorp.

We wish to thank Dr. Steven W. Rhodes for providing the review of emerging technologies applicable to the treatment of hazardous waste in biomedical research institutions (Appendix IV).

Contents

DESTRUCTION OF HAZARDOUS CHEMICALS IN THE LABORATORY

INTRODUCTION

Biological agents can be completely inactivated by treating them with formaldehyde, ethylene oxide or moist heat, and radioactive materials will decay with the passage of sufficient time, but there are no general destruction techniques that are applicable to chemical agents. The availability of destruction techniques for hazardous chemical agents would be particularly helpful because of the dangers associated with their handling and disposal. In addition, being able to destroy or inactivate the hazardous materials where they are used is advantageous because the user should be familiar with the hazards of these materials and the precautions required in their handling.

Here we present summaries of destruction procedures for a variety of hazardous chemicals. Most of the procedures have been validated, many by international collaborative testing. We have drawn on information available in the literature[1–13] and on our own published and unpublished work.

About This Book

This book is a collection of techniques for destroying a variety of hazardous chemicals. It is intended for those whose knowledge of the chemistry of the compounds covered is rather sophisticated; that is, for those who are aware not only of the obvious dangers, such as the toxic effects of the compounds themselves and of some of the reagents used in the methods, but also of the potential hazards represented, for example, by the possible formation of diazoalkanes when *N*-nitrosamides are treated with base. If you are not thoroughly familiar with the potential hazards and the chemistry of the materials to be destroyed and the reagents to be used, do not proceed.

The destruction methods are organized in what we believe to be rational categories. These categories are listed in the Table of Contents. It is quite likely, however, that others would have categorized these methods differently, so we have provided three indexes. We have assembled many synonyms of the compounds covered into a Name Index. In each case, the page number given is the first page of the monograph in which the destruction of that compound is discussed. In some cases, the compound itself may not have been studied; it may have been referred to in the Related Compounds section. Since it is not possible to cite every synonym and every variation in spelling, we have also provided a CAS Registry Number Index and a Molecular Formula Index. With these aids one should be able to find the appropriate destruction method for the compound in question. As a further aid, recognizing the fact that frequently a destruction method is sought only after an accident occurs, we have added an appendix that lists the solvents recommended for use with wipes that are used to sample the area where a spill has occurred in order to determine whether the cleanup has been complete.

One of the difficulties in preparing a book such as this is deciding what should be included and what should be excluded from the text. We have tried to make the method descriptions and the supporting references complete, but at the same time not include unnecessary details. We also tried to eliminate ambiguity wherever possible, going so far as to repeat almost verbatim certain procedures for some compounds rather than noting a minor change and referring to another section and risk a wrong page number or a misinterpretation. Some general safety precautions are given below. These are not repeated for each group of compounds; in some cases, unusual hazards are noted. For many of the destruction procedures we use the word "discard" in connection with the final reaction mixture. This *always* means "discard in compliance with all applicable regulations."

Although we have included all the validated destruction procedures known to us, we realize that there may be other procedures in the literature. Thus, we would be pleased to hear from readers who have any information or suggestions. This work is continuing and we would also be pleased to hear from readers who have suggestions for future work or other comments.

Properties of a Destruction Technique

We have already indicated the advantages of destroying hazardous chemicals at the place where they were generated. It is also useful to consider the desirable properties of a destruction technique.

- Destruction of the hazardous chemical should be complete.
- A substantially complete material accountance should be available, with the detectable products being innocuous materials. (This accountance is often difficult to accomplish. In the absence of a complete material accountance, an assessment of the mutagenic activity of the reaction mixture may provide useful information concerning the potential biological hazards associated with the decomposition products.)
- The effectiveness of the technique should be easy to verify analytically.
- The equipment and reagents required should be readily available, inexpensive, and easy and safe to use. The reagents should have no shelf-life limitations.
- The destruction technique should require no elaborate operations (such as distillation or extraction) that might be difficult to contain; it must be easy to perform reliably and should require little time.
- The method should be applicable to the real world; that is, it should be capable of destroying the compound itself, solutions in various solvents, and spills.

These properties characterize an ideal destruction technique. Most techniques cannot meet all of these criteria, but they represent a goal toward which one should strive.

Contents of a Monograph

Each monograph usually contains the following information:

- An introduction describes the various properties of the compound or class of compounds being considered.
- The principles of destruction section details, in general terms, the chemistry of the destruction procedures, the products, and the efficiency of destruction.

- The destruction procedures section may be subdivided into procedures for bulk quantities, solutions in water, organic solvents, and so on.
- The analytical procedures section describes one or more procedures that may be used to test the final reaction mixtures to ensure that the compound has been completely degraded. The techniques usually involve packed column gas chromatography (GC) or reverse phase high-performance liquid chromatography (HPLC), but colorimetric procedures and thin-layer chromatography (TLC) are also used in some cases.
- The mutagenicity assays section describes the data available on the mutagenic activity of the starting materials, possible degradation products, and final reaction mixtures. The data were generally obtained from the plate incorporation technique of the *Salmonella*/mammalian microsome mutagenicity assay (see below).
- The related compounds section describes other compounds to which the destruction procedures should be applicable. The destruction procedures have not usually been validated for these materials, however; they should be fully investigated before adopting them.
- References identify the sources of the information given in the monograph.

Mutagenicity Assays

The residues produced by the destruction methods were tested for mutagenicity. Unless otherwise specified, the reaction mixtures from the destruction procedures and some of the starting materials and products were tested for mutagenicity using the plate incorporation technique of the *Salmonella*/mammalian microsome assay essentially as recommended by Ames et al.[14] with the modifications of Andrews et al.[15] Some or all of the tester strains TA98, TA100, TA1530, TA1535, TA1537, and TA1538 of *Salmonella typhimurium* were used with and without S9 rat liver microsomal activation. The reaction mixtures were neutralized before testing. In general, basic reaction mixtures were neutralized by adding acetic acid. Acidic reaction mixtures were neutralized by adding solid sodium bicarbonate. Reaction mixtures containing potassium permanganate were decolorized with sodium ascorbate before neutralization. A 100 μL aliquot of the solution (corresponding to varying amounts of undegraded material) was used per plate. Pure compounds were generally tested at a level of 1 mg per plate in either dimethyl sulfoxide (DMSO) or aqueous solution. To each plate were added 100 μL of these solutions. The criterion for significant mutagenicity was set at more than twice the level of the control value. The control value was the average of the cells only and cells plus solvent runs. Unless otherwise specified, residues did not exhibit mutagenic activity. The absence of mutagenic activity in the residual solutions, however, does not necessarily imply that they are nontoxic or have no other adverse biological or environmental effects.

Analytical Procedures

Unless otherwise specified the analytical equipment used in our work consisted of the following. For HPLC a dual pump computer-controlled solvent delivery system (Rainin Instrument Co., Woburn, MA) was used with ultraviolet (UV) detection using either a Knauer Model 87 variable wavelength detector (Rainin) or an ABI 1000S diode array detector (Applied Biosystems, Foster City, CA). The injection volume was 20 μL and the flow rate was 1 mL/min. The column was a 250 × 4.6-mm i.d. column of Microsorb 5 μm C8 fitted with a 15 × 4.6-mm guard column of the same material. For GC a Hewlett Packard HP 5880A instrument was fitted with a 1.8-m × 2-mm i.d. × 0.25-in. o.d. packed silanized glass column. The column was fitted with a guard column packed with the same material. The guard column was changed periodically. Thc injector temperature was 200°C and the flame ionization detector temperature was 300°C. The carrier gas was nitrogen flowing at 30 mL/min. Injection was by syringe and sample volumes were in the 1–5-μL range. For each instrument an electronic integrator was used to determine peak areas automatically.

In some cases we found that injecting unneutralized reaction mixtures onto the hot GC column caused degradation of the material for which we were analyzing. Thus it might be that degradation was incomplete but the appropriate peak was not observed in the chromatogram because the compound was degraded on the GC column. Spiking experiments can be used to determine if this is a problem. In a spiking experiment a small amount of the original compound is added to the final reaction mixture and this spiked mixture is analyzed. If an appropriate peak is observed, compound degradation on the GC column is not a problem. If an appropriate peak is not observed, it may be necessary to neutralize the reaction mixture before analysis and/or use a different GC column. Similar problems may be encountered when using HPLC because of the formation of salts or the influence of the sample solvent; again, spiking experiments should be employed. We have indicated in the monographs some instances where problems such as these were encountered (see, e.g., Halogenated Compounds monograph) but spiking experiments should be used routinely to test the efficacy of the analytical techniques.

Spills

The initial step in dealing with a spill should be the removal of as much of the spill as possible by using a high efficiency particulate air (HEPA) filter equipped vacuum cleaner for solids and absorbents for liquids or solutions. The residue should be decontaminated as described in the monographs.

Whereas solutions or bulk quantities may be treated with heterogeneous [e.g., nickel–aluminum (Ni–Al) alloy reduction] or homogeneous methods [e.g., potassium permanganate/sulfuric acid ($KMnO_4/H_2SO_4$) oxidation], decontamination of glassware, surfaces, and equipment and the treatment of spills is best accomplished with homogenous methods. These methods allow the reagent, which is in solution, to contact all parts of the surface to be decontaminated. At the end of the cleanup it is frequently useful to rub the surface with a wipe moistened with a suitable solvent and analyze the wipe for the spilled compound. A list of suitable solvents is given in Appendix I.

Applicability of Procedures

Methods that successfully degrade some compounds may not affect other compounds of the same class or other classes of compounds. For example, oxidation with $KMnO_4$ in H_2SO_4 solution has been successfully applied to the destruction of several classes of compounds such as aromatic amines[8] and polycyclic aromatic hydrocarbons.[4] This method gave satisfactory results with some of the antineoplastic agents but not with others, including most of the *N*-nitrosourea drugs.[9] Sodium hypochlorite treatment, often recommended as a general destruction technique, failed to give satisfactory results with doxorubicin and daunorubicin[9] and polycyclic aromatic hydrocarbons,[4] whereas it did work for aflatoxins.[2] Nickel–aluminum alloy in dilute base worked well for *N*-nitrosamines[3] but was unsatisfactory for the destruction of polycyclic aromatic hydrocarbons.[4]

Chromic acid is an attractive oxidizing agent and has been used successfully to degrade many compounds, but the spent chromium compounds are potentially carcinogenic. These compounds are also environmentally hazardous and may not be discharged into the sewer. For this reason, we have not recommended the use of chromic acid for degrading any of the compounds we have covered. Potassium permanganate/sulfuric acid degradation appears to be as efficient and has fewer hazards.

Safety Considerations

A first step in minimizing risks associated with hazardous chemicals is to prepare a set of guidelines regulating such work. Many organizations have produced such guidelines and many texts have been written on the subject.[16–33]

Such documents will provide many useful suggestions when preparing guidelines for any laboratory situation. It is important that the guidelines "fit" the management and administrative structure of the institution and that any particular work requirements be taken into account.

To ensure the safety of those working with hazardous materials of any kind, policies, responsibility, and authority must be clearly defined. The responsibilities of the laboratory director, the supervisor, the employee, and the safety committee should be clearly spelled out.

It is important that potentially hazardous materials are handled only by those workers who have received the appropriate training. For that reason glassware and equipment should be decontaminated in the laboratory before they are transferred to any central washing system.

Obviously, it is important to consider the waste disposal aspects of one's work before the work begins. Experiments should always be designed to use the minimum quantities of potentially hazardous materials, and plans should be made in advance to minimize the wastes generated by any experimentation. Although we concentrate here on laboratory methods for destroying or decontaminating hazardous chemicals, it is valuable to briefly discuss some other approaches to handling chemical wastes. Regardless of the disposal approach selected, only completely decontaminated wastes producing no adverse biological effects should be discarded. Procedures for disposing of hazardous chemicals must comply with all applicable regulations. It is obviously undesirable to deliberately dispose of hazardous chemicals through the sewage system or by evaporation into the atmosphere, unless one has solid evidence that their subsequent degradation is extremely rapid, irreversible, complete, and produces safe degradation products.

It is impossible to provide a concise summary of safety practices for handling hazardous chemicals in the laboratory. For a complete discussion the reader is advised to consult readily available references.[16–33] Each institution and facility should tailor its program to meet its needs. It is important that the safety program include procedures for working with chemicals, biological materials, compressed gases, high-voltage power supplies, radioisotopes, and so on.

The following descriptions are designed to give a sufficiently complete guide to the destruction methods available in order to allow one to implement them successfully. The user may wish to consult the sources cited in order to determine the exact reaction conditions, limitations, and hazards that we have not been able to list because of space limitations. In some cases more than one procedure is listed. In these instances all the procedures should be regarded as equally valid unless restrictions on applicability are noted. In the course of collaborative testing we have occasionally found that the efficacy of the same technique varies between laboratories and may also depend on the batch of reagents being used. Thus, we strongly recommend that these methods be periodically validated to ensure that the chemicals are actually being destroyed. These methods have been tested on a limited number of compounds. The efficiency of the destruction techniques must be confirmed when they are applied to a new compound.

The details of analytical techniques are also included. It should be noted that even if 99.5% of a compound is destroyed, the remaining amount may still pose a considerable hazard, particularly if the original reaction was performed on a large scale. The efficiency of degradation is generally indicated by giving the limit of detection, for example, <0.5% of the original compound remained. This means that **none** of the original compound could be detected in the final reaction mixture. However, because of the limitations of the analytical techniques used, it is possible that traces of the original compound, which were below the limit of detection, remained. If this is the case, to use the example given above, the quantity that remained was less than 0.5% of the original amount.

The reactions described were generally performed on the scale specified. If the scale is greatly increased unforeseen hazards may be introduced, particularly with respect to the production of large amounts of heat, which may not be apparent in a small scale reaction. Extra care should therefore be exercised when these reactions are performed on a large scale.

In addition to the potential hazards posed by the compounds themselves, many of the reagents used in degradation procedures are hazardous. *Acids and bases are corrosive and should be prepared and used carefully. As noted below, the dilution of concentrated H_2SO_4 is a very exothermic process, which can result in splattering if carried out incorrectly.* All reactions should be carried out in a properly functioning chemical fume hood, which is vented to the outside. Laminar flow cabinets or other recirculating hoods with or without filters are not appropriate. The performance of the hood should be checked by qualified personnel at regular intervals. Hoods should be equipped with an alarm that sounds if the airflow drops below a preset value.

Dissolving concentrated H_2SO_4 in H_2O is a very exothermic process and appropriate protective clothing, including eye protection, should be worn. Concentrated H_2SO_4 should **always** be added to H_2O and **never** the other way around (otherwise splashing of hot concentrated H_2SO_4 may occur). To prepare H_2SO_4 solutions the appropriate quantity of concentrated H_2SO_4 is slowly and cautiously added to about 500 mL of H_2O, which is stirred in a 1-L flask. When addition is complete H_2O is added to bring the volume up to 1 L and the mixture is allowed to cool to room temperature before use. To prepare a 1 *M* H_2SO_4 solution use 53 mL of concentrated H_2SO_4 and to prepare a 3 *M* H_2SO_4 solution use 160 mL of concentrated H_2SO_4.

Appropriate protective clothing should be worn. This clothing includes, but is not limited to, eye protection (safety glasses or face shield), lab coat, and gloves. Rubber gloves generally allow the passage of organic liquids and solutions in organic solvents; they should not be allowed to routinely come into contact with them. Protective clothing should be regarded as the last line of defense and should be changed immediately if it becomes contaminated.

Wastes should be segregated into solid, aqueous, nonchlorinated organic, and chlorinated organic material and disposed of in accordance with local regulations.

In the introductions to the monographs we did not try to give an exhaustive listing of the toxicity data [e.g., LD_{50} (the dose that is lethal for 50% of the animals tested) or TLV (threshold limit values) data] or other hazards associated with the compounds under consideration. Instead, we attempted to give some indication of the main hazards associated with each compound or class of compounds. Extensive listings of all the *known* hazards associated with these compounds can be found elsewhere.[16–18]

All organic compounds discussed in this book should be regarded as flammable and all volatile compounds should be regarded as having the capacity of forming explosive mixtures in confined spaces. In many cases the toxic properties of many of these compounds have simply not been adequately investigated. Prudence dictates that, unless there is good reason for believing otherwise, all of the compounds discussed in this book should be regarded as volatile, highly toxic, flammable, human carcinogens, and should be handled with great care.

Other hazards are introduced by the reagents needed to perform the destruction procedures. Examples are the use of Ni–Al alloy and the use of $KMnO_4$.

Safety Considerations With Nickel–Aluminum Alloy

In the course of the reaction Ni–Al alloy reacts with base to produce hydrogen, a flammable gas that forms explosive mixtures with air. Providing the reactions are done in a fume hood this should not be a problem. It has been found that this reaction frequently exhibits an induction period.[34] There is an initial temperature rise when the Ni–Al alloy is first added but the temperature soon declines to ambient levels. Typically, after about 3 h, a much larger temperature rise occurs and the reaction mixture has frequently been observed to boil at this stage. For this reason the reaction should be carried out in a flask that is certainly no more than one-half full. In some cases we have observed that considerable foaming occurs and that an even larger flask is required. These instances are mentioned in the monographs (see, e.g., Antineoplastic Alkylating Agents). We have found it convenient to perform these reactions in a round-bottom flask fitted with an air condenser. The reaction also produces finely divided nickel, which is potentially pyrophoric. This product does not appear to be a problem, however, as long as it is allowed to dry on a metal tray away from flammable solvents for 24 h before being discarded.

Safety Considerations with Potassium Permanganate

A recent paper[35] pointed out that when $KMnO_4$ in H_2SO_4 was used to degrade hazardous compounds mutagenic reaction mixtures were produced because manganese was left in solution. The Ames test was used with tester strain *Salmonella typhimurium* TA102 (which was most sensitive to manganese) to assess mutagenic activity. Mutagenic activity was also detected with strain TA100 but at a lower level. Manganese is also known to be a carcinogen.[36,37] Thus disposal of reaction mixtures that contain manganese is not desirable. However, by manipulating the workup conditions, $KMnO_4$ can be used to degrade hazardous reagents and the manganese can subsequently be removed from solution.[38] These procedures have been incorporated into the monographs. A fuller account of the procedures that can be used to remove manganese from solution can be found in Appendix III.

A number of potential hazards have been identified. We have made no attempt to provide comprehensive guidelines for safe work, however, and it is essential that workers follow a code of good practice.

References

1. National Research Council Committee on Hazardous Substances in the Laboratory. *Prudent Practices for Disposal of Chemicals from Laboratories*; National Academy Press: Washington, DC, 1983.
2. Castegnaro, M.; Hunt, D.C.; Sansone, E.B.; Schuller, P.L.; Siriwardana, M.G.; Telling, G.M.; van Egmond, H.P.; Walker, E.A., Eds., *Laboratory Decontamination and Destruction of Aflatoxins B_1, B_2, G_1, and G_2 in Laboratory Wastes*; International Agency for Research on Cancer: Lyon, 1980 (IARC Scientific Publications No. 37).
3. Castegnaro, M.; Eisenbrand, G.; Ellen, G.; Keefer, L.; Klein, D.; Sansone, E. B.; Spincer, D.; Telling, G.; Webb K., Eds., *Laboratory Decontamination and Destruction of Carcinogens in Laboratory Wastes: Some* N-*Nitrosamines*; International Agency for Research on Cancer: Lyon, 1982 (IARC Scientific Publications No. 43).
4. Castegnaro, M.; Grimmer,G.; Hutzinger,O.; Karcher,W.; Kunte, H.; Lafontaine,M.; Sansone, E. B.; Telling, G.; Tucker, S.P., Eds., *Laboratory Decontamination and Destruction of Carcinogens in Laboratory Wastes: Some Polycyclic Aromatic Hydrocarbons*; International Agency for Research on Cancer: Lyon, 1983 (IARC Scientific Publications No. 49).
5. Castegnaro, M.; Ellen, G.; Lafontaine, M.; van der Plas, H.C.; Sansone, E. B.; Tucker, S.P., Eds., *Laboratory Decontamination and Destruction of Carcinogens in Laboratory Wastes: Some Hydrazines*; International Agency for Research on Cancer: Lyon, 1983 (IARC Scientific Publications No. 54).
6. Castegnaro, M.; Benard, M.; van Broekhoven, L. W.; Fine, D.; Massey, R.; Sansone, E.B.; Smith, P.L.R.; Spiegelhalder, B.; Stacchini, A.; Telling, G.; Vallon, J.J., Eds., *Laboratory Decontamination and Destruction of Carcinogens in Laboratory Wastes: Some* N-*Nitrosamides*; International Agency for Research on Cancer: Lyon, 1983 (IARC Scientific Publications No. 55).
7. Castegnaro, M.; Alvarez, M.; Iovu, M.; Sansone, E. B.; Telling, G.M.; Williams, D.T., Eds., *Laboratory Decontamination and Destruction of Carcinogens in Laboratory Wastes: Some Haloethers*; International Agency for Research on Cancer: Lyon, 1984 (IARC Scientific Publications No. 61).
8. Castegnaro, M.; Barek, J.; Dennis, J.; Ellen, G.; Klibanov, M.; Lafontaine, M.; Mit-

chum, R.; van Roosmalen, P.; Sansone, E.B.; Sternson, L.A.; Vahl, M., Eds., *Laboratory Decontamination and Destruction of Carcinogens in Laboratory Wastes: Some Aromatic Amines and 4-Nitrobiphenyl*; International Agency for Research on Cancer: Lyon, 1985 (IARC Scientific Publications No. 64).

9. Castegnaro, M.; Adams, J.; Armour, M-. A.; Barek, J.; Benvenuto, J.; Confalonieri, C.; Goff, U.; Ludeman, S.; Reed, D.; Sansone, E. B.; Telling, G., Eds., *Laboratory Decontamination and Destruction of Carcinogens in Laboratory Wastes: Some Antineoplastic Agents*; International Agency for Research on Cancer: Lyon, 1985 (IARC Scientific Publications No. 73).
10. Castegnaro, M.; Barek, J.; Frémy, J-.M.; Lafontaine, M.; Miraglia, M.; Sansone, E.B.; Telling, G.M., Eds., *Laboratory Decontamination and Destruction of Carcinogens in Laboratory Wastes: Some Mycotoxins*; International Agency for Research on Cancer: Lyon, 1991 (IARC Scientific Publications No. 113).
11. Castegnaro, M.; Barek, J.; Jacob, J.; Kirso, U.; Lafontaine, M.; Sansone, E.B.; Telling, G.M.; Vu Duc, T., Eds., *Laboratory Decontamination and Destruction of Carcinogens in Laboratory Wastes: Some Polycyclic Heterocyclic Hydrocarbons*; International Agency for Research on Cancer: Lyon, 1991 (IARC Scientific Publications No. 114).
12. Armour, M-. A. *Hazardous Laboratory Chemicals Disposal Guide*; CRC Press: Boca Raton, FL, 1991.
13. Armour, M-.A.; Browne, L.M.; McKenzie, P.A.; Renecker, D.M.; Bacovsky,R.A., Eds., *Potentially Carcinogenic Chemicals, Information and Disposal Guide*; University of Alberta: Edmonton, Alberta, 1986.
14. Ames, B.N.; McCann, J.; Yamasaki, E. Methods for detecting carcinogens and mutagens with the *Salmonella*/mammalian-microsome mutagenicity test. *Mutat. Res.* **1975**, *31*, 347–364.
15. Andrews, A.W.; Thibault, L.H.; Lijinsky, W. The relationship between carcinogenicity and mutagenicity of some polynuclear hydrocarbons. *Mutat. Res.* **1978**, *51* , 311–318.
16. Lewis, R.J.,Sr. *Sax's Dangerous Properties of Industrial Materials,* 8th ed.; Van Nostrand-Reinhold: New York, 1992.
17. Bretherick, L. *Bretherick's Handbook of Reactive Chemical Hazards*, 4th ed.; Butterworths: London, 1990.
18. Bretherick, L., Ed., *Hazards in the Chemical Laboratory*, 4th ed.; Royal Society of Chemistry: London, 1986.
19. National Research Council, Committee on Hazardous Substances in the Laboratory. *Prudent Practices for Handling Hazardous Chemicals in Laboratories*; National Academy Press: Washington, DC, 1981.
20. Manufacturing Chemists Association. *Guide for Safety in the Chemical Laboratory*, 2nd ed.; Van Nostrand-Reinhold: New York, 1972.
21. Montesano, R.; Bartsch, H.; Boyland, E.; Della Porta, G.; Fishbein, L.; Griesemer, R.A.; Swan, A.B.; Tomatis, L., Eds., *Handling Chemical Carcinogens in the Laboratory: Problems of Safety*; International Agency for Research on Cancer: Lyon, 1979 (IARC Scientific Publications No. 33).
22. Castegnaro, M.; Sansone, E.B. *Chemical Carcinogens*; Springer-Verlag: New York, 1986.

23. Furr, A.K., Ed., *CRC Handbook of Laboratory Safety*, 3rd ed.; CRC Press: Boca Raton, FL, 1989.
24. Young, J.A., Ed., *Improving Safety in the Chemical Laboratory: A Practical Guide*; Wiley: New York, 1987.
25. American Chemical Society, Committee on Chemical Safety. *Safety in Academic Chemistry Laboratories*, 5th ed.; American Chemical Society: Washington, DC, 1990.
26. Pal, S.B., Ed., *Handbook of Laboratory Health and Safety Measures*; Kluwer Academic Publishers: Hingham, MA, 1985.
27. Freeman, N.T.; Whitehead, J. *Introduction to Safety in the Chemical Laboratory*; Academic Press: New York, 1982.
28. *Occupational Health and Safety*, 2nd ed.; National Safety Council: Chicago, 1993.
29. Miller, B.M., Ed., *Laboratory Safety: Principles and Practices*; American Society for Microbiology: Washington, DC, 1986.
30. Fuscaldo, A.A.; Erlick, B.J.; Hindman, B., Eds., *Laboratory Safety: Theory and Practice*; Academic Press: New York, 1980.
31. Rosenlund, S.J. *The Chemical Laboratory: Its Design and Operation: A Practical Guide for Planners of Industrial, Medical, or Educational Facilities*; Noyes Publishers: Park Ridge, NJ, 1987.
32. Lees, R.; Smith, A.F., Eds., *Design, Construction, and Refurbishment of Laboratories*; Ellis Horwood: Chichester, 1984.
33. DiBerardinis, L.J.; Baum, J.S.; First, M.W.; Gatwood, G.T.; Groden, E.; Seth, A.K. *Guidelines for Laboratory Design: Health and Safety Considerations*, 2nd ed.; Wiley: New York, 1993.
34. Lunn, G. Reduction of heterocycles with nickel–aluminum alloy. *J. Org. Chem.* **1987**, *52*, 1043–1046.
35. De Méo, M.; Laget, M.; Castegnaro, M.; Duménil, G. Genotoxic activity of potassium permanganate in acidic solutions. *Mutat. Res.* **1991**, *260*, 295–306.
36. Stoner, G.D.; Shimkin, M.B.; Troxell, M.C.; Thompson, T.L.; Terry, L.S. Test for carcinogenicity of metallic compounds by the pulmonary tumor response in Strain A mice. *Cancer Res.* **1976**, *36*, 1744–1747.
37. DiPaolo, J.A. The potentiation of lymphosarcomas in the mouse by manganous chloride. *Fed. Proc.* **1964**, *23*, 393(Abstract).
38. Lunn, G.; Sansone, E.B.; De Méo, M.; Laget, M.; Castegnaro, M. Potassium permanganate can be used for degrading hazardous compounds. *Am. Ind. Hyg. Assoc. J.* **1994**, *55*, 167–171.

MONOGRAPHS

ACID HALIDES AND ANHYDRIDES

CAUTION! Refer to safety considerations section on page 7 before starting any of these procedures.

Acid halides, sulfonyl chlorides, and anhydrides are widely used in organic chemistry. The safe disposal of a number of these compounds has been investigated. All of these compounds are corrosive, can cause burns, and some may be lachrymators. In general these compounds react violently with dimethyl sulfoxide (DMSO).[11] A number of other incompatibilities have been noted, for example, acetyl chloride reacts violently with ethanol,[12] propionyl chloride reacts violently with diisopropyl ether,[13] thionyl chloride reacts violently with a variety of reagents including ammonia, *N,N*-dimethylformamide (DMF), tetrahydrofuran (THF), and ethanol,[14] sulfuryl chloride is incompatible with lead(IV) oxide, ether, red phosphorus, dinitrogen pentaoxide, and alkalies;[15] benzenesulfonyl chloride reacts violently with methylformamide;[16] and acetic anhydride reacts violently with a variety of compounds including boric acid, chromium triox-

References 1–10 are listed in table on page 18.

Compound	Reference	Formula	bp or mp	Registry Number
Acetyl chloride	1	$CH_3C(O)Cl$	bp 52°C	[75-36-5]
Propionyl chloride	2	$CH_3CH_2C(O)Cl$	bp 77–79°C	[79-03-8]
Dimethylcarbamoyl chloride	3	$(CH_3)_2NC(O)Cl$	bp 167–8°C	[79-44-7]
Benzoyl chloride	4	$PhC(O)Cl$	bp 198°C	[98-88-4]
Thionyl chloride	5	$SOCl_2$	bp 79°C	[7719-09-7]
Sulfuryl chloride	6	SO_2Cl_2	bp 68–70°C	[7791-25-5]
Methanesulfonyl chloride	7	CH_3SO_2Cl	bp 60°C/21 mm Hg	[124-63-0]
Benzenesulfonyl chloride	8	$PhSO_2Cl$	bp 251–2°C	[98-09-9]
p-Toluenesulfonyl chloride	9	*p*-$CH_3C_6H_4SO_2Cl$	mp 67–9°C	[98-59-9]
Acetic anhydride	10	$(CH_3C(O))_2O$	bp 138–140°C	[108-24-7]

ide, ethanol, nitric acid, and perchloric acid.[17] This list is not exhaustive and standard reference works should be consulted before proceeding.[18] Benzenesulfonyl chloride may explode on storage.[16] These compounds all react readily, and sometimes violently, with H_2O, alcohols, and amines. Dimethylcarbamoyl chloride is carcinogenic in experimental animals[19] and sulfuryl chloride may be carcinogenic.[20]

Principle of Destruction

Under controlled conditions these compounds are readily hydrolyzed to the corresponding acids. Highly reactive compounds (e.g., acetyl chloride, propionyl chloride, dimethylcarbamoyl chloride, benzoyl chloride, thionyl chloride, sulfuryl chloride, methanesulfonyl chloride, and acetic anhydride) are simply added to a 2.5 *M* sodium hydroxide (NaOH) solution at room temperature while compounds of lesser reactivity (e.g., benzenesulfonyl chloride and *p*-toluenesulfonyl chloride) require prolonged stirring or refluxing with a 2.5 *M* NaOH solution. For these compounds destruction was greater than 99.98%. Chlorosulfonic acid is too reactive to be degraded using any of these procedures (see the monograph for Chlorosulfonic Acid).

Destruction Procedures

Destruction Procedure for Highly Reactive Compounds (e.g., Acetyl Chloride, Propionyl Chloride, Dimethylcarbamoyl Chloride, Benzoyl Chloride, Thionyl Chloride, Sulfuryl Chloride, Methanesulfonyl Chloride, and Acetic Anhydride)[21]

Cautiously add 5 mL or 5 g of the compound to 100 mL of a 2.5 *M* NaOH solution. Stir the reaction at room temperature until it is over (it may be useful to monitor the temperature), neutralize, and discard it.

Destruction Procedure for Compounds of Lesser Reactivity (e.g., Benzenesulfonyl Chloride and p-Toluenesulfonyl Chloride)[21]

1. Add 5 mL or 5 g of the compound to 100 mL of a 2.5 *M* NaOH solution. Cover and stir the reaction at room temperature for 3 h (benzenesulfonyl chloride) or 24 h (*p*-toluenesulfonyl chloride), analyze for completeness of destruction, neutralize the reaction mixture, and discard it.
2. Add 5 mL or 5 g of the compound to 100 mL of a 2.5 *M* NaOH solution. Reflux the reaction mixture for 1 h, cool, analyze for completeness of destruction, neutralize the reaction mixture, and discard it.

Destruction Procedure for Compounds of Unknown Reactivity[22]

To degrade 0.5 mol of the compound stir a NaOH solution (2.5 *M*, 600 mL) in a 1-L flask and add a few milliliters of the compound. If the compound dissolves and heat is generated, add the rest of the compound at such a rate that the reaction remains under control. If the reaction is slow (e.g., with *p*-toluenesulfonyl chloride), heat the mixture to about 90°C (e.g., with a steam bath) and, when the compound has dissolved, add the rest of the compound dropwise. When a clear solution is obtained, allow it to cool. Neutralize the final, cooled, reaction mixture, analyze for completeness of destruction, and discard it.

Analytical Procedures[21]

The following procedure has been found useful for the analysis of benzenesulfonyl chloride and *p*-toluenesulfonyl chloride. A 100-μL aliquot of the reaction mixture is neutralized by adding it to 1 mL of a 20 μL/mL solution of acetic acid in methanol. Analyze by reverse phase HPLC using acetonitrile:water 60:40 flowing at 1 mL/min and a UV detector set at 254 nm. The approximate retention times are 6.5 min for benzenesulfonyl chloride and 7.6 min for *p*-toluenesulfonyl chloride. Impurities in the *p*-toluenesulfonyl chloride may interfere with the analysis.

Related Compounds

This procedure should be generally applicable to acid halides, sulfonyl halides, and acid anhydrides. Chlorosulfonic acid is, however, too reactive to be treated by any of these methods. See the monograph on Chlorosulfonic Acid.

References

1. Other names for this compound are ethanoyl chloride, acetic acid chloride, and acetic chloride.
2. Other names for this compound are propanoyl chloride and propionic chloride.
3. Other names for this compound are chloroformic acid dimethylamide, DDC, (dimethylamino)carbonyl chloride, dimethylcarbamic acid chloride, dimethylcarbamic chloride, dimethylcarbamidoyl chloride, *N,N*-dimethylcarbamoyl chloride, dimethylcarbamyl chloride, *N,N*-dimethylcarbamyl chloride, and DMCC.
4. Other names for this compound are benzenecarbonyl chloride, benzoic acid chloride, or α-chlorobenzaldehyde.
5. Other names for this compound are sulfinyl chloride, sulfurous oxychloride, sulfur chloride oxide, sulfurous dichloride, and thionyl dichloride.
6. Other names for this compound are sulfonyl chloride and sulfuric oxychloride.
7. Another name for this compound is mesyl chloride.
8. Other names for this compound are benzene sulfone chloride, benzenesulfonic acid chloride, and benzenosulphochloride.
9. Other names for this compound are 4-methylbenzenesulfonyl chloride, tosyl chloride, and toluenesulfonic acid chloride.
10. Other names for this compound are acetic oxide, acetyl oxide, ethanoic anhydrate, acetyl ether, acetyl anhydride, and acetic acid anhydride.
11. Bretherick, L. *Bretherick's Handbook of Reactive Chemical Hazards*, 4th ed.; Butterworths: London, 1990; pp. 299–300.
12. Lewis, R.J., Sr. *Sax's Dangerous Properties of Industrial Materials*, 8th ed.; Van Nostrand-Reinhold: New York, 1992; p. 41.
13. Reference 11, pp. 364–365.
14. Reference 11, pp. 1023–1025.

15. Reference 11, pp. 1026, 1357, 1413, and 1438.
16. Reference 12, p. 363.
17. Reference 11, pp. 449–452 and 1147–1148.
18. For example, references 11 and 12.
19. International Agency for Research on Cancer. Volume 12, *IARC Monographs on the Evaluation of Carcinogenic Risk of Chemicals to Man. Some Carbamates, Thiocarbamates and Carbazides*; International Agency for Research on Cancer: Lyon, 1976; pp. 77–84.
20. Reference 12, p. 3168.
21. Lunn, G. Unpublished observations.
22. National Research Council, Committee on Hazardous Substances in the Laboratory. *Prudent Practices for Disposal of Chemicals from Laboratories;* National Academy Press: Washington, DC, 1983; p. 67

AFLATOXINS

CAUTION! Refer to safety considerations section on page 7 before starting any of these procedures.

Aflatoxins are fungal metabolites produced by *Aspergillus parasiticus* and *Aspergillus flavus*. In hot humid areas peanuts, beans, and corn may be contaminated with aflatoxins. A variety of aflatoxins are known and they are all high-melting (> 180°C) crystalline solids. The most commonly encountered aflatoxins are B_1 [1162-65-8], B_2 [7220-81-7], G_1 [1165-39-5], G_2 [7241-98-7], and M_1 [6795-23-9] (which is the major metabolite of aflatoxin B_1 in milk). Other aflatoxins are known. These aflatoxins are all chemically very similar[1] and the structure of aflatoxin B_1 is shown below.

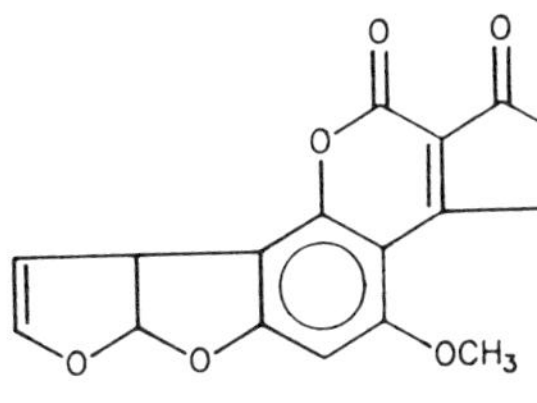

Aflatoxin B_1

Aflatoxins are carcinogenic in humans and laboratory animals.[2] These compounds are also acutely poisonous by ingestion,[3] are used in the laboratory in cancer research, and may also be found as analytical standards in laboratories doing surveillance of foodstuffs. Solid aflatoxins may become electrostatically charged and cling to glassware or protective clothing.

Principles of Destruction

Aflatoxins may be degraded using ammonia (NH_3),[4] potassium permanganate in sulfuric acid ($KMnO_4$ in H_2SO_4),[4] potassium permanganate in 2 *M* sodium hydroxide solution ($KMnO_4$ in NaOH),[5] or 5.25% sodium hypochlorite (NaOCl) solution followed by the addition of acetone.[4] The acetone is required to destroy any 2,3-dichloroaflatoxin B_1 that may have been formed by the action of the NaOCl. Before the addition of the acetone the NaOCl concentration should be reduced to 1.3% or less so that the haloform reaction does not occur.[6] When $KMnO_4$ is used the final reaction mixtures should be made strongly basic and filtered to remove manganese compounds.[7] Animal carcasses may be decontaminated by burying them in quicklime (calcium oxide).[4]

Destruction Procedures

Destruction of Stock Quantities

1. Add sufficient methanol (~ 1 mL or more if required) to solubilize the aflatoxins and wet the glassware, then add 2 mL of 5.25% NaOCl solution (see below for assay procedure) for each microgram (μg) of aflatoxin. Allow this to stand overnight, then add three volumes of H_2O and add a volume of acetone equal to 5% of the total diluted volume. After 30 min check for completeness of destruction and discard it.

2. Add sufficient H_2O so that the aflatoxins are dissolved and their concentration does not exceed 2 μg/mL. Then, for each 100 mL of this solution, **cautiously** add 10 mL of concentrated H_2SO_4 with stirring (**exothermic reaction!**). Add 16 g of $KMnO_4$ per liter of the resulting solution. The purple color should remain for at least 3 h. If it does not, add more $KMnO_4$. Leave it to react for a further 3 h, then decolorize it with sodium metabisulfite, make it strongly basic by adding 10 *M* KOH solution (**Caution!** Exothermic), dilute with H_2O, filter, test the filtrate for completeness of destruction, and discard it.

3. Prepare a 0.3 *M* solution of $KMnO_4$ in 2 *M* NaOH solution by stirring the mixture for at least 30 min but no more than 2 h. Dissolve 300 μg of

aflatoxins in 5 mL of acetonitrile and add 10 mL of $KMnO_4$ in NaOH. Stir for at least 3 h. The color should be either green or purple. If it is not, add more $KMnO_4$ in NaOH until the green or purple color persists for at least 1 h. For each 10 mL of $KMnO_4$ in NaOH add 0.8 g of sodium metabisulfite (more if necessary for complete decolorization), dilute with an equal volume of water, filter to remove the manganese salts, check for completeness of destruction, and discard the solid and filtrate appropriately.

Destruction of Aflatoxins in Aqueous Solution

1. For each microgram of aflatoxin add 2 mL of 5.25% NaOCl solution (see below for assay procedure). Allow it to stand overnight, then add three volumes of H_2O and a volume of acetone equal to 5% of the total diluted volume. After 30 min check for completeness of destruction and discard it.

2. For each 100 mL of solution **cautiously** add 10 mL of concentrated H_2SO_4 with stirring (**exothermic reaction!**). Add 16 g of $KMnO_4$ per liter of the resulting solution. The purple color should remain for at least 3 h. If it does not, add more $KMnO_4$. Leave it to react for a further 3 h, then decolorize it with sodium metabisulfite, make it strongly basic by adding 10 *M* KOH solution (**Caution!** Exothermic), dilute with H_2O, filter, test the filtrate for completeness of destruction, and discard it.

3. Dilute with H_2O, if necessary, so that the concentration of aflatoxins does not exceed 200 μg/mL. Add sufficient NaOH, with stirring, to make the concentration 2 *M*, then add sufficient solid $KMnO_4$ to make the concentration 0.3 *M*. Stir for at least 3 h. The color should be either green or purple. If it is not, add more $KMnO_4$ in NaOH until the green or purple color persists for at least 1 h. For each 10 mL of $KMnO_4$ in NaOH add 0.8 g of sodium metabisulfite (more if necessary for complete decolorization), dilute with an equal volume of water, filter to remove the manganese salts, check for completeness of destruction, and discard the solid and filtrate appropriately.

Destruction of Aflatoxins in Volatile Organic Solvents

1. Evaporate to dryness under reduced pressure using a rotary evaporator (add an equal volume of dichloromethane to dimethyl sulfoxide (DMSO) solutions before evaporation), then solubilize the residual aflatoxins in a little methanol (~ 1 mL). For each microgram of aflatoxin add 2 mL of 5.25% NaOCl solution (see below for assay procedure). Allow it to stand

overnight, then add three volumes of H_2O and a volume of acetone equal to 5% of the total diluted volume. After 30 min check for completeness of destruction and discard it.

2. Evaporate to dryness under reduced pressure using a rotary evaporator (add an equal volume of dichloromethane to DMSO solutions before evaporation), then dissolve the residual aflatoxins in H_2O (~ 10 mL for each 20 μg of aflatoxins; more if required). For each 100 mL of this solution **cautiously** add 10 mL of concentrated H_2SO_4 with stirring (**exothermic reaction!**). Add 16 g of $KMnO_4$ per liter of the resulting solution. The purple color should remain for at least 3 h. If it does not, add more $KMnO_4$. Leave it to react for a further 3 h, then decolorize it with sodium metabisulfite, make it strongly basic by adding 10 *M* KOH solution (**Caution!** Exothermic), dilute with H_2O, filter, test the filtrate for completeness of destruction, and discard it.

3. Prepare a 0.3 *M* solution of $KMnO_4$ in 2 *M* NaOH solution by stirring the mixture for at least 30 min but no more than 2 h. Remove the organic solvent under reduced pressure using a rotary evaporator (add an equal volume of dichloromethane to DMSO solutions before evaporation). Dissolve 300 μg of aflatoxins in 5 mL of acetonitrile and add 10 mL of $KMnO_4$ in NaOH. Stir for at least 3 h. The color should be either green or purple. If it is not, add more $KMnO_4$ in NaOH until the green or purple color persists for at least 1 h. For each 10 mL of $KMnO_4$ in NaOH add 0.8 g of sodium metabisulfite (more if necessary for complete decolorization), dilute with an equal volume of H_2O, filter to remove the manganese salts, check for completeness of destruction, and discard the solid and filtrate appropriately.

Destruction of Aflatoxins in Oil

Add 2 mL of a 5.25% NaOCl solution (see below for assay procedure) for each microgram of aflatoxin, shake the mixture on a mechanical shaker for at least 2 h, add three volumes of H_2O for each volume of NaOCl used, then add a volume of acetone equal to 5% of the total diluted volume. After 30 min check for completeness of destruction and discard it.

Decontamination of Equipment and Thin-Layer Chromatography Plates

First rinse equipment with a little methanol to solubilize the aflatoxins. Immerse equipment, thin-layer chromatography (TLC) plates, protective clothing, and absorbent paper in a 1:3 mixture of 5.25% NaOCl solution (see below for assay procedure) and H_2O for at least 2 h, then add an amount of acetone equal to 5% of the total volume, allow the mixture to react for at least 30 min, and discard it.

Treatment of Spills

1. First remove as much of the spill as possible by high efficiency particulate air (HEPA) vacuuming (not sweeping), then rinse the area with a little methanol to solubilize the aflatoxins. Take up the rinse with absorbent paper. Immerse the absorbent paper in a 1:3 mixture of a 5.25% NaOCl solution (see below for assay procedure) and H_2O for at least 2 h, then add an amount of acetone equal to 5% of the total volume, allow the mixture to react for at least 30 min, and discard it. Wash the surface from which the spill has been removed with a 5.25% NaOCl solution and leave it for 10 min before adding a 5% aqueous solution of acetone.

2. Prepare a 0.3 *M* solution of $KMnO_4$ in 2 *M* NaOH solution by stirring the mixture for at least 30 min but no more than 2 h. Collect spills of liquid with a dry tissue and spills of solid with a tissue wetted with dichloromethane. Immerse all tissues in the $KMnO_4$ in NaOH solution. Allow to react for at least 3 h. The color should be either green or purple. If it is not, add more $KMnO_4$ in NaOH until the green or purple color persists for at least 1 h. For each 10 mL of $KMnO_4$ in NaOH add 0.8 g of sodium metabisulfite (more if necessary for complete decolorization), dilute with an equal volume of H_2O, filter, check for completeness of destruction, and discard the solid and filtrate appropriately. Cover the spill area with an excess of the $KMnO_4$ in NaOH solution and allow to react for 3 h. Collect the solution on a tissue and immerse the tissue in 2 *M* sodium metabisulfite solution. If the pH of this solution is acidic, make it alkaline with NaOH. Rinse the spill area with a 2 *M* solution of sodium metabisulfite. Check the surface for completeness of decontamination by using a wipe moistened with methanol and analyzing the wipe for the presence of aflatoxins.

Destruction of Aflatoxins in Animal Litter

Spread the litter on a metal tray to a maximum depth of about 5 cm, then sprinkle it with a 5% NH_3 solution (30–40 mL per 25 g of litter). Autoclave the tray for 20 min at 128–130°C, then discard the litter. **Do not** preevacuate the autoclave as this would remove the NH_3.

Destruction of Aflatoxins in Animal Carcasses

Bury carcasses in quicklime and cover to a depth of about 1 cm.

Analytical Procedures

1. Extract 200 mL of decontaminated waste solution three times with 50-mL portions of chloroform and combine the extracts.[4] Concentrate the extracts to about 3 mL using a rotary evaporator and add this solution to a graduated tube. Wash the flask twice with 2-mL portions of chloroform and add these washes to the tube. Concentrate the contents of the tube at about 60°C to 0.5 mL under a stream of nitrogen. Spot a TLC plate with 10 μL of this solution and with 5 μL of a 0.2 mg/L standard solution of aflatoxins and develop with a mixture of chloroform:acetone (9:1) in subdued light. Determine the presence or absence of aflatoxins by visualizing under ultraviolet (UV) light (365 nm). (The TLC plates used were Kieselgel 60 Merck.) More cleanup of the sample may be required, prior to TLC, if the sample is highly colored or if the aflatoxins were initially dissolved in oil. The cleaned sample may also be analyzed by HPLC as described below.

2. For reaction mixtures obtained using the $KMnO_4$ in NaOH procedure, acidify an aliquot to pH 2–3 using concentrated hydrochloric acid (HCl). Extract this mixture three times with an equal volume of dichloromethane, pool the extracts, and dry them over anhydrous sodium sulfate.[5] Remove the sodium sulfate by filtration, evaporate to dryness and take up the residue in 0.5 mL of water:methanol:acetonitrile 2:1:1. Analyze by reverse phase HPLC using water:methanol:acetonitrile 2:1:1 flowing at 1 mL/min with spectrofluorometric detection (excitation 360 nm, emission 440 nm).

3. Other analytical techniques have been reviewed[8,9] and techniques using reverse phase HPLC with a 10-μm Spherisorb ODS column, water:acetonitrile: methanol 15:3:2, fluorescence detection with excitation 365 nm, emission about 450 nm[10] or a C_{18} μ Bondapak column, methanol:water 40:60, fluorescence detection with excitation 360 nm and a 417 nm cut-off emission filter[11] have been described.

Mutagenicity Assays

Aflatoxins B_1 and G_1 have been shown to be mutagenic in *Salmonella typhimurium* and other species[12] but specific studies of possible mutagenic products from the degradation procedures involving NH_3, $KMnO_4$ in H_2SO_4, and 5.25% NaOCl solution followed by the addition of acetone have not been carried out. The residues from the degradation reactions involving $KMnO_4$ in NaOH were tested for mutagenicity using tester strains

TA97, TA98, TA100, and TA102 of *S. typhimurium*.[5] No mutagenic activity was found.

Related Compounds

The above techniques were investigated for aflatoxins B_1, B_2, G_1, and G_2 but they should also be applicable to other aflatoxins.

Assay of Sodium Hypochlorite Solution

Sodium hypochlorite solutions tend to deteriorate with time, so they should be periodically checked for the amount of active chlorine they contain. Pipette 10 mL of the NaOCl solution into a 100-mL volumetric flask and fill it to the mark with distilled H_2O. Pipette 10 mL of this solution into a conical flask containing 50 mL of distilled H_2O, 1 g of potassium iodide (KI), and 12.5 mL of 2 *M* acetic acid. Titrate this solution against a 0.1 *N* sodium thiosulfate solution using starch as an indicator. Each 1 mL of the sodium thiosulfate solution corresponds to 3.545 mg of active chlorine. Commercially available NaOCl solution (Clorox bleach) contains 5.25% NaOCl and should contain 45–50 g of active chlorine per liter.

References

1. The systematic names are aflatoxin B_1: 2,3,6aα,9aα-tetrahydro-4-methoxycyclopenta-[*c*]furo[3′,2′:4,5]furo[2,3-*h*][1]benzopyran-1,11-dione, aflatoxin B_2: 2,3,6aα,8,9,9aα-hexahydro-4-methoxycyclopenta[*c*]furo[3′,2′:4,5]furo-[2,3-*h*][1]benzopyran-1,11-dione, aflatoxin G_1:3,4,7aα,10aα-tetrahydro-5-methoxy-1*H*,12*H*-furo[3′,2′:4,5]furo[2,3-*h*]pyrano-[3,4-*c*][1]benzopyran-1,12-dione, aflatoxin G_2: 3,4,7aα,9,10,10aα-hexahydro-5-methoxy-1*H*,12*H*-furo[3′,2′:4,5]furo[2,3-*h*]pyrano[3,4-*c*][1]benzopyran-1,12-dione, and aflatoxin M_1: 2,3,6a,9a-tetrahydro-9a-hydroxy-4-methoxycyclopenta[*c*]furo-[3′,2′:4,5]furo[2,3-*h*]-[1]benzopyran-1,11-dione (4-hydroxyaflatoxin B_1).
2. International Agency for Research on Cancer. *IARC Monographs on the Evaluation of the Carcinogenic Risk of Chemicals to Humans, Supplement No. 7, Overall Evaluations of Carcinogenicity: An Updating of* IARC Monographs *Volumes 1 to 42*; International Agency for Research on Cancer: Lyon, 1987; pp. 83–87.
3. Lewis, R.J., Sr. *Sax's Dangerous Properties of Industrial Materials*; 8th ed.; Van Nostrand-Reinhold: New York, 1992; pp. 85–87.
4. Castegnaro, M.; Hunt, D.C.; Sansone, E.B.; Schuller, P.L.; Siriwardana, M.G.; Telling, G.M.; van Egmond, H.P.; Walker, E.A., Eds., *Laboratory Decontamination and Destruction of Aflatoxins B_1, B_2, G_1, and G_2 in Laboratory Wastes*; International Agency for Research on Cancer: Lyon, 1980 (IARC Scientific Publications No. 37).

5. Castegnaro, M.; Barek, J.; Frémy, J-.M.; Lafontaine, M.; Miraglia, M.; Sansone, E.B.; Telling, G.M., Eds., *Laboratory Decontamination and Destruction of Carcinogens in Laboratory Wastes: Some Mycotoxins*; International Agency for Research on Cancer: Lyon, 1991 (IARC Scientific Publications No. 113).

6. Castegnaro, M.; Friesen, M.; Michelon, J.; Walker, E.A. Problems related to the use of sodium hypochlorite in the detoxification of aflatoxin B_1. *Am. Ind. Hyg. Assoc. J.* **1981**, *42*, 398–401.

7. Lunn, G.; Sansone, E.B.; De Méo, M.; Laget, M.; Castegnaro, M. Potassium permanganate can be used for degrading hazardous compounds. *Am. Ind. Hyg. Assoc. J.* **1994**, *55*, 167–171.

8. Schuller, P.L.; Horwitz, W.; Stoloff, L. A review of sampling plans and collaboratively studied methods of analysis for aflatoxins. *J. Assoc. Off. Anal. Chem.* **1976**, *59*, 1315–1343.

9. Pohland, A.E.; Thorpe, C.W.; Neshein, S. Newer developments in mycotoxin methodology. *Pure Appl. Chem.* **1979**, *52*, 213–223.

10. Beebe, R.M. Reverse phase high pressure liquid chromatographic determination of aflatoxins in foods. *J. Assoc. Off. Anal. Chem.* **1978**, *61*, 1347–1352.

11. Vazquez, M.; Franco, C.M.; Cepeda, A.; Prognon, P.; Mahuzier, G. Liquid chromatographic study of the interaction between aflatoxins and β-cyclodextrin. *Anal. Chim. Acta* **1992**, *269*, 239–247.

12. Wong, J.J.; Hsieh, D.P.H. Mutagenicity of aflatoxins related to their metabolism and carcinogenic potential. *Proc. Natl. Acad. Sci. USA* **1976**, *73*, 2241–2244.

ALKALI AND ALKALINE EARTH METALS

CAUTION! Refer to safety considerations section on page 7 before starting any of these procedures.

The alkali metals sodium (Na) [7440-23-5],[1,2] potassium (K) [7440-09-7],[3,4] and lithium (Li) [7439-93-2][5,6] react violently with H_2O or even moist air to generate hydrogen, which can then be ignited by the heat of the reaction. These metals are corrosive to the skin and incompatible with many organic and inorganic compounds, including halocarbons. Potassium may oxidize on storage and oxidized metal may explode violently when handled or cut.[4] These metals are used in organic synthesis and as drying agents. Alkali metals require special fire extinguishing procedures. The alkaline earth metals magnesium (Mg) [7439-95-4],[7,8] calcium (Ca) [7440-70-2],[9,10] strontium (Sr) [7440-25-7],[11,12] and barium (Ba) [7440-39-3][13,14] are less reactive to water but they are incompatible with many organic and inorganic compounds. Magnesium and barium have been reported to be incompatible with halocarbons.

Principles of Destruction

The alkali metals are allowed to react with an alcohol in a slow and controlled fashion to generate the metal alkoxide and hydrogen. The metal alkoxide is subsequently hydrolyzed with H_2O to give the metal hydroxide and alcohol. A method involving dropping lumps of K into a hole in the ground partially filled with H_2O[15] would probably not be acceptable in today's regulatory climate. Barium, calcium, and strontium are allowed to react with H_2O to generate the metal hydroxide and hydrogen. It has been reported[9] that when calcium reacts with H_2O the heat of the reaction may ignite the hydrogen that is evolved. However, we have experienced no problems with this procedure. Magnesium is allowed to react with dilute hydrochloric acid (HCl) to generate magnesium chloride and hydrogen. In all cases the hydrogen is vented into the fume hood. A procedure for recycling scrap pieces of sodium has recently been published.[16]

Destruction Procedures

Caution! These procedures present a high fire hazard and should be conducted in a properly functioning chemical fume hood away from flammable solvents. The presence of a nonflammable board or cloth for smothering the reaction, as well as an appropriate fire extinguisher, may be advisable. If possible, do the reaction in batches to minimize the risk.

Sodium and Lithium

Add 1 g of Na or Li to 100 mL of cold ethanol at such a rate that the reaction does not become violent.[17] Stir the reaction mixture. If the reaction mixture becomes viscous and the rate of reaction slows, add more ethanol. When all the metal has been added stir the reaction mixture until all reaction ceases, then examine carefully for the presence of unreacted metal. If none is found, dilute the mixture with H_2O, neutralize, and discard it.

Potassium

Potassium is the most treacherous of the alkali metals and fires during its destruction are not infrequent. Precautions for its safe handling have been described.[18,19]

Add the K to *tert*-butyl alcohol[18] at a rate so that the reaction does not become violent. If the reaction mixture becomes viscous and the rate of reaction slows, add more *tert*-butyl alcohol. When all the K has been added, stir the reaction mixture until all reaction ceases, then examine *carefully* for the presence of unreacted metal. If none is found, dilute the mixture with H_2O, neutralize, and discard. *tert*-Amyl alcohol may also be used.[19] Whichever alcohol is used it is important to use an anhydrous grade. If necessary, the alcohol should be dried before use. Powdered 3-Å molecular sieve has been recommended.[20]

Magnesium

Add 1 g of Mg to 100 mL of 1 *M* HCl and stir the mixture.[17] When the reaction has ceased, neutralize the reaction mixture and discard it.

Barium, Calcium, and Strontium

Add 1 g of Ba, Ca, or Sr metal to 100 mL of H_2O in portions and stir the mixture.[17] When the reaction has ceased, neutralize the reaction mixture and discard it.

References

1. Bretherick, L. *Bretherick's Handbook of Reactive Chemical Hazards*; 4th ed.; Butterworths: London, 1990; pp. 1370–1379.
2. Lewis, R.J., Sr. *Sax's Dangerous Properties of Industrial Materials*; 8th ed.; Van Nostrand-Reinhold: New York, 1992; pp. 3056–3057.
3. Reference 1, pp. 1286–1292.
4. Reference 2, pp. 2855–2856.
5. Reference 1, pp. 1312–1318.
6. Reference 2, pp. 2123–2124.
7. Reference 1, pp. 1320–1325.
8. Reference 2, p. 2147.
9. Reference 1, pp. 916–918.
10. Reference 2, pp. 660–661.
11. Reference 1, p. 1460.
12. Reference 2, p. 3139
13. Reference 1, pp. 78–79.
14. Reference 2, p. 333.
15. Burfield, D.R.; Smithers, R.H. Safe handling and disposal of potassium. *Chem. Ind. (London)* **1979**, 89.

16. Hübler-Blank, B.; Witt, M.; Roesky, H.W. Recycling of sodium waste. *J. Chem. Educ.* **1993**, *70*, 408–409.

17. Lunn, G. Unpublished observations.

18. Johnson, W.S.; Schneider, W.P. β-Carbethoxy-γ,γ-diphenylvinylacetic acid. In *Organic Syntheses, Coll. Vol. 4*; Rabjohn, N., Ed., Wiley: New York, 1963; pp. 132–135.

19. Fieser, L.F.; Fieser, M. *Reagents for Organic Synthesis, Volume 1*; Wiley: New York, 1967; pp. 905–906.

20. Burfield, D.R.; Smithers, R.H. Desiccant efficiency in solvent and reagent drying. 7. Alcohols. *J. Org. Chem.* **1983**, *48*, 2420–2422.

ALKALI METAL ALKOXIDES

CAUTION! Refer to safety considerations section on page 7 before starting any of these procedures.

The alkali metal alkoxides sodium methoxide (sodium methylate, CH_3ONa) [124-41-4], sodium ethoxide (sodium ethylate, CH_3CH_2ONa) [141-52-6], and potassium *tert*-butoxide [$(CH_3)_3COK$] [865-47-4] are corrosive. Sodium methoxide[1] and sodium ethoxide[2] may ignite in moist air. Sodium methoxide reacts violently with chloroform[3] and $(CH_3)_3COK$ ignites on contact with acids or reactive solvents.[4] These compounds are used in organic synthesis.

Principle of Destruction

The alkali metal alkoxides are hydrolyzed with H_2O to sodium or potassium hydroxide (NaOH or KOH) and the corresponding alcohol.

Destruction Procedure

Add 5 g of the alkoxide to 100 mL of H_2O and stir the mixture. When all the alkoxide has dissolved and the reaction appears to be over discard the mixture.[5]

References

1. Bretherick, L. *Bretherick's Handbook of Reactive Chemical Hazards*; 4th ed.; Butterworths: London, 1990, p. 169.
2. Reference 1, p. 289.
3. Reference 1, p. 133.
4. Reference 1, pp. 474–475.
5. Lunn, G. Unpublished observations.

ANTINEOPLASTIC ALKYLATING AGENTS

CAUTION! Refer to safety considerations section on page 7 before starting any of these procedures.

The drugs considered in this monograph are all antineoplastic agents and have an N—CH_2CH_2Cl functionality in common. These compounds are all crystalline solids and are generally moderately soluble in alcohols. Because these agents are basic they are generally soluble in acid and those compounds having a carboxylic acid group are soluble in base. The H_2O solubility varies.

The following compounds are considered:

Mechlorethamine [51-75-2][1] [mp of hydrochloride 108–110°C]	$(ClCH_2CH_2)_2NCH_3.HCl$ Hydrochloride soluble in H_2O
Melphalan [148-82-3][2] [mp 182–183°C]	**I** Almost insoluble in H_2O

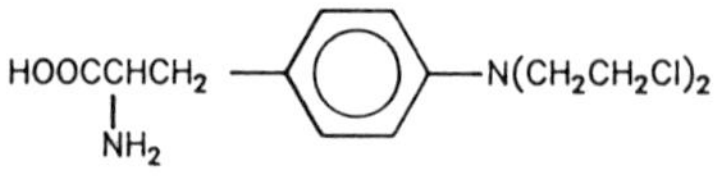

I

Chlorambucil [305-03-3][3] **II**
[mp 64–66°C]

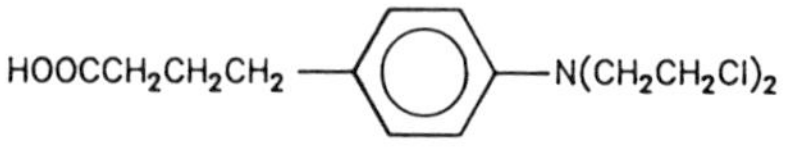

II

Cyclophosphamide [50-18-0][4] **III**
[mp 41–45°C (monohydrate)] Soluble in H_2O (40 mg/mL)

Ifosfamide [3778-73-2][5] **IV**
[mp 39–41°C] Soluble in H_2O (1 in 10)

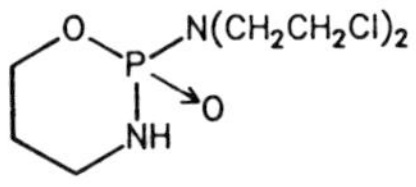

III

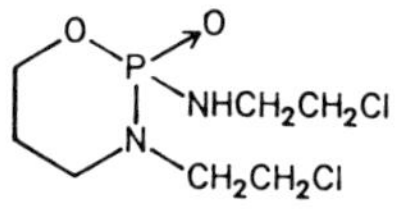

IV

Uracil mustard [66-75-1][6] **V**
[mp 206°C (decomposes)] Sparingly soluble in H_2O

Spirohydantoin mustard **VI**
[56605-16-4][7]
[mp 127–129°C][8]

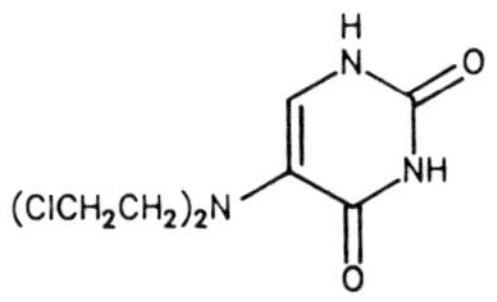

V

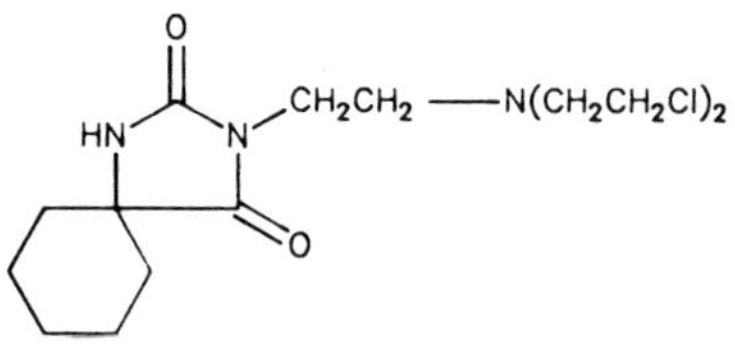

VI

Mechlorethamine,[9–11] melphalan,[12–14] chlorambucil,[15–17] cyclophosphamide,[18–21] ifosfamide,[22] and uracil mustard[23,24] are carcinogenic in experimental animals; chlorambucil,[16,17] cyclophosphamide,[19–21] and melphalan[12–14] are human carcinogens; there is limited evidence that mechlorethamine[11] is a human carcinogen. All of these compounds are mutagenic.[25] These compounds are used as antineoplastic drugs.

Principles of Destruction

All the drugs were degraded by reduction with nickel–aluminum (Ni–Al) alloy in potassium hydroxide (KOH) solution.[25] The products detected were ethanol from cyclophosphamide and ifosfamide, and ethylmethylamine and diethylmethylamine from mechlorethamine. Mechlorethamine and chlorambucil were also degraded by reaction with saturated sodium bicarbonate solution ($NaHCO_3$). Pharmaceutical preparations were also degraded using these procedures,[25] but cyclophosphamide tablets first had to be refluxed in 1 *M* hydrochloric acid (HCl) before the Ni–Al reduction. If this step was omitted the destruction was incomplete. Degradation of mechlorethamine by sodium thiosulfate and $NaHCO_3$ has been recommended,[26] but when the reaction was carried out for the recommended time (45 min) mutagenic products were observed.[25] If the reaction was allowed to proceed for 18 h, however, no mutagenic activity was seen.[25] The International Agency for Research on Cancer (IARC) has recommended alkaline hydrolysis for the degradation of cyclophosphamide and ifosfamide and acid hydrolysis for the degradation of cyclophosphamide,[27] but we found[8] that these reactions gave incomplete degradation and mutagenic products. Cyclophosphamide in urine was degraded using alkaline potassium permanganate ($KMnO_4$) followed by addition of sodium thiosulfate ($Na_2S_2O_3$)[28] and this procedure has also been reported to work for pharmaceutical preparations of cyclophosphamide and ifosfamide.[29] Melphalan was degraded by oxidation with $KMnO_4$ in basic solution.[30] In all cases destruction was greater than 99.8% except for pharmaceutical preparations of mechlorethamine ($>$ 90% degradation) and cyclophosphamide in urine (degradation efficiency not reported).[28]

Destruction Procedures

Destruction of Bulk Quantities of Melphalan, Uracil Mustard, and Spirohydantoin Mustard

Dissolve the drug in **methanol** (CH_3OH) so that the concentration does not exceed 10 mg/mL, then add an equal volume of 2 *M* KOH solution.

For every 20 mL of this basified solution add 1 g of Ni–Al alloy. Add quantities of more than 5 g in portions to prevent the reaction from becoming too violent. Some foaming may occur, so the reaction should be done in a vessel at least five times larger than the final volume. Stir the mixture overnight, then filter through a pad of Celite®. Test the filtrate for completeness of destruction, neutralize, and discard it. Allow the spent nickel, which is filtered off, to dry on a metal tray away from flammable solvents for 24 h, then discard it with the solid waste.

Destruction of Bulk Quantities of Melphalan

Dissolve 20 mg of melphalan in 20 mL of a 2 *M* NaOH solution and add 0.2 g of $KMnO_4$. Stir for 1 h, decolorize with sodium bisulfite, dilute with an equal volume of H_2O, filter to remove manganese compounds,[31] neutralize the filtrate, check for completeness of destruction, and discard it.

Destruction of Bulk Quantities of Mechlorethamine, Chlorambucil, Cyclophosphamide, and Ifosfamide

Dissolve the drug in **water** so that the concentration does not exceed 10 mg/mL, then add an equal volume of 2 *M* KOH solution. For every 20 mL of this basified solution add 1 g of Ni–Al alloy. Add quantities of more than 5 g in portions to prevent the reaction from becoming too violent. Some foaming may occur, so the reaction should be done in a vessel at least five times larger than the final volume. Stir the mixture overnight, then filter through a pad of Celite®. Test the filtrate for completeness of destruction, neutralize, and discard it. Allow the spent nickel, which is filtered off, to dry on a metal tray away from flammable solvents for 24 h, then discard it with the solid waste.

Destruction of Pharmaceutical Preparations of Mechlorethamine, Melphalan, Chlorambucil, and Cyclophosphamide

Melphalan. Dissolve the pharmaceutical preparation (100 mg) in the supplied diluent (10 mL), as directed, and add an equal volume of 2 *M* KOH solution.

Cyclophosphamide. The pharmaceutical preparation consists of 100 mg of drug in 10 mL of saline solution. Add an equal volume of 2 *M* KOH solution.

Chlorambucil. Dissolve each 2-mg tablet in 10 mL of 1 *M* KOH solution.

For every 20 mL of basified solution add 1 g of Ni–Al alloy. Add quantities of more than 5 g in portions to prevent the reaction from becoming too violent. Some foaming may occur, so the reaction should be done in a vessel at least five times larger than the final volume. Stir the mixture overnight, then filter through a pad of Celite®. Test the filtrate for completeness of destruction, neutralize, and discard it. Allow the spent nickel, which is filtered off, to dry on a metal tray away from flammable solvents for 24 h, then discard it with the solid waste.

Destruction of Mechlorethamine and Chlorambucil

Take up bulk quantities in H_2O so that the concentration does not exceed 10 mg/mL. If necessary, dilute pharmaceutical preparations so that their concentrations do not exceed 10 mg/mL. The chlorambucil may not be completely soluble in the H_2O but it will dissolve when the base is added. Add five volumes of a saturated $NaHCO_3$ solution for each volume of aqueous solution and allow the mixture to stand overnight, check for completeness of destruction, and discard it. Prepare a saturated $NaHCO_3$ solution by mixing $NaHCO_3$ and H_2O in a container. Shake the container occasionally. If solid persists, the solution is saturated, if not, add more $NaHCO_3$.

Destruction of Cyclophosphamide Tablets

For each 50 mg tablet add 10 mL of 1 *M* HCl and reflux the mixture for 1 h, cool, and place in a vessel whose volume is at least 10 times the final solution volume. Add an equal volume of 2 *M* KOH solution and stir. For every 20 mL of this basified solution, add 1 g of Ni–Al alloy. Add quantities of more than 5 g in portions to prevent the reaction from becoming too violent. Considerable foaming will occur but the reaction should stay in the flask. Stir the mixture overnight, then filter through a pad of Celite®. Test the filtrate for completeness of destruction, neutralize, and discard it. Allow the spent nickel, which is filtered off, to dry on a metal tray away from flammable solvents for 24 h, then discard it with the solid waste.

Destruction of Cyclophosphamide in Urine[28]

For each 20 mL of urine add 0.5 mL of 5 *M* KOH solution followed by 1.2 g of $KMnO_4$. After 2 h add sodium bisulfite until the color of the

$KMnO_4$ disappears, then add 1 mL of 5 *M* KOH solution and 0.66 g of sodium thiosulfate. After 20 min neutralize this mixture by the addition of acid, test for completeness of destruction, and discard it.

Analytical Procedures

Analysis was by high-performance liquid chromatography (HPLC) using a 250 × 4.6-mm i.d. column of Microsorb C8. The injection volume was 20 μL and the mobile phase flowed at 1 mL/min. Ultraviolet (UV) detection, at 254 nm unless otherwise stated, was used. For melphalan, chlorambucil, uracil mustard, and spirohydantoin mustard the mobile phase was a mixture of methanol and 20 m*M* potassium phosphate, monobasic (KH_2PO_4) buffer: melphalan (58:42); chlorambucil (65:35); uracil mustard (40:60) (UV 200 nm); and spirohydantoin mustard (65:35). It was frequently advantageous to add a little KH_2PO_4 buffer to an aliquot of the neutralized reaction mixture and to centrifuge before analysis. This technique removed salts that could clog the chromatograph. For cyclophosphamide the mobile phase was acetonitrile : 20 m*M* KH_2PO_4 (25:75) and the UV detector was set at 190 nm. Although direct injection could be employed for cyclophosphamide, a massive early eluting peak made detection difficult. It was found advantageous to add solid sodium chloride (NaCl) to 0.5 mL of the reaction mixture and stir. If necessary, more NaCl was added until solid persisted. Acetonitrile (0.25 mL) was then added and the mixture was stirred for 5 min. The acetonitrile layer, which now contained any traces of the drug, was analyzed and a much cleaner chromatogram resulted. On our equipment these mobile phase combinations were found to give reasonable retention times (8–16 min).

Mechlorethamine was determined using a colorimetric procedure. Thus 100 μL of the reaction mixture was mixed with 1 mL of a solution of 2 mL of glacial acetic acid in 98 mL of 2-methoxyethanol and 1 mL of a 5% (w/v) solution of 4-(4-nitrobenzyl)pyridine was added. This mixture was heated at 100°C for 10 min, then cooled in ice for 5 min. Piperidine (0.5 mL) was then added, the mixture centrifuged, and the absorbance determined at 560 nm against an appropriate blank using 10-mm cuvettes. Under these conditions the limit of detection was about 50 μg/mL.

Gas chromatography (GC) using a 1.8-m × 2-mm i.d. glass column packed with 28% Pennwalt 223 + 4% KOH on 80/100 Gas Chrom R was used to determine the products of these reactions. The injection temperature was 200°C and the flame ionization detector operated at 300°C. The oven temperature was 60°C and the approximate retention times were ethanol (2.7 min), ethylmethylamine (1.5 min), and diethylmethylamine (4.9 min).

Mutagenicity Assays[25]

The mutagenicity assays were carried out as described on page 4 using tester strains TA98, TA100, TA1530, and TA1535. The final reaction mixtures from the Ni–Al alloy reductions were tested at a level corresponding to 0.25 mg (0.125 mg for cyclophosphamide tablets) of undegraded material per plate. To avoid toxicity problems it was generally necessary to mix the neutralized reaction mixtures with an equal volume of pH 7 buffer before testing. The reaction mixtures from $NaHCO_3$ degradation were tested without using buffer and the degradation products from about 0.17 mg of drug were applied to each plate. None of the reaction mixtures were found to be mutagenic. All of the drugs were tested at a level of 0.5 mg per plate and were found to be mutagenic. None of the products identified were found to be mutagenic (1 mg per plate). After decontamination with alkaline $KMnO_4$ and sodium thiosulfate, urine that contained cyclophosphamide was not mutagenic to TA98, TA100, UTH8414, and UTH8413.[28] Residues obtained from the alkaline $KMnO_4$ oxidation of melphalan were not mutagenic to TA98, TA100, and TA1535.[30]

Related Compounds

The Ni–Al alloy technique described above should be applicable to compounds of the general form $RR'N—CH_2CH_2Cl$ but the procedure should be thoroughly validated.

References

1. Other names for this compound are *N,N*-bis(2-chloroethyl)methylamine, chloramin, di(2-chloroethyl)methylamine, 2,2′-dichlorodiethyl-*N*-methylamine, 2,2′-dichloro-*N*-methyldiethylamine, *N*-methylbis(2-chloroethyl)amine, *N*-methyl-2,2′-dichlorodiethylamine, methyldi(2-chloroethyl)amine, *N*-methyl-lost, mustine, nitrogen mustard, 2-chloro-*N*-(2-chloroethyl)-*N*-methylethanamine, chlormethine, MBA, HN2, Caryolysine, Cloramin, Dichloren, Embichen, Embikhine, Erasol, Mustargen hydrochloride, nitrogranulogen, Dichlor Amine, ENT 25,294, Mutagen, NSC-762, TL 146, Antimit, Azotoyperite, C 6866, Carolysine, Chloramin, Chloramine, Chlorethamine, Chlorethazine, Chlormethinum, Dema, Dimitan, Embechine, Erasol, Erasol-Ido, Kloramin, N-Lost, Mebichloramine, Merchlorethanamine, Mitoxine, NCI-C56382, Nitol, Nitol "Takeda", Pliva, stickstofflost, and Zagreb. The compound is generally supplied as the hydrochloride.

2. Other names for this compound are 4-[bis(2-chloroethyl)amino]phenylalanine, *p*-di(2-chloroethyl)aminophenylalanine, phenylalanine mustard, sarcolysine, L-PAM, melfalan, Alkeran, Sarcoclorin, alanine nitrogen mustard, medphalan, merphalan, phenylalanine nitrogen mustard, NSC-8806, CB 3025, AT 2901, 3-(*p*-s(2-chloroethyl)amino]-phenyl)alanine, NCI-C04853, and SK 15673 as well as numerous variants based on the D, L, or DL forms of the phenylalanine.

3. Other names for this compound are 4-[bis(2-chloroethyl)amino]benzenebutanoic acid, γ-[*p*-bis(2-chloroethyl)aminophenyl]butyric acid, 4-(*p*-[bis(2-chloroethyl)amino]phenyl)-butyric acid, chloraminophene, chloroambucil, chlorobutine, *N,N*-di-2-chloroethyl-γ-*p*-aminophenylbutyric acid, *p*-(*N,N*-di-2-chloroethyl)aminophenylbutyric acid, γ-[*p*-di(2-chloroethyl)aminophenyl]butyric acid, Amboclorin, Leukeran, phenylbutyric acid nitrogen mustard, CB 1348, Ambochlorin, chloraminophen, chlorobutin, Ecloril, Elcoril, Leukersan, Leukoran, Linfolizin, Linfolysin, NCI-C03485, and NSC-3088.

4. Other names for this compound are bis(2-chloroethyl)phosphoramide cyclic propanolamide ester, bis(2-chloroethyl)phosphamide cyclic propanolamide ester, cyclophosphoramide, 1-bis(2-chloroethyl)amino-1-oxo-2-aza-5-oxaphosphoridine, 2-[bis(2-chloroethyl)-amino]-2*H*-1,3,2-oxazaphosphorine 2-oxide, 2-[bis(2-chloroethyl)amino]tetrahydro-(2*H*)-1,3,2-oxazaphosphorine 2-oxide, *N,N*-bis(2-chloroethyl)-*N*'-(3-hydroxypropyl)phosphorodiamidic acid intramol ester, *N,N*-bis(2-chloroethyl)-*N*',*O*-propylenephosphoric acid ester diamide, *N,N*-bis(2-chloroethyl)tetrahydro-2*H*-1,3,2-oxazaphosphorin-2-amine 2-oxide (Chemical Abstracts name), *N,N*-bis(2-chloroethyl)-*N*',*O*-trimethylenephosphoric acid ester diamide, *N,N*-di(2-chloroethyl)amino-*N,O*-propylenephosphoric acid ester diamide, cyclophosphane, cytophosphane, Cytoxan, Endoxan, Procytox, Sendoxan, CB-4564, Clafen, cyclophosphamidum, cyclophosphan, cyclophosphanum, cytophosphan, Endoxana, Endoxan-Asta, Endoxan R, Enduxan, Genoxal, Mitoxan, NSC-26271, Semdoxan, Senduxan, Asta, Asta B518, B 518, Claphene, CP, CPA, CTX, CY, Cyclostin, Endoxanal, Hexadrin, NCI-C04900, Neosar, 2-*H*-1,3,2-oxazaphosphorinane, and SK 20501. The compound is generally supplied as the monohydrate.

5. Other names for this compound are *N*,3-bis(2-chloroethyl)tetrahydro-2*H*-1,3,2-oxaphosphorin-2-amine 2-oxide, 3-(2-chloroethyl)-2-[(2-chloroethyl)amino]tetrahydro-2*H*-1,3,2-oxaphosphorin-2-oxide, isophosphamide, iphosphamide, isoendoxan, Cyfos, Holoxan, Mitoxana, Naxamide, A 4942, Asta Z 4942, MJF 9325, NSC-109724, Z 4942, ifosfamid, isofosfamide, and NCI-C01638.

6. Other names for this compound are chlorethaminacil, aminouracil mustard, 5-[bis(2-chloroethyl)amino]-2,4(1*H*,3*H*)-pyrimidinedione, 5-[bis(2-chloroethyl)amino]uracil, 5-[di(2-chloroethyl)amino]uracil, 2,6-dihydroxy-5-bis(2-chloroethyl)aminopyrimidine, demethyldopan, desmethyldopan, uramustine, aminouracil mustard, CB-4835, chlorethaminacil, 2,6-dihydroxy-5-bis(2-chloroethyl)aminopyramidine, ENT 50,439, NCI-C04820, Nordopan, NSC-34462, SK 19849, U-8344, Uracillost, Uracilmostaza, and Uramustin.

7. Other names for this compound are 3-(2-[bis(2-chloroethyl)amino]ethyl)-1,3-diazaspiro[4,5]decane-2,4-dione, spiromustine, 3-(2-[bis(2-(chloroethyl)amino]ethyl)-5,5-pentamethylenehydantoin, NSC-172112, and SHM.

8. Lunn, G. Unpublished observations.

9. International Agency for Research on Cancer. *IARC Monographs on the Evaluation of the Carcinogenic Risk of Chemicals to Man.* Volume 9, *Some Aziridines,* N, S- *and* O-*Mustards and Selenium*; International Agency for Research on Cancer: Lyon, 1975; pp. 193–207.
10. International Agency for Research on Cancer. *IARC Monographs on the Evaluation of the Carcinogenic Risk of Chemicals to Humans, Supplement No. 4, Chemicals, Industrial Processes and Industries Associated with Cancer in Humans. IARC Monographs, Volumes 1 to 29*; International Agency for Research on Cancer: Lyon, 1982; pp. 170–172.
11. International Agency for Research on Cancer. *IARC Monographs on the Evaluation of the Carcinogenic Risk of Chemicals to Humans, Supplement No. 7, Overall Evaluations of Carcinogenicity: An Updating of* IARC Monographs *Volumes 1 to 42*; International Agency for Research on Cancer: Lyon, 1987; pp. 269–271.
12. Reference 9, pp. 167–180.
13. Reference 10, pp. 154–155.
14. Reference 11, pp. 239–240.
15. Reference 9, pp. 125–134.
16. International Agency for Research on Cancer. *IARC Monographs on the Evaluation of the Carcinogenic Risk of Chemicals to Humans.* Volume 26, *Some Antineoplastic and Immunosuppressive Agents*; International Agency for Research on Cancer: Lyon, 1981; pp. 115–136.
17. Reference 11, pp. 144–145.
18. Reference 9, pp. 135–156.
19. Reference 16, pp. 165–202.
20. Reference 10, pp. 99–100.
21. Reference 11, pp. 182–184.
22. Reference 16, pp. 237–247.
23. Reference 11, pp. 370–371.
24. Reference 9, pp. 235–241.
25. Lunn, G.; Sansone, E.B.; Andrews, A.W.; Hellwig, L.C. Degradation and disposal of some antineoplastic drugs. *J. Pharm. Sci.* **1989**, *78*, 652–659.
26. *Physician's Desk Reference*, 47th ed.; Medical Economics Co.:Oradell, NJ, 1993; p. 1568.
27. Castegnaro, M.; Adams, J.; Armour, M-. A.; Barek, J.; Benvenuto, J.; Confalonieri, C.; Goff, U.; Ludeman, S.; Reed, D.; Sansone, E. B.; Telling, G.; Eds. *Laboratory Decontamination and Destruction of Carcinogens in Laboratory Wastes: Some Antineoplastic Agents*; International Agency for Research on Cancer: Lyon, 1985 (IARC Scientific Publications No. 73).
28. Monteith, D.K.; Connor, T.H.; Benvenuto, J.A.; Fairchild, E.J.; Theiss, J.C. Stability and inactivation of mutagenic drugs and their metabolites in the urine of patients administered antineoplastic therapy. *Environ. Mol. Mutagenesis* **1987**, *10*, 341–356.
29. Benvenuto, J.A.; Connor, T.H.; Monteith, D.K.; Laidlaw, J.L.; Adams, S.C.; Matney,

T.S.; Theiss, J.C. Degradation and inactivation of antitumor drugs. *J. Pharm. Sci.* **1993**, *82*, 988–991.

30. Barek, J.; Castegnaro, M.; Malaveille, C.; Brouet, I.; Zima, J. A method for the efficient degradation of melphalan into nonmutagenic products. *Microchem. J.* **1987**, *36*, 192–197.

31. Lunn, G.; Sansone, E.B.; De Méo, M.; Laget, M.; Castegnaro, M. Potassium permanganate can be used for degrading hazardous compounds. *Am. Ind. Hyg. Assoc. J.* **1994**, *55*, 167–171.

AROMATIC AMINES

CAUTION! Refer to safety considerations section on page 7 before starting any of these procedures.

Aromatic amines constitute a group of widely used synthetic organic chemicals. Many have been shown to be carcinogens in experimental animals and a number are thought to be human carcinogens. 4-Aminobiphenyl [92-67-1],[1,2] benzidine [92-87-5],[3–5] and 2-naphthylamine [91-59-8][6,7] are human and animal carcinogens and 3,3′-dichlorobenzidine [91-94-1],[8–10] 3,3′-dimethoxybenzidine [119-90-4],[11,12] di-(4-amino-3-chlorophenyl)methane [101-14-4],[13] 3,3′-dimethylbenzidine [119-93-7],[14] 2,4-diaminotoluene [95-80-7],[15] and 2-aminoanthracene [613-13-8][16,17] cause cancer in laboratory animals. Diaminobenzidine [91-95-2] may cause cancer in experimental animals.[18] Benzidine may cause damage to the blood.[19]

In a collaborative study organized by the International Agency for Research on Cancer (IARC) on the laboratory destruction of aromatic amines[20] the following aromatic amines were considered: 4-aminobiphenyl (4-ABP),[21] benzidine (Bz; **I**, R = H),[22] 3,3′-dichlorobenzidine (DClB; **I**, R = Cl),[23] 3,3′-dimethoxybenzidine (DMoB; **I**, R = OCH_3),[24] 3,3′-dimethylbenzidine (DMB; **I**, R = CH_3),[25] di(4-amino-3-chlorophenyl)methane (MOCA),[26] 1-naphthylamine (1-NAP) [134-32-7],[27] 2-naphthylamine (2-NAP),[28] and 2,4-diaminotoluene (TOL).[29] Procedures for the destruction of diaminobenzidine[30] (DAB; **I**, R = NH_2)[31,32] and 2-aminoanthracene[33] (2-AA, **II**)[34] have also been published.

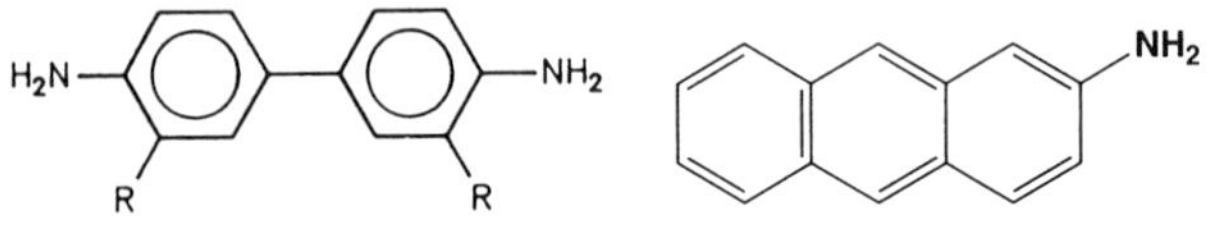

H_2N — — NH_2 R R NH_2

I **II**

All of these compounds are crystalline solids and are generally very sparingly soluble in cold H_2O, more soluble in hot H_2O, and very soluble in acid and in organic solvents. Some of these compounds are used in the chemical industry. Aromatic amines are also used in chemical laboratories and in biomedical research, for example, as stains and analytical reagents. *Note:* Unless otherwise stated all of these procedures can be used for all of the amines mentioned above except for 2-aminoanthracene.

Principles of Destruction

Aromatic amines may be oxidized with potassium permanganate in sulfuric acid ($KMnO_4$ in H_2SO_4).[20,32] The products of this reaction have not been identified. Aromatic amines may also be removed from solution using horseradish peroxidase in the presence of hydrogen peroxide (H_2O_2).[20,32,35] The enzyme catalyzes the oxidation of the aromatic amine to a radical. These radicals diffuse into solution and polymerize. The polymers are insoluble and fall out of solution. Although the resulting solution is nonmutagenic, the precipitated polymer is mutagenic so this method was only recommended for the treatment of large quantities of aqueous solution containing small amounts of aromatic amines.[20] In all cases destruction was greater than 99%. Recently, we showed that the above procedures may also be applied to diaminobenzidine.[32] We also found that residual amounts of H_2O_2 in the horseradish peroxidase procedure produced a mutagenic response and that the mutagenicity could be removed by adding ascorbic acid solution (to reduce the H_2O_2) to the final reaction mixtures.[32] Various procedures involving diazotization followed by decomposition of the diazo compound have been investigated but the results seem to depend on the nature of the aromatic amine.[20] These procedures are not discussed here.

Destruction Procedures

Destruction of Aromatic Amines in Bulk and in Organic Solvents[20,32]

Evaporate solutions of aromatic amines in organic solvents to dryness using a rotary evaporator. Dissolve the aromatic amines as follows: For each 9 mg of diaminobenzidine tetrahydrochloride dihydrate dissolve in 10 mL of H_2O, for each 9 mg of Bz, DAB, DClB, DMB, DMoB, 1-NAP, 2-NAP, and TOL dissolve in 10 mL of 0.1 *M* HCl; for each 2.5 mg of MOCA dissolve in 10 mL of 1 *M* H_2SO_4; for each 2 mg of 4-ABP dissolve in 10 mL of glacial acetic acid; for each 2 mg of mixtures of the above amines add 10 mL of glacial acetic acid. Stir these solutions until the aromatic amines have completely dissolved, then for each 10 mL of the solution so formed add 5 mL of 0.2 *M* $KMnO_4$ solution and 5 mL of 2 *M* H_2SO_4. Allow the mixture to stand for at least 10 h, then analyze for completeness of destruction. Decolorize the mixture by the addition of sodium metabisulfite, make strongly basic by the addition of 10 *M* KOH solution (**Caution!** Exothermic), dilute with H_2O, filter to remove manganese compounds,[36] check the filtrate for completeness of destruction, neutralize, and discard it.

Destruction of Aromatic Amines in Aqueous Solution[20,32]

Dilute with H_2O, if necessary, so that the concentration of MOCA does not exceed 0.25 mg/mL, the concentration of 4-ABP does not exceed 0.2 mg/mL, and the concentration of the other amines does not exceed 0.9 mg/mL. For each 10 mL of solution add 5 mL of 0.2 *M* $KMnO_4$ solution and 5 mL of 2 *M* H_2SO_4 solution. Allow the mixture to stand for at least 10 h, then analyze for completeness of destruction. Decolorize the mixture by the addition of sodium metabisulfite, make strongly basic by the addition of 10 *M* KOH solution (**Caution!** Exothermic), dilute with H_2O, filter to remove manganese compounds,[36] check the filtrate for completeness of destruction, neutralize, and discard it.

Destruction of Aromatic Amines in Oil[20]

Extract the oil solution with 0.1 *M* HCl until all the amines are removed (at least 2 mL of HCl will be required for each micromole of amine). For each 10 mL of HCl solution add 5 mL of 0.2 *M* $KMnO_4$ solution and 5 mL of 2 *M* H_2SO_4 solution. Allow the mixture to stand for at least 10 h, then analyze for completeness of destruction. Decolorize the mixture by

the addition of sodium metabisulfite, make strongly basic by the addition of 10 *M* KOH solution (**Caution!** Exothermic), dilute with H_2O, filter to remove manganese compounds,[36] check the filtrate for completeness of destruction, neutralize, and discard it.

Destruction of 2-Aminoanthracene[34]

Evaporate solutions in organic solvents to dryness using a rotary evaporator. Prepare a 0.3 *M* solution of $KMnO_4$ in 3 *M* H_2SO_4 by stirring $KMnO_4$ in 3 *M* H_2SO_4 (47.4 g $KMnO_4$ for each liter of acid) for at least 15 min but no longer than 1 h. Dissolve the 2-aminoanthracene in glacial acetic acid so that the concentration does not exceed 10 mg/mL. Stir until the 2-aminoanthracene has completely dissolved, then for each 1 mL of the solution so formed add 40 mL of 0.3 *M* $KMnO_4$ in 3 *M* H_2SO_4 and stir the mixture for at least 18 h. Decolorize the mixture by the addition of sodium metabisulfite, make strongly basic by the addition of 10 *M* KOH solution (**Caution!** Exothermic), dilute with H_2O, filter to remove manganese compounds,[36] check the filtrate for completeness of destruction, neutralize, and discard it.

Decontamination of Spills[20]

Remove as much of the spill as possible by the use of absorbents and high efficiency particulate air (HEPA) vacuuming, then wet the surface with glacial acetic acid until all the amines are dissolved. Add an excess of a mixture of equal volumes of 0.2 *M* $KMnO_4$ solution and 2 *M* H_2SO_4 to the spill area. Allow the mixture to stand for at least 10 h, decolorize with sodium metabisulfite (while ventilating the area), and mop up the liquid with paper towels. Squeeze the solution out of the towels, basify, dilute with H_2O, filter to remove manganese compounds,[36] and test the filtrate for completeness of destruction. Test for completeness of decontamination by wiping the surface with a wipe moistened with an appropriate solvent.

Decontamination of Glassware[20]

Immerse the glassware in a mixture of equal volumes of 0.2 *M* $KMnO_4$ and 2 *M* H_2SO_4. Allow the glassware to stand in the bath for at least 10 h, then decolorize the mixture by the addition of sodium metabisulfite, make strongly basic by the addition of 10 *M* KOH solution (**Caution!** Exothermic), dilute with H_2O, filter to remove manganese compounds,[36] check the filtrate for completeness of destruction, neutralize, and discard it.

Decontamination of Large Quantities of Solutions Containing Aromatic Amines[20,32]

Note: This method is recommended for all the amines listed above except 4-ABP and 2-NAP where destruction was found to be incomplete and 2-aminoanthracene (not tested).

Adjust the pH of aqueous solutions to 5–7 using acid or base as appropriate and dilute so that the concentration of aromatic amines does not exceed 100 mg/L. Dilute solutions in methanol, ethanol, dimethyl sulfoxide (DMSO), or *N,N*-dimethylformamide (DMF) with sodium acetate solution (1 g/L) so that the concentration of organic solvent does not exceed 20% and the concentration of aromatic amines does not exceed 100 mg/L. For each liter of solution add 3 mL of a 3% solution of H_2O_2 and 300 units of horseradish peroxidase. Allow the mixture to stand for 3 h, then remove the precipitate by filtration or centrifugation. For each liter of filtrate add 100 mL of a 5% (w/v) ascorbic acid solution. Check the solution for completeness of degradation and discard. The residue is mutagenic and should be treated as hazardous. It has been reported that further oxidation of this residue by $KMnO_4$ in H_2SO_4 produces nonmutagenic residues.[20] Filter the reaction mixture through a porous glass filter and immerse this filter in a 1:1 mixture of 0.2 *M* aqueous $KMnO_4$ solution and 2 *M* H_2SO_4 solution.[32,37]

The horseradish peroxidase used was Donor: hydrogen peroxide oxidoreductase; EC 1.11.1.7 (Type II) having a specific activity of 150–200 purpurogallin units per milligram obtained from Sigma. An appropriate amount was dissolved in sodium acetate solution (1 g/L), then an aliquot of this solution was used to obtain the requisite number of units.

Analytical Procedures

There are many publications on the analysis of aromatic amines. The following HPLC analysis was recommended by the IARC.[20] A 250 × 4.6-mm i.d. reverse phase column was used and the mobile phase was acetonitrile : methanol : buffer (10:30:20) flowing at 1.5 mL/min. The buffer was 1.5 m*M* in potassium phosphate, dibasic (K_2HPO_4) and 1.5 m*M* in potassium phosphate, monobasic (KH_2PO_4). If a variable wavelength UV detector is available the following wavelengths can be used: Bz = 285 nm, DMoB = 305 nm, DMB = 285 nm, DClB = 285 nm, 1-NAP = 240 nm, 2-NAP = 235 nm, 4-ABP = 275 nm, MOCA = 245 nm, and TOL = 235 nm. If only a fixed wavelength detector is available, use 280 nm for the first four; 254 nm for the rest. Diaminobenzidine can be determined using the above buffer : methanol (75:25) flowing at 1 mL/min with the UV detector set at 300 nm.[32] 2-Aminoanthracene was determined using methanol:water 70:30 flowing at 1 mL/min with the UV detector set at 254 nm.[34] Greater sensitivity can be obtained by extracting the basified reaction mixtures with cyclohexane, drying these extracts over anhydrous sodium sulfate, evaporating to dryness, and taking up the residue in a little methanol.

Mutagenicity Assays[20,32]

Reaction mixtures obtained when the aromatic amines (except for diaminobenzidine) were degraded with $KMnO_4$ in H_2SO_4 were tested for mutagenicity using *Salmonella typhimurium* strains TA97, TA98, and TA100. No mutagenic activity was seen. Using the same strains, the supernatants from the horseradish peroxidase–H_2O_2 reactions were tested. Again, no mutagenic activity was seen but the solid residues from Bz, DClB, DMoB, and 2-NAP were mutagenic. Reaction mixtures obtained from the degradation of diaminobenzidine and 2-aminoanthracene were tested using strains TA98, TA100, TA1530, and TA1535 and no mutagenic activity was found.

Related Compounds

The procedures described above should be generally applicable to other aromatic amines but thorough validation is essential in each case as some variability may be observed, particularly for the horseradish peroxidase method.[20]

References

1. International Agency for Research on Cancer. *IARC Monographs on the Evaluation of the Carcinogenic Risk of Chemicals to Humans, Supplement No. 7, Overall Evaluations of Carcinogenicity: An Updating of* IARC Monographs *Volumes 1 to 42*; International Agency for Research on Cancer: Lyon, 1987; pp. 91–92.
2. International Agency for Research on Cancer. *IARC Monographs on the Evaluation of*

Carcinogenic Risk of Chemicals to Man. Volume 1; International Agency for Research on Cancer: Lyon, 1971; pp. 74–79.

3. Reference 1, pp. 123–125.
4. Reference 2, pp. 80–86.
5. International Agency for Research on Cancer. *IARC Monographs on the Evaluation of the Carcinogenic Risk of Chemicals to Humans.* Volume 29, *Some Industrial Chemicals and Dyestuffs*; International Agency for Research on Cancer: Lyon, 1982; pp. 149–183.
6. Reference 1, pp. 261–263.
7. International Agency for Research on Cancer. *IARC Monographs on the Evaluation of Carcinogenic Risk of Chemicals to Man.* Volume 4, *Some Aromatic Amines, Hydrazine and Related Substances,* N-*Nitroso Compounds and Miscellaneous Alkylating Agents*; International Agency for Research on Cancer: Lyon, 1974; pp. 97–111.
8. Reference 1, pp. 193–194.
9. Reference 7, pp. 49–55.
10. Reference 5, pp. 239–256.
11. Reference 1, pp. 198–199.
12. Reference 7, pp. 41–47.
13. Reference 1, pp. 246–247.
14. Reference 2, pp. 87–91.
15. International Agency for Research on Cancer. *IARC Monographs on the Evaluation of Carcinogenic Risk of Chemicals to Man.* Volume 16, *Some Aromatic Amines and Related Nitro Compounds—Hair Dyes, Colouring Agents and Miscellaneous Industrial Chemicals*; International Agency for Research on Cancer: Lyon, 1978; pp. 83–95.
16. Griswold, Jr., D.P.; Casey, A.E.; Weisburger, E.K.; Weisburger, J.H.; Schabel, F.M. On the carcinogenicity of a single intragastric dose of hydrocarbons, nitrosamines, aromatic amines, dyes, coumarins, and miscellaneous chemicals in female Sprague–Dawley rats. *Cancer Res.* **1966**, *26*, 619–625.
17. Griswold, Jr., D.P.; Casey, A.E.; Weisburger, E.K.; Weisburger, J.H. The carcinogenicity of multiple intragastric doses of aromatic and heterocyclic nitro or amino derivatives in young female Sprague–Dawley rats. *Cancer Res.* **1968**, *28*, 924–933.
18. Weisburger, E.K.; Russfield, A.B.; Homburger, F.; Weisburger, J.H.; Boger, E.; Van Dongen, C.G.; Chu, K.C. Testing of twenty-one environmental aromatic amines or derivatives for long-term toxicity or carcinogenicity. *J. Environ. Pathol. Toxicol.* **1978**, *2*, 325–356.
19. Lewis, R.J., Sr. *Sax's Dangerous Properties of Industrial Materials*, 8th ed.; Van Nostrand-Reinhold: New York, 1992; p. 367.
20. Castegnaro, M.; Barek, J.; Dennis, J.; Ellen, G.; Klibanov, M.; Lafontaine, M.; Mitchum, R.; van Roosmalen, P.; Sansone, E.B.; Sternson, L.A.; Vahl, M., Eds., *Laboratory Decontamination and Destruction of Carcinogens in Laboratory Wastes: Some Aromatic Amines and 4-Nitrobiphenyl*; International Agency for Research on Cancer: Lyon, 1985 (IARC Scientific Publications No. 64).
21. Other names for this compound are (1,1′-biphenyl)-4-amine, 4-aminodiphenyl, anilino-

benzene, xenylamine, 4-biphenylamine, *p*-phenylaniline, paraaminodiphenyl, and *p*-phenylaniline.

22. Other names for this compound are (1,1′-biphenyl)-4,4′-diamine, 4,4′-biphenyldiamine, 4,4′-diaminobiphenyl, 4,4′-diaminodiphenyl, *p,p*′-dianiline, *p,p*′-bianiline, 4,4′-biphenylenediamine, C.I. 37225, C.I. azoic dye component 112, *p,p*′-dianiline, 4,4′-diphenylenediamine, Fast Corinth Base B, and NCI-C03361.
23. Other names for this compound are 4,4′-diamino-3,3′-dichlorobiphenyl, 3,3′-dichloro-4,4′-biphenyldiamine, 3,3′-dichlorobiphenyl-4,4′-diamine, DCB, 3,3′-dichloro-4,4′-diaminobiphenyl, C.I. 23060, Curithane C126, and 3,3′-dichlorobenzidene.
24. Other names for this compound are 3,3′-dimethoxy(1,1′-biphenyl)-4,4′-diamine, 3,3′-dimethoxy-4,4′-diaminobiphenyl, *o*-dianisidine *o,o*′-dianisidine, Acetamine Diazo Black RD, Amacel Developed Navy SD, Azoene Fast Blue Base, Azofix Blue B Salt, Azogne Fast Blue B, Blue BN Balse, Brentamine Fast Blue Base, Cellitazol B, C.I. 24110, C.I. Azoic Diazo Component 48, Cibacete Diazo Navy Blue 2B, C.I. Disperse Black 6, Diacelliton Fast Grey G, Diacel Navy DC, Diato Blue Base B, Fast Blue B Base, Hiltonil Fast Blue B Base, Hiltosal Fast Blue B Salt, Hindasol Blue B Salt, Kako Blue B Salt, Kayaku Blue B Base, Lake Blue B Base, Meisei Teryl Diazo Blue HR, Mitsui Blue B Base, Napthanil Blue B Base, Neutrosel Navy BN, Sanyo Fast Blue Salt B, Setacyl Diazo Navy R, and Spectrolene Blue B.
25. Other names for this compound are 3,3′-dimethyl(1,1′-biphenyl)-4,4′-diamine, 3,3′-tolidine, bianisidine, 4,4′-bi-*o*-toluidine, 4,4′-diamino-3,3′-dimethylbiphenyl, 4,4′-diamino-3,3′-dimethyldiphenyl, 3,3′-dimethyl-4,4′-biphenyldiamine, 3,3′-dimethyl-4,4′-diphenyldiamine, 3,3′-dimethylbiphenyl-4,4′-diamine, 3,3′-dimethyldiphenyl-4,4′-diamine, *o*-tolidine, *o,o*′-tolidine, C.I. 37230, C.I. Azoic Diazo Component 113, diaminoditolyl, 4,4′-di-*o*-toluidine, Fast Dark Blue Base R, and *o*-tolidin.
26. Other names for this compound are 4,4′-methylenebis(2-chlorobenzenamine), 4,4′-diamino-3,3′-dichlorodiphenylmethane, 4,4′-methylene(bis)chloroaniline, methylene-4,4′-bis(*o*-chloroaniline), *p,p*′-methylenebis(α-chloroaniline), 4,4′-methylenebis(*o*-chloroaniline), *p,p*′-methylenebis(*o*-chloroaniline), 4,4′-methylenebis(2-chloroaniline), DACPM, Bis amine, Curalin M, Curene 442, Cyanaset, 3,3′-dichloro-4,4′-diaminodiphenylmethane, and MBOCA.
27. Other names for this compound are 1-naphthalenamine, α-naphthylamine, 1-aminonaphthalene, naphthalidine, and C.I. Azoic Dye Component 114
28. Other names for this compound are 2-naphthalenamine, β-naphthylamine, 2-aminonaphthalene, 6-naphthylamine, C.I. 37270, Fast Scarlet Base B, NA, 2-naphthalamine, 2-naphthalenamine, and USAF CB-22.
29. Other names for this compound are toluene-2,4-diamine, *m*-tolylenediamine, 3-amino-*p*-toluidine, 5-amino-*o*-toluidine, 1,3-diamino-4-methylbenzene, 2,4-diamino-1-methylbenzene, 4-methyl-1,3-benzenediamine, 4-methyl-*m*-phenylenediamine, 2,4-tolamine, *m*-toluenediamine, 2,4-toluenediamine, *m*-toluylenediamine, 2,4-toluylenediamine, 2,4-tolylenediamine, tolylene-2,4-diamine, Azogen Developer H, Benzofur MT, C.I. 76035, C.I. Oxidation Base, Developer H, Eucanine GB, Fouramine, Fourrine M, MTD, Nako TMT, NCI-C02302, Pelagol Grey J, Pontamine Developer TN, Renal MD, TDA, Zoba GKE, and Zogen Developer H.

30. Other names for this compound are 3,3′,4,4′-biphenyltetramine, 3,3′,4,4′-diphenyltetramine, 3,3′,4,4′-tetraaminobiphenyl, and 3,3′-diaminobenzidine.

31. Barek, J. Destruction of carcinogens in laboratory wastes. II. Destruction of 3,3′-dichlorobenzidine, 3,3′-diaminobenzidine, 1-naphthylamine, 2-naphthylamine, and 2,4-diaminotoluene by permanganate. *Microchem. J.* **1986**, *33*, 97–101.

32. Lunn, G.; Sansone, E.B. The safe disposal of diaminobenzidine. *Appl. Occup. Environ. Hyg.* **1991**, *6*, 49–53.

33. Other names for this compound are 2-anthracenamine, β-aminoanthracene, 2-anthracylamine, 2-anthramine, and 2-anthrylamine.

34. Lunn, G.; Sansone, E.B. Destruction of azo and azoxy compounds and 2-aminoanthracene. *Appl. Occup. Environ. Hyg.* **1991**, *6*, 1020–1026.

35. Klibanov, A.M.; Morris, E.D. Horseradish peroxidase for the removal of carcinogenic aromatic amines from water. *Enzyme Microb. Technol.* **1981**, *3*, 119–122.

36. Lunn, G.; Sansone, E.B.; De Méo, M.; Laget, M.; Castegnaro, M. Potassium permanganate can be used for degrading hazardous compounds. *Am. Ind. Hyg. Assoc. J.* **1994**, *55*, 167–171.

37. Barek, J.; Pacáková, V.; Štulík, K.; Zima, J. Monitoring of aromatic amines by HPLC with electrochemical detection. Comparison of methods for destruction of carcinogenic aromatic amines in laboratory wastes. *Talanta* **1985**, *32*, 279–283.

AZIDES

CAUTION! Refer to safety considerations section on page 7 before starting any of these procedures.

Sodium azide (Smite, Azium, Kazoe, NaN_3) [26628-22-8] can decompose explosively on heating and can form shock sensitive and highly explosive azides when it comes in contact with heavy metals.[1] For this reason solutions of NaN_3 should **never** be poured down the sink. Sodium azide is also incompatible with a number of other reagents and it is acutely toxic and may be carcinogenic.[2] A death has been reported when H_2O preserved with NaN_3 was ingested in the laboratory.[3] Sodium azide is a powerful mutagen. Treatment with acid liberates explosive, toxic volatile (bp 37°C) hydrazoic acid (HN_3).[4]

Organic azides vary greatly in stability but a number are known to decompose explosively with shock or heating. For example, phenyl azide ($C_6H_5N_3$, azidobenzene) [622-37-7] explodes when heated and when mixed with Lewis acids and benzyl azide ($C_6H_5CH_2N_3$) [622-79-7] is a heat-sensitive explosive.[5] These compounds should be handled very carefully. An explosion has been reported when diazidomethane formed in a reaction mixture containing dichloromethane and sodium azide.[6]

Table 1 Destruction of Azide in Buffer Solution with Nitrite

Solvent	Final pH	$NaNO_2$ Required (g)	Molar Ratio NO_2^-/N_3^-	NaN_3 Left (%)
1 *M* HCl	1.2	0.25	4.7	<0.02
pH 3 buffer	4.2	0.25	4.6	<0.02
pH 7 buffer	7.1	2.5	45	<0.02
H_2O	8.0	5	95	<0.02
0.1 *M* Na_2CO_3	10.3	5	95	<0.02
1 *M* KOH	12.9	5	95	<0.02

In each case 50 mL of a 1 mg/mL solution of NaN_3 was stirred with the indicated amount of sodium azide for 18 h before analysis.

Principles of Destruction

Sodium azide can be oxidized by ceric ammonium nitrate[7] to nitrogen[8] or by nitrous acid[9] to nitrous oxide.[8] Some toxic nitrogen dioxide may be produced as a by-product of these reactions so they should always be done in a chemical fume hood. Destruction was greater than 99.996%. Sodium azide in buffer solution may also be degraded by the addition of sodium nitrite.[10] The reaction proceeds much more readily at low pH but if sufficient sodium nitrite is added it will proceed to completion even at high pH (Table 1). At low pH it may be possible to completely degrade the azide present in the buffer with less than the amount of sodium nitrite indicated below. However, the reaction mixture must be carefully checked to make sure that no azide remains. At high pH it is possible for unreacted azide to remain in the presence of excess nitrite.

Organic azides (e.g., phenyl azide, PhN_3) can be reduced to the corresponding amines with tin in hydrochloric acid (HCl) or with stannous chloride in methanol.

Destruction Procedures

Sodium Azide

1. For each gram of NaN_3, stir 9 g of ceric ammonium nitrate in 30 mL of H_2O until it has dissolved. Dissolve the NaN_3 in 5 mL of H_2O and add it in 1-mL portions over 5 min. On a larger scale an ice bath may be required for cooling. Stir the reaction mixture for 1 h, then check that it is still oxidizing. Add a few drops of the reaction mixture to an equal volume of

10% (w/v) potassium iodide (KI) solution, then acidify with a drop of 1 *M* HCl and add a drop of starch solution as an indicator. The deep blue color of the starch–iodine complex indicates that excess oxidant is present. Analyze for completeness of destruction and discard the reaction mixture. If excess oxidant is not present, add more ceric ammonium nitrate.

2. Dissolve NaN_3 (5 g) in 100 mL of H_2O. Stir the reaction mixture and add 7.5 g of sodium nitrite dissolved in 38 mL of H_2O.[8] Slowly add dilute H_2SO_4 (4 *M*) until the reaction mixture is acidic to litmus. On a larger scale an ice bath may be required for cooling. Stir for 1 h. Add a few drops of the reaction mixture to an equal volume of 10% (w/v) KI solution, then acidify with a drop of 1 *M* HCl and add a drop of starch solution as an indicator. The deep blue color of the starch–iodine complex indicates that excess nitrous acid is present. If excess nitrous acid is present, analyze for completeness of destruction and discard the reaction mixture. If excess nitrous acid is not present, add more sodium nitrite.

Note: It is important to add the sodium nitrite, **then** the H_2SO_4. Adding these reagents in the reverse order will generate explosive, volatile, toxic, hydrazoic acid.

3. If necessary, dilute with water so that the concentration of sodium azide in the buffer solution does not exceed 1 mg/mL. For each 50 mL of buffer solution add 5 g of sodium nitrite, stir the reaction for 18 h, check for completeness of destruction, and discard it.

Organic Azides

1. Suspend 10.9 g of stannous chloride with stirring in 40 mL of methanol and add 4 g (0.03 mol) of benzyl azide dropwise.[11] When addition is complete stir the reaction mixture at room temperature for 30 min, then dilute with H_2O, neutralize, and discard it.

2. Slowly add 1 g (0.0084 mol) phenyl azide to a stirred mixture of 6 g of granular tin in 100 mL of concentrated HCl.[12] Stir the mixture for 30 min after addition is complete, then pour it into a large quantity of cold H_2O, neutralize, and discard it. Unreacted tin may be recycled or discarded with the solid waste.

Analytical Procedures

Analysis for Sodium Azide

Sodium azide is analyzed by reacting azide ion with 3,5-dinitrobenzoyl chloride to form 3,5-dinitrobenzoyl azide, which can be determined by high-performance liquid chromatography (HPLC).[13] Prepare the following solutions. Sulfamic acid (20% w/v in H_2O), sodium azide 100 μg/mL in H_2O), potassium hydroxide (KOH) (1 *M*), and HCl 0.2 *M*. To prepare the indicator dissolve 0.1 g bromocresol purple in 18.5 mL 0.01 *M* KOH and make up to 25 mL with H_2O. The analysis was performed by reverse phase HPLC with a mobile phase of water : acetonitrile (50:50) flowing at 1 mL/min. The ultraviolet (UV) detector was set at 254 nm and the peak for 3,5-dinitrobenzoyl azide came at about 9 min. The limit of detection of sodium azide was 0.2 μg/mL.

1. *Analysis for azide in the presence of nitrite.* Remove a 5-mL aliquot of the reaction mixture and remove excess nitrite by adding at least 1 mL of the sulfamic acid solution. More sulfamic acid solution may be required for strongly basic reaction mixtures or those containing high concentrations of nitrite. [Complete destruction of nitrite can be checked by using a modified Griess reagent (see below).] After standing for at least 3 min the analytical solution may be spiked, if desired, by adding a small quantity of NaN_3 solution (10–20 μL). Add 1 drop of indicator and basify the mixture by adding KOH solution until it turns purple (typically, 3–10 mL are required). Add 2 mL of acetonitrile (4 mL if more than 1 mL of sulfamic acid was used), then add HCl dropwise until the mixture is acidic (yellow color) followed by 1 more drop. Prepare a 10% (w/v) solution of 3,5-dinitrobenzoyl chloride in acetonitrile and add 50 μL. Shake the mixture and allow it to stand for at least 3 min before analysis by HPLC. Further standing times have no effect on the analyses. In this analytical procedure the crucial part is the use of freshly prepared 3,5-dinitrobenzoyl azide solution. This solution should be used within minutes of preparation. It is generally most convenient to prepare all the analytical samples (spiked and unspiked) with the fresh solution at the beginning of the day and then analyze them in the course of the day.

2. *Analysis for azide in the presence of ceric salts.* Remove a 10 mL aliquot of the reaction mixture and dilute with 40 mL of H_2O. Add 5 mL of this solution to 3 mL of 1 *M* KOH. (If less than 3 mL of 1 *M* KOH is used precipitation of ceric salts is not complete.) Shake and centrifuge the mixture. Remove 2 mL of the supernatant and add to 1 mL of acetonitrile. At this point the analytical solution may be spiked, if desired, by adding a small quantity of sodium azide solution (~10–20 μL). Add 1 drop of indicator and then add HCl dropwise until the mixture is acidic (yellow color) followed by 1 more drop. Prepare a 10% (w/v) solution of 3,5-dinitrobenzoyl chloride in acetonitrile and add 50 μL. Shake the mixture

and allow it to stand for at least 3 min before analysis by HPLC. Further standing times have no effect on the analyses. In this analytical procedure the crucial part is the use of freshly prepared 3,5-dinitrobenzoyl azide solution. This solution should be used within minutes of preparation. It is generally most convenient to prepare all the analytical samples (spiked and unspiked) with the fresh solution at the beginning of the day and then analyze them in the course of the day.

Analysis for Nitrite[14]

Boil 0.1 g α-naphthylamine in 20 mL of H_2O until it dissolves. Pour this solution, while hot, into 150 mL of 15% v/v aqueous acetic acid. To this solution add a solution of 0.5 g of sulfanilic acid in 150 mL of 15% v/v aqueous acetic acid. Store the reagent in a brown bottle. Add 3 mL of the mixture to be tested to 1 mL of the reagent and allow to stand at room temperature for 6 min. Measure the absorbance at 520 nm against a suitable blank. Limit of detection was 0.06 μg/mL of sodium nitrite. Note that at high pH the reaction between azide and nitrite is quite slow, and so the presence of excess nitrite does not mean that all the azide has been degraded. α-Naphthylamine may be a carcinogen.[15,16] A procedure using *N*-(1-naphthyl)ethylenediamine. 2HCl instead has been described.[17]

Related Compounds

The procedures described above for NaN_3 should not be used for heavy metal azides, many of which are shock sensitive explosives. Professional help should be sought in these cases. The procedures for organic azides should be generally applicable although in all cases the reactions should be thoroughly validated to ensure that the azides are completely destroyed. The products of these reactions are the corresponding amines, which may themselves be hazardous compounds.

References

1. Bretherick, L. *Bretherick's Handbook of Reactive Chemical Hazards*; 4th ed.; Butterworths: London, 1990; pp. 1360–1362.
2. Lewis, R.J., Sr. *Sax's Dangerous Properties of Industrial Materials*, 8th ed.; Van Nostrand-Reinhold: New York, 1992; pp. 3063–3064.
3. *CHAS Notes*, Newsletter of the American Chemical Society Division of Chemical Health and Safety, Vol. 6, No. 3, July–Sept, 1988.

4. Reference 2, p. 1890.
5. Reference 1, pp. 609 and 712.
6. Peet, N.P.; Weintraub, P.M. Explosion with sodium azide in DMSO–CH_2Cl_2. *Chem. Eng. News* **1993,** *April 19*, 4.
7. Manufacturing Chemists Association. *Laboratory Waste Disposal Manual*; Manufacturing Chemists Association: Washington, DC, 1973; p. 136.
8. Mason, K.G. Hydrogen azide. In *Mellor's Comprehensive Treatise on Inorganic and Theoretical Chemistry, Vol. VIII, Suppl. II*; Wiley: New York, 1967; pp. 1–15.
9. National Research Council, Committee on Hazardous Substances in the Laboratory. *Prudent Practices for Disposal of Chemicals from Laboratories*; National Academy Press: Washington, DC, 1983; p. 88.
10. Lunn, G. Unpublished results.
11. Maiti, S.N.; Singh, M.P.; Micetich, R.G. Facile conversion of azides to amines. *Tetrahedron Lett.* **1986**, *27*, 1423–1424.
12. Armour, M-.A.; Browne, L.M.; Weir, G.L., Eds., *Hazardous Chemicals. Information and Disposal Guide*, 3rd ed.; University of Alberta: Edmonton, Alberta, 1987; p. 291.
13. Swarin, S.J.; Waldo, R.A. Liquid chromatographic determination of azide as the 3,5-dinitrobenzoyl derivative. *J. Liquid Chromatogr.* **1982**, *5*, 597–604.
14. Horwitz, W., Ed., *Official Methods of Analysis of the Association of Official Analytical Chemists*; 12th ed.; Association of Official Analytical Chemists: Washington, D.C., 1975; p. 422.
15. International Agency for Research on Cancer. *IARC Monographs on the Evaluation of the Carcinogenic Risk of Chemicals to Man:* Volume 4, *Some Aromatic Amines, Hydrazine and Related Substances,* N-*Nitroso Compounds and Miscellaneous Alkylating Agents*; International Agency for Research on Cancer: Lyon, 1973; pp. 87–96.
16. International Agency for Research on Cancer. *IARC Monographs on the Evaluation of the Carcinogenic Risk of Chemicals to Humans, Supplement No. 7, Overall Evaluations of Carcinogenicity: An Updating of IARC Monographs Volumes 1 to 42*; International Agency for Research on Cancer: Lyon, 1987; pp. 260–261.
17. Helrich, K., Ed., *Official Methods of Analysis of the Association of Official Analytical Chemists*; 15th ed.; Association of Official Analytical Chemists: Arlington, VA, 1990; p. 938.

AZO AND AZOXY COMPOUNDS AND TETRAZENES

> **CAUTION!** Refer to safety considerations section on page 7 before starting any of these procedures.

Azobenzene (**I**, R = R′ = H, mp 68°C) [103-33-3],[1] azoxybenzene (**II**, R = R′ = H, mp 36°C) [495-48-7],[2] *N,N*-dimethyl-4-amino-4′-hydroxyazobenzene [**I**, R = $N(CH_3)_2$, R′ = OH, mp 201–202°C] [2496-15-3], azoxyanisole (**II**, R = R′ = OCH_3, 4,4′-dimethoxyazoxybenzene, mp 118°C) [13620-57-0], phenylazophenol (**I**, R = OH, R′ = H, mp 150–152°C) [1689-82-3],[3] phenylazoaniline (**I**, R = NH_2, R′ = H, mp 123–126°C) [60-09-3],[4] and fast garnet (**III**, mp 101–102°C) [97-56-3][5] are solids and azoxymethane (AOM, $CH_3—N{=}N(O)—CH_3$, bp 98°C) [25843-45-2] and tetramethyltetrazene [$(CH_3)_2N—N{=}N—N(CH_3)_2$, bp 130°C] [6130-87-6] are liquids. Generally, these compounds are only slightly soluble in H_2O (5–50 ppm), but they are soluble in alcohols and organic solvents. Azoxymethane,[6] phenylazoaniline,[7] and fast garnet[8] are carcinogenic in animals and azobenzene may be carcinogenic in animals.[9,10] Azoxybenzene is toxic by ingestion and is a skin and eye irritant;[11] tetramethyltetrazene explodes when heated to its boiling point.[12] The mutagenicity of azo dyes has been reviewed.[13] Azoxymethane is used in cancer research and the other compounds are used in organic synthesis and as intermediates in the chemical industry.

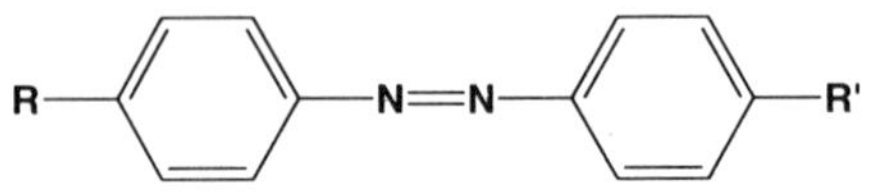

I

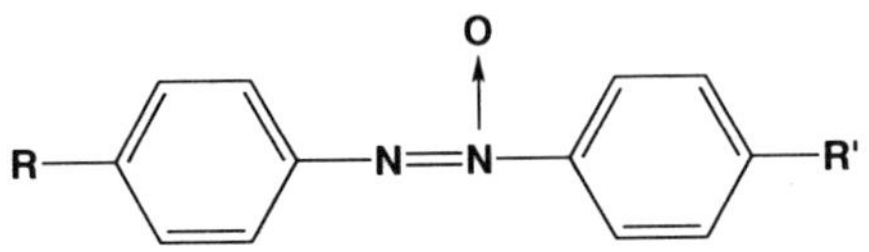

II

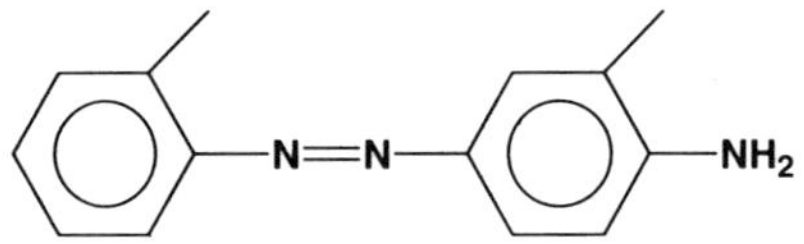

III

Principles of Destruction

All of the compounds except azoxybenzene (destruction was incomplete) and azoxymethane and tetramethyltetrazene (not tested) can be oxidized by potassium permanganate in sulfuric acid ($KMnO_4$ in H_2SO_4). Destruction is greater than 99.5%; the products have not been identified.[14,15] The other compounds can be reduced to their parent amines by using nickel–aluminum (Ni–Al) alloy in potassium hydroxide (KOH) solution.[16] Destruction is greater than 99% for tetramethyltetrazene and greater than 99.5% for the other compounds.[15] When performing these reactions it should be noted that in some cases the products are aromatic amines, which may also be hazardous. For example, the product of the reduction of azobenzene and azoxybenzene is aniline for which limited evidence of carcinogenicity in animals exists.[17] In fact this procedure is not recommended for phenylazoaniline or fast garnet because the final reaction mixtures and *p*-phenylenediamine (from phenylazoaniline) and 2,5-diaminotoluene (from fast garnet) are mutagenic.

Destruction Procedures

Destruction of Azobenzene, Azoxybenzene, Azoxyanisole, Phenylazophenol, Azoxymethane, and Tetramethyltetrazene

Dissolve the compound in H_2O (azoxymethane and tetramethyltetrazene) or methanol (others) so that the concentration does not exceed 5 mg/mL and add an equal volume of 2 *M* KOH solution. For each 100 mL of basified solution add 5 g of Ni–Al alloy. Add quantities of more than 5 g in portions over approximately 1 h to avoid excessive frothing. Stir the reaction mixture overnight, then filter through a pad of Celite®. Neutralize the filtrate, test for completeness of destruction, and discard it. Place the nickel that is filtered off on a metal tray away from flammable solvents for 24 h, then discard it with the solid waste.

Table 1 Destruction of Azo and Azoxy Compounds Using Nickel–Aluminum Alloy in Potassium Hydroxide Solution

Compound	% Remaining	Recovery	Mutagenicity of Reaction Mixture[a]	Products (%)
Azobenzene	<0.26	101	–	Aniline (84)
Azoxybenzene	<0.12	92	–	Aniline (95)
Azoxyanisole	<0.22	92	–	*p*-Anisidine (69)
Phenylazophenol	<0.38	67	–	Aniline (99); *p*-Aminophenol (34)
Phenylazoaniline	<0.14	97	+	Aniline (93); *p*-Phenylenediamine (81)
Fast Garnet	<0.05	104	+	*o*-Toluidine (92); 2,5-Diaminotoluene (98)
Azoxymethane	<0.5		nt	Methylamine (100)
Tetramethyltetrazene	<1		nt	Dimethylamine (94)

[a] The – = nonmutagenic; + = mutagenic; nt = not tested.

Destruction of Azobenzene, Azoxyanisole, Phenylazophenol, Phenylazoaniline, and Fast Garnet

Prepare a 0.3 *M* solution of $KMnO_4$ in 3 *M* H_2SO_4 by stirring 47.4 g of $KMnO_4$ per liter of 3 *M* H_2SO_4 for at least 15 min but no more than 1 h. Take up the compound in glacial acetic acid so that the concentration does not exceed 10 mg/mL (5 mg/mL for phenylazoaniline) and add 40 mL (80 mL for phenylazoaniline) of the $KMnO_4/H_2SO_4$ mixture for each 1 mL of this solution. After 18 h decolorize with sodium metabisulfite, make strongly basic with 10 *M* KOH solution (**Caution!** Exothermic), dilute with H_2O, filter to remove manganese compounds,[18] test the filtrate for completeness of destruction, neutralize, and discard it.

Table 2 Destruction of Azo and Azoxy Compounds Using Potassium Permanganate in Sulfuric Acid

Compound	% Remaining	Recovery	Mutagenicity of Reaction Mixture[a]
Azobenzene	<0.13	70	–
Azoxybenzene	0.46	69	–
Azoxyanisole	<0.17	57	–
Phenylazophenol	<0.033	110	–
Phenylazoaniline	<0.07	84	–
Fast garnet	<0.063	69	–

[a] The – = nonmutagenic.

Destruction of N,N-Dimethyl-4-amino-4′-hydroxyazobenzene

Take up 0.24 mg of *N,N*-dimethyl-4-amino-4′-hydroxyazobenzene in 1 mL of 50% (v/v) acetic acid and add 1 mL of 2 *M* H_2SO_4 and 1 mL of 0.2 *M* $KMnO_4$ solution. After 2 h decolorize with oxalic acid, test for completeness of destruction, neutralize and discard it.

Analytical Procedures

Basify 20 mL aliquots of the $KMnO_4$ reaction mixtures with 20 mL 10 *M* KOH solution, dilute with 50 mL water, cool and extract three times with 10-mL portions of cyclohexane. Dry the extracts over anhydrous sodium sulfate, evaporate, and take up the residue in about 2 mL of methanol. Analyze this solution by high-performance liquid chromatography (HPLC) for the presence of the starting material.

Filter 20 mL aliquots of the Ni–Al alloy reaction mixtures through a pad of Celite® with 50 mL of cyclohexane. Separate the layers and extract the aqueous layer three times with 10-mL portions of cyclohexane. Combine all the organic layers, dry over anhydrous sodium sulfate, evaporate, and take up the residue in about 2 mL of methanol. Analyze this solution by HPLC for the presence of the starting material. Nickel–aluminum alloy reaction mixtures can also be analyzed directly by gas chromatography (GC).

Because of the presence of the phenolic group the procedures for phenylazophenol were a little different. The $KMnO_4$ reaction mixture was not basified before extraction and the Ni–Al alloy reaction mixture was processed as follows. A 20-mL aliquot was filtered through a pad of Celite® with 50 mL of cyclohexane and the layers of the filtrate were separated. The aqueous layer was extracted once with 20 mL of cyclohexane then acidified with 2 mL of 6 *M* H_2SO_4 and extracted five times with 20-mL portions of cyclohexane. All the organic layers were combined and dried over anhydrous sodium sulfate, evaporated, and the residue taken up in about 2 mL of methanol.

In all cases control experiments were performed to ascertain that the recoveries of the starting materials were satisfactory.

Some of the starting materials were determined by HPLC using a 250 × 4.6-mm i.d. column of Microsorb C8 (Rainin Instrument Co., Woburn, MA) with methanol-water mixtures flowing at 1 mL/min as the mobile phases. The ultraviolet (UV) detector was set at 254 nm. The methanol:water ratios, with retention times in minutes in parentheses, were azobenzene 70:30 (13.9), azoxybenzene 70:30 (12.3), azoxyanisole 70:30 (14.3), phenylazophenol 60:40 (16.9), phenylazoaniline 60:40 (12.4), and fast garnet 70:30 (11.4). *p*-Aminophenol was also determined by HPLC using methanol:20 g/L ammonium acetate solution 7:93 as the mobile phase.[19] An aliquot of the centrifuged reaction mixture was acidified with an equal volume of 2 *M* HCl, neutralized with sodium bicarbonate, centrifuged, and an aliquot of the supernatant diluted 10-fold with 20 g/L ammonium acetate solution. The retention time was 6.9 min.

Some of the starting materials and products were determined by GC, using a 1.8-m × 2-mm i.d. packed column with flame ionization detection.[15] For methylamine (retention time 0.9 min at 60°C), dimethylamine (1.3 min at 100°C), azoxymethane (2.6 min at 120°C), and tetramethyltetrazene (3.4 min at 150°C) the packing was 28% Pennwalt 223 + 4% KOH on 80/100 Gas Chrom R, for aniline (2.9 min at 80°C), azobenzene (1.5 min at 180°C), and azoxybenzene (3.2 min at 180°C) the packing was 3% SP 2401-DB on 100/120 Supelcoport, and for *p*-anisidine (2.7 min at 100 °C), *p*-phenylenediamine (1.2 min at 130 °C), *o*-toluidine (2.7 min at 80 °C), and 2,5-diaminotoluene (1.7 min at 130 °C) the packing was 3% OV1 on 80/100 Supelcoport. These chromatographic conditions are only a guide; the exact conditions would have to be determined experimentally.

N,N-Dimethyl-4-amino-4′-hydroxyazobenzene was determined by differential pulse polarography of the reaction mixture.[14]

Mutagenicity Assays

Reaction mixtures from the degradation of azobenzene, azoxybenzene, azoxyanisole. phenylazophenol, phenylazoaniline, and fast garnet were tested for mutagenicity as described on page 4. Tester strains TA98, TA100, TA1530, and TA1535 were used. Before testing $KMnO_4$ reaction mixtures were decolorized with sodium ascorbate then neutralized with sodium bicarbonate and Ni–Al alloy reaction mixtures were centrifuged and neutralized with acetic acid. The only reaction mixtures that were mutagenic were those obtained when phenylazoaniline and fast garnet were degraded with Ni–Al alloy. This is probably because the products of these reactions, *p*-phenylenediamine and 2,5-diaminotoluene, were mutagenic. Azobenzene, azoxybenzene, phenylazophenol, phenylazoaniline, and fast garnet were all mutagenic when tested as dimethyl sulfoxide (DMSO) solutions.

Related Compounds

Nickel–aluminum alloy in KOH solution appears to be a general method for the cleavage of N—N and N—O bonds[16] and so it should be applicable to other azo and azoxy compounds. Potassium permanganate in H_2SO_4 is a general oxidative procedure and it should be applicable to many azo compounds.[14] In each case full validation should be carried out before using the procedure.

References

1. Other names for this compound are azobenzide, azobenzol, azobisbenzene, azodibenzene, azodibenzeneazofume, benzeneazobenzene, 1,2-diphenyldiazene, diphenyl diimide, diazobenzene, ENT 14,611, NCI-C02926, and USAF EK-704.
2. Other names for this compound are diphenyldiazene 1-oxide, azobenzene oxide, azoxybenzide, and azoxydibenzene.
3. Other names for this compound are 4-hydroxyazobenzene, Solvent Yellow 7, *p*-benzeneazophenol, C.I. Solvent Yellow 7, C.I. 11800, and 4-phenylazophenol.

4. Other names for this compound are 4-aminoazobenzene, 4-(phenylazo)benzenamine, C.I. Solvent Yellow I, *p*-aminodiphenylimide, aniline yellow, C.I. 11000, AAB, 4-aminoazobenzol, 4-benzeneazoaniline, Brasilazina Oil Yellow G, Ceres Yellow R, C.I. Solvent Blue 7, Fast Spirit Yellow AAB, Oil Soluble Aniline Yellow, Oil Yellow AAB, Organol Yellow, paraphenolazo aniline, *p*-phenylazophenylamine, Sudan Yellow R, and USAF EK-1375.
5. Other names for this compound are C.I. 11160, 4′-amino-2,3′-dimethylazobenzene, *o*-aminoazotoluene, Solvent Yellow 3, 2-amino-5-azotoluene, AAT, *o*-AAT, *o*-aminoazotoluol, *o*-AT, Brasilazina Oil Yellow R, Butter Yellow, C.I. 11160, C.I. 11160B, C.I. Solvent Yellow 3, 2′,3-dimethyl-4-aminoazobenzene, Fast Garnet GBC Base, Fast Oil Yellow, Fast Yellow AT, Fast Yellow B, Hidaco Oil Yellow, 2-methyl-4-[(2-methylphenyl)azo]benzenamine, OAAT, Oil Yellow, Oil Yellow 21, Oil Yellow 2681, Oil Yellow AT, Oil Yellow A, Oil Yellow C, Oil Yellow I, Oil Yellow 2R, Oil Yellow T, Organol Yellow 25, Somalia Yellow R, Sudan Yellow RRA, *o*-tolueneazo-*o*-toluidine, 5-(*o*-tolylazo)-2-aminotoluene, 4-(*o*-tolylazo)-*o*-toluidine, Tulabase Fast Garnet GB, Tulabase Fast Garnet GBC, and Waxacol Yellow NL.
6. Narisawa, T.; Wong, C-.Q.; Weisburger, J.H. Azoxymethane-induced liver hemangiosarcomas in inbred strain-2 guinea pigs. *J. Natl. Cancer Inst.* **1976**, *56*, 653–654.
7. International Agency for Research on Cancer. *IARC Monographs on the Evaluation of Carcinogenic Risk of Chemicals to Man.* Volume 8, *Some Aromatic Azo Compounds*; International Agency for Research on Cancer: Lyon, 1975, pp. 53–60.
8. Reference 6, pp. 61–74.
9. Reference 6, pp. 75–81.
10. Lewis, R.J., Sr. *Sax's Dangerous Properties of Industrial Materials*, 8th ed.; Van Nostrand-Reinhold: New York, 1992; p. 325.
11. Reference 10, pp. 327–328.
12. Reference 10, p. 3249.
13. Chung, K.T.; Cerniglia, C.E. Mutagenicity of azo dyes: Structure-activity relationships. *Mutat. Res.* **1992**, *277*, 201–220.
14. Barek, J.; Kelnar, L. Destruction of carcinogens in laboratory wastes. IV. Destruction of *N,N*-dimethyl-4-amino-4′-hydroxyazobenzene by permanganate. *Microchem. J.* **1986**, *33*, 239–242.
15. Lunn, G.; Sansone, E.B. Destruction of azo and azoxy compounds and 2-aminoanthracene. *Appl. Occup. Environ. Hyg.* **1991**, *6*, 1020–1026.
16. Lunn, G.; Sansone, E.B.; Keefer, L.K. General cleavage of N—N and N—O bonds using nickel/aluminum alloy. *Synthesis* **1985**, 1104–1108.
17. International Agency for Research on Cancer. *IARC Monographs on the Evaluation of the Carcinogenic Risk of Chemicals to Humans, Supplement No. 7, Overall Evaluations of Carcinogenicity: An Updating of* IARC Monographs *Volumes 1 to 42*; International Agency for Research on Cancer: Lyon, 1987; pp. 99–100.
18. Lunn, G.; Sansone, E.B.; De Méo, M.; Laget, M.; Castegnaro, M. Potassium permanganate can be used for degrading hazardous compounds. *Am. Ind. Hyg. Assoc. J.* **1994**, *55*, 167–171.
19. Sternson, L.A.; DeWitte, W.J. High-pressure liquid chromatographic analysis of aniline and its metabolites. *J. Chromatogr.* **1977**, *137*, 305–314.

BIOLOGICAL STAINS

CAUTION! Refer to safety considerations section on page 7 before starting any of these procedures.

Biological stains are widely used in biomedical research laboratories and also for diagnostic purposes. A number of these stains are known to be toxic. Trypan blue is carcinogenic in experimental animals[1] and erythrosin B,[2] Janus green,[3] and rose Bengal[4] may be carcinogenic. Methylene blue,[5] and toluidine blue O[6] are toxic to humans. Congo red[7] and rose Bengal[4] have been reported to be teratogens and acridine orange,[8] Congo red,[9] erythrosin B,[10] methylene blue,[8] neutral red,[11] propidium iodide,[12] toluidine blue O,[13] and trypan blue[9] have been reported to be mutagens. In our work we have found that azure A,[14] azure B,[14] brilliant blue R,[15] Coomassie brilliant blue G,[15] cresyl violet acetate,[14] ethidium bromide,[14] Janus green B,[14] and safranine O,[14] are mutagenic. The destruction of the stains diaminobenzidine, ethidium bromide, and propidium iodide are described in other monographs. The decontamination of solutions containing the following stains (Table 1), except for brilliant blue R and Coomassie brilliant blue G,[15] was investigated in a recent paper.[14] In this work acridine orange biological stain (**I**), the zinc chloride double salt, was used. The chemical structures of these stains vary considerably. The structures of eosin B (**II**), methylene blue (**III**), propidium iodide (**IV**), Congo red (**V**), and trypan blue (**VI**) are shown below.

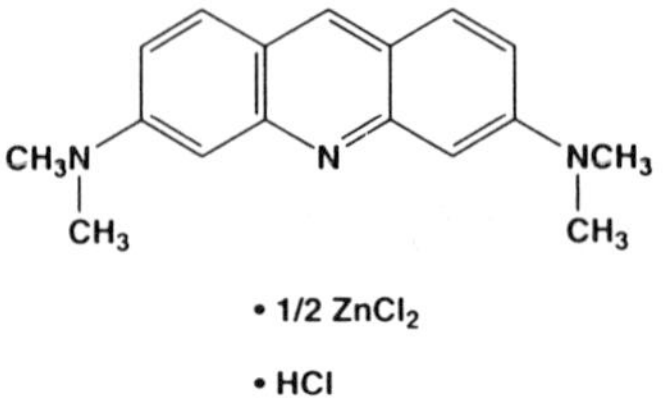

$\cdot$ 1/2 $ZnCl_2$
$\cdot$ HCl

I

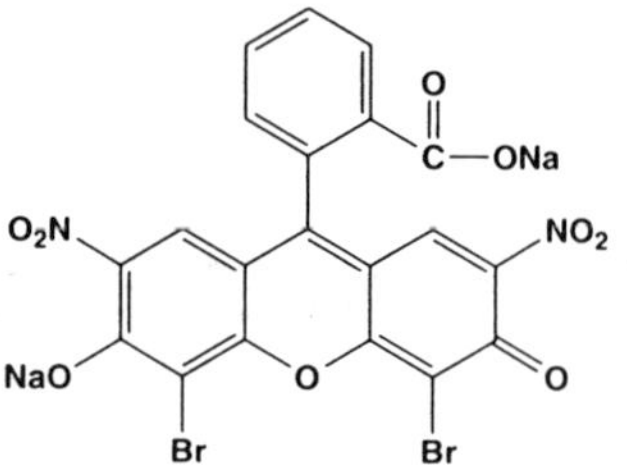

II

Cl^-

$2I^-$

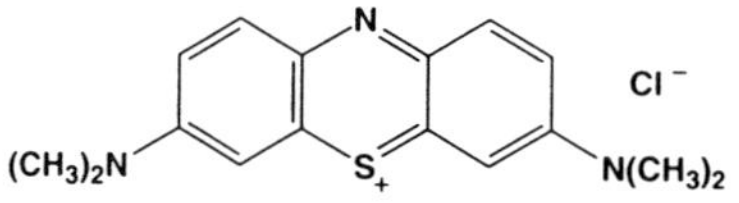

III

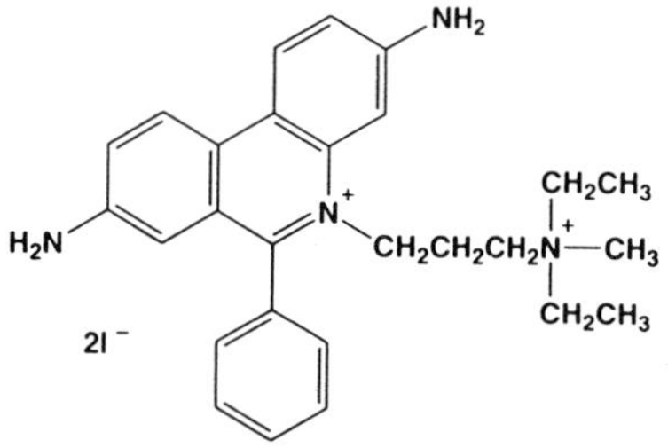

IV

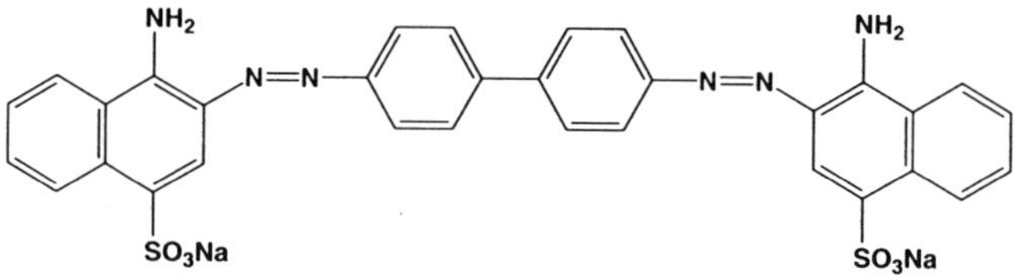

V

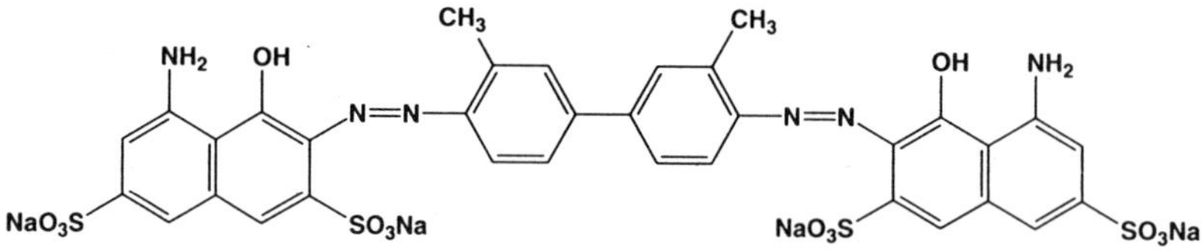

VI

Table 1 Stains Considered

Stain	Reference	Colour Index Number	Chemical Abstracts Registry Number
Acridine orange (biological stain)	16	46005	[10127-02-3]
Alcian blue 8GX	17	74240	[33864-99-2]
Alizarin red S (monohydrate)	18	58005	[130-22-3]
Azure A	19	52005	[531-53-3]
Azure B	20	52010	[531-55-5]
Brilliant blue R	21	42660	[6104-59-2]
Congo red	22	22120	[573-58-0]
Coomassie brilliant blue G	23	42655	[6104-58-1]
Cresyl violet acetate	24		[10510-54-0]
Crystal violet	25	42555	[548-62-9]
Eosin B	26	45400	[548-24-3]
Erythrosin B	27	45430	[568-63-8]
Ethidium bromide	28		[1239-45-8]
Janus green B	29	11050	[2869-83-2]
Methylene blue	30	52015	[7220-79-3]
Neutral red	31	50040	[553-24-2]
Nigrosin, water soluble	32	50240	[8005-03-6]
Orcein, synthetic	33		[1400-62-0]
Propidium iodide	34		[25535-16-4]
Rose Bengal	35	45440	[632-69-9]
Safranine O	36	50240	[477-73-6]
Toluidine blue O	37	52040	[92-31-9]
Trypan blue	38	23850	[72-57-1]

Principle of Decontamination

The stains can be removed from solution using the polymeric resin Amberlite XAD-16.[14,15] In some cases the resins Amberlite XAD-2, XAD-4, and XAD-7 can also be used. The reader is referred to the original publication[14] for these details. When a batch procedure is used the resin is stirred in the solution to be decontaminated and, at the end of the reaction time, removed by filtration (Tables 2 and 3). The decontaminated solution may be disposed of with the nonhazardous aqueous waste and the resin with the hazardous solid waste. The volume of contaminated resin gen-

Table 2 Time Required for Complete Decontamination

Compound	Time Required for Complete Decontamination
Acridine orange	18 h
Alcian blue 8GX	10 min
Alizarin red S	18 h
Azure A	10 min
Azure B	10 min
Brilliant blue R	2 h
Congo red	2 h
Coomassie brilliant blue G	2 h
Cresyl violet acetate	2 h
Crystal violet	30 min
Eosin B	30 min
Erythrosin B	18 h
Ethidium bromide	4 h
Janus green B	30 min
Methylene blue	30 min
Neutral red	10 min
Nigrosin	2 h
Orcein	2 h
Propidium iodide	2 h
Rose Bengal	3 h
Safranine O	1 h
Toluidine blue O	30 min
Trypan blue	2 h

erated is much smaller than the original volume of the solution of biological stain so the waste disposal problem is greatly reduced. When complete decontamination was achieved decontamination was greater than 99% except for Orcein when it was greater than 98.85%. Preliminary results indicate that some stains can be removed from the resin using methanol and that the clean resin can then be reused.[39] For treating large volumes of dilute aqueous solutions it is possible to put the resin in a column and run the contaminated solution through the column in a continuous flow system. Good results were obtained using 10 g of Amberlite XAD-16 in an 11-mm i.d. column. Aqueous solutions of biological stains having a molecular

Table 3 Volume Decontaminated per Gram of Resin

Compound	Volume of Solution Decontaminated per Gram of Resin (mL)
Acridine orange	20
Alcian blue 8GX	500
Alizarin red S	5
Azure A	80
Azure B	80
Brilliant blue R	80
Congo red	40
Coomassie brilliant blue G	80
Cresyl violet acetate	40
Crystal violet	200
Eosin B	40
Erythrosin B	10
Ethidium bromide	20
Janus green B	80
Methylene blue	80
Neutral red	500
Nigrosin	80
Orcein	200
Propidium iodide	20
Rose bengal	20
Safranine O	20
Toluidine blue O	80
Trypan blue	40

weight of less than 600 can be completely decontaminated using this procedure (Table 4). The initial concentration of each stain was 25 μg/mL and the final effluents had stain concentrations of less than 0.25 μg/mL. Thus decontamination was greater than 99%. Better results were obtained by grinding the Amberlite XAD-16 with H_2O in a Waring Blendor for 10 s. In general, the most useful system for decontaminating aqueous solutions of biological stains is a 15-mm i.d. column packed with 20 g of Amberlite XAD-16 ground for 10 s in a Waring Blendor. When a 25 μg/mL solution of methylene blue was decontaminated the breakthrough volume (at the

Table 4 Decontamination of Aqueous Solutions of Biological Stains Using a Flow-Through System

	Breakthrough Volumes (mL)		
Compound	Limit of Detection	1 ppm	5 ppm
Neutral red	>2480	>2480	>2480
Crystal violet	1020	>1630	>1630
Cresyl violet acetate	706	>1396	>1396
Azure B	630	882	>1209
Azure A	615	810	>975
Acridine orange	465	>990	>990
Methylene blue	420	645	1050
Safranine O	365	438	584
Toluidine blue O	353	494	606
Ethidium bromide	260	312	416
Janus green B	170	650	>870
Alizarin red S	120	150	240

limit of detection) was 420 mL for 10 g of unground resin, 2595 mL for 10 g of ground resin, and 4800 mL for 20 g of ground resin. It was also found that decontamination became more efficient as the flow rate decreased. However, decontamination times at low flow rates became inconveniently long and it was found that 2 mL/min was a good compromise between speed and efficiency. It was found that methylene blue could be removed from a column of Amberlite XAD-16 with methanol and the column could then be reused. Adsorption efficiency appears to be dependent on the properties of the stain and the adsorption system operating parameters. Compounds with formula weights greater than about 600 were poorly retained and thus had relatively low breakthrough volumes, whereas compounds with formula weights less than 600 were more efficiently retained on the column and thus had greater breakthrough volumes. A possible explanation for the dependence of adsorption efficiency on formula weight may be that the high formula weight compounds are larger and cannot penetrate pores in the resin. A batch procedure may be more appropriate for decontaminating solutions of compounds with low breakthrough volumes. There was no correlation between molecular weight and efficiency of adsorption in the batch experiments. In effect the batch system results in high contact times compared with low contact times for a flow through system. On the other hand, the flow through system is considerably more convenient as filtration steps are not required and the Amberlite XAD-16 is readily regenerated. In some cases it may be desired to decontaminate very dilute solutions of stains and in these cases a flow through system may be appropriate even for compounds having molecular weight greater than 600.[15]

Decontamination Procedures

1. Add 1 g of Amberlite XAD-16 (4 g for alizarin red S and 2 g for erythrosin B) to 20 mL of a 100 μg/mL solution of the biological stain in H_2O. Stir the mixture for at least the time indicated in Table 2 and then remove the resin by filtration. Discard the resin and the filtrate appropriately.

2. Add 1 g of Amberlite XAD-16 to the volume indicated in Table 3 of a 100 μg/mL solution of the biological stain in H_2O. Stir the mixture for at least 18 h and then remove the resin by filtration. Discard the resin and the filtrate appropriately.

3. Use an 11-mm i.d. × 300-mm long glass chromatography column (Cat. No. 5820-04), Ace Glass (Vineland, NJ) fitted with threaded adapters with flow regulating valves (Cat. No. 5838-45) at top and bottom. The adapters should accommodate 1.5-mm i.d. × 0.3-mm wall Teflon tubing (Cat. No. 12684-17) using nut connectors (Cat. No. 5855-08) and insert connectors (Cat. No. 5855-71). Attach the tubing using an insertion tool (Cat. No. 5855-80). To prevent clogging of the column outlet, place a small plug of glass wool at the bottom of the chromatography column. Magnetically stir 10 g of Amberlite XAD-16 and 25 mL of H_2O in a beaker for 5 min to facilitate wetting and pour the resin slurry into the column through a funnel. As the resin settles, tap the column with a rubber stopper fitted over a pencil to encourage even packing. Pump the solution to be decontaminated through the column using a model QG 20 lab pump (Fluid Metering, Inc., Oyster Bay, NY) at 2 mL/min. Alternatively, gravity flow coupled with periodic adjustment of the flow regulating valve can be used. Check the effluent from the column for completeness of decontamination and stop the pump when stain is detected. Discard the decontaminated effluent and the contaminated resin appropriately. In many cases the stain can be washed off the resin with methanol and the resin reused. Discard the methanol solution of stain appropriately. Results obtained when decontaminating 25 μg/mL solutions of various stains are shown in Table 4.

A 25 μg/mL solution of each stain was pumped at 2 mL/min through

10 g of unground Amberlite XAD-16 in an 11-mm i.d. glass column. The numbers in Table 4 are the volumes (mL) that had to be collected to obtain a fraction containing the specified stain concentration.

4. Use a 15-mm i.d. × 300-mm long glass chromatography column (Spectrum 124010, Fisher Scientific Co., Pittsburgh, PA, Cat. No. 06-651-2B). To prevent clogging of the column outlet, place a small plug of glass wool at the bottom of the chromatography column. Grind 20 g of Amberlite XAD-16 with 200 mL of H_2O for 10 s in a Waring Blendor, then pour the resin slurry into the column through a funnel. As the resin settles, tap the column with a rubber stopper fitted over a pencil to encourage even packing. Pump the solution to be decontaminated through the column using a model QG 20 lab pump (Fluid Metering, Inc., Oyster Bay, NY) at 2 mL/min. Alternatively, gravity flow coupled with periodic adjustment of the flow regulating valve can be used. Check the effluent from the column for completeness of decontamination and stop the pump when stain is detected. Discard the decontaminated effluent and the contaminated resin appropriately. In many cases the stain can be washed off the resin with methanol and the resin reused. Discard the methanol solution of stain appropriately.

Analytical Procedures

The filtrates from the decontamination procedures were analyzed using one of two procedures.[14,15] The absorbance of the solution could be measured at a fixed wavelength using a spectrophotometer. Details of the wavelengths used are given in Table 5. Using disposable 10-mm plastic cuvettes and assigning 0.015 as the minimum reliable absorbance reading the limits of detection shown in Table 5 were obtained. In the second procedure the fluorescence of the solution was measured with a spectrophotofluorometer using disposable 10 × 75-mm glass culture tubes as cuvettes. The excitation and emission wavelengths are shown in Table 5. In most cases a reading of 0.005 was assigned as the minimum reliable reading. In some cases, however, a rather high reading was observed for blanks and for these compounds the minimum reliable reading was set at twice the blank reading. These readings were used to calculate the limits of detection. In some cases traces of acid or base on the resins induced color changes in the stains and for cresyl violet acetate and neutral red aliquots of the filtrates were mixed with equal volumes of pH 5 buffer solution before analysis. For alizarin red and orcein an aliquot of the filtrate was taken and a volume of 1 *M* potassium hydroxide (KOH) solution equal to

Table 5 Analytical Procedures

Compound	Reagent	Procedure[a]	Wavelength(s) (nm)	Limit of Detection (ppm)
Acridine orange	DNA solution	F	ex.[b] 492 em.[c] 528	0.0032
Alcian blue 8GX		A	615	0.9
Alizarin red S	1 *M* KOH	A	556	0.46
Azure A		A	633	0.15
Azure B		A	648	0.13
Brilliant blue R		A	585	1.0
Congo red		A	497	0.25
Coomassie brilliant blue G		A	610	1.7
Cresyl violet acetate	pH 5 buffer	F	ex. 588 em. 618	0.021
Crystal violet		A	588	0.1
Eosin B		A	514	0.21
Erythrosin B		F	ex. 488 em. 556	0.025
Ethidium bromide		F	ex. 540 em. 590	0.5
Janus green B		A	660	0.6
Methylene blue		A	661	0.13
Neutral red	pH 5 buffer	A	540	0.6
Nigrosin		A	570	0.8
Orcein	1 *M* KOH	A	579	1.15
Propidium iodide	DNA solution	F	ex. 350 em. 600	0.1
Rose Bengal		F	ex. 520 em. 576	0.04
Safranine O		F	ex. 460 em. 582	0.03
Toluidine blue O		A	626	0.2
Trypan blue		A	607	0.22

[a] F = The fluorescence of the sample was measured; A = the absorbance of the sample was measured.

[b] ex. = excitation wavelength

[c] em. = emission wavelength

one tenth the volume of the aliquot was added. The fluorescence of solutions of acridine orange and propidium iodide was increased by mixing an aliquot of the filtrate with an equal volume of a 20 μg/mL solution of calf thymus DNA (Sigma) in TBE buffer. After standing for 15 min the fluorescence of this solution was measured. The TBE buffer (pH 8.1) contained tris(hydroxymethyl)aminomethane (0.089 *M*), boric acid (0.089 *M*), and ethylenediaminetetraacetic acid (EDTA) (0.002 *M*). The pH 5 buffer consisted of 2.04 g of potassium hydrogen phthalate and 38 mL of 0.1 *M* KOH solution made up to 200 mL.

Mutagenicity Assays[14,15]

The mutagenicity assays were carried out as described on page 4 using tester strains TA98, TA100, TA1530, and TA1535. Reaction mixtures were tested by applying 100 μL to each plate. None of the completely decontaminated reaction mixtures was mutagenic. Acridine orange, azure A, azure B, brilliant blue R, Coomassie brilliant blue G, cresyl violet acetate, ethidium bromide, Janus green B, methylene blue, neutral red, propidium iodide, safranine O, toluidine blue O, and trypan blue were all mutagenic to TA98 with activation and they were also mutagenic to some other strains. Crystal violet was extremely toxic to the cells but it was not mutagenic. Congo red[9] and erythrosin B[10] have been found to be mutagenic in other assays.

Related Compounds

These procedures should be generally applicable to aqueous solutions of other biological stains but thorough testing is required before routine use. It was found that aqueous solutions of Eosin Y[40] and Giemsa stain[41] could not be completely decontaminated even with high ratios of resin–solution. However, the residual concentrations of these stains were very low, less than 25 ppb. Eosin Y may be carcinogenic.[42] See also the monograph on Ethidium Bromide for decontamination and chemical degradation procedures for ethidium bromide and propidium iodide and the monograph on Aromatic Amines for degradation procedures for the stain diaminobenzidine.

References

1. International Agency for Research on Cancer. *IARC Monographs on the Evaluation of Carcinogenic Risk of Chemicals to Man.* Volume 8, *Some Aromatic Azo Compounds*; International Agency for Research on Cancer: Lyon, 1975; pp. 267–278.
2. Lewis, R.J., Sr. *Sax's Dangerous Properties of Industrial Materials*; 8th ed.; Van Nostrand-Reinhold: New York, 1992; pp. 1689–1690.
3. Reference 2, p. 1191.
4. Reference 2, p. 2999.
5. Reference 2, p. 480.

6. Reference 2, p. 166.
7. Reference 2, p. 3080.
8. Webb, R.B.; Hass, B.S. Biological effects of dyes on bacteria. VI. Mutation induction by acridine orange and methylene blue in the dark with special reference to Escherichia coli WP6 (polA1). *Mutat. Res.* **1984**, *137*, 1–6.
9. Kornbrust, D.J.; Barfknecht, T.R. Comparison of 7 azo dyes and their reduction products in the rat and hamster hepatocyte primary culture/DNA-repair assays. *Mutat. Res.* **1984**, *136*, 255–266.
10. Lakdawalla, A.A.; Netrawali, M.S. Erythrosine, a permitted food dye, is mutagenic in the Bacillus subtilis multigene sporulation assay. *Mutat. Res.* **1988**, *206*, 171–176.
11. Longnecker, D.S.; Curphey, T.J.; Daniel, D.S. Mutagenicity of neutral red. *Mutat. Res.* **1977**, *48*, 109–112.
12. Fukunaga, M.; Cox, B.A.; von Sprecken, R.S.; Yielding, L.W. Production of frameshift mutagens in Salmonella by phenanthridinium derivatives: enzymatic activation and photoaffinity labeling. *Mutat. Res.* **1984**, *127*, 31–37.
13. Dunipace, A.J.; Beaven, R.; Noblitt, T.; Li, Y.; Zunt, S.; Stookey, G. Mutagenic potential of toluidine blue evaluated in the Ames test. *Mutat. Res.* **1992**, *279*, 255–259.
14. Lunn, G.; Sansone, E.B. Decontamination of aqueous solutions of biological stains. *Biotechnic Histochem.* **1991**, *66*, 307–315.
15. Lunn, G.; Klausmeyer, P.K.; Sansone, E.B. Removal of biological stains from aqueous solution using a flow-through decontamination procedure. *Biotechnic Histochem.* **1994**, *69*, 45–54.
16. Other names for this compound are 3,6-bis(dimethylamino)acridine zinc chloride double salt Solvent Orange 15, Basic Orange 3RN, acridine orange NO, acridine orange R, C.I. Basic Orange 14, rhoduline orange NO, *N,N,N′,N′*-tetramethyl-3,6-acridinediamine, 2,8-bisdimethylaminoacridine, brilliant acridine orange E, 3,6-di(dimethylamino)acridine, euchrysine, solvent orange 15, waxoline orange A, and 3,6-bis(dimethylamino)acridine.
17. Other names for this compound are Ingrain Blue 1, Michrome No. 24, Alcian Blue 8GS, and (*N,N′,N″,N‴*-[29*H*,31*H*-phthalocyaninetetrayltetrakis(methylenethio-[(dimethylamino)methylidyne])]tetrakis(*N*-methylmethanaminiumato)(2-)-*N*(29),*N*(30),*N*(31), *N*(32)copper(4^+) tetrahydrochloride.
18. Another name for this compound is 3,4-dihydroxy-9,10-dioxo-2-anthracenesulfonic acid, sodium salt.
19. Other names for this compound are 3-amino-7-(dimethylamino)phenothiazin-5-ium chloride, 7-(dimethylamino)-3-imino-3*H*-phenothiazine hydrochloride, 3-amino-7-dimethylaminophenazathionium chloride, *asym*-dimethyl-3,7-diaminophenazathionium chloride, *asym*-dimethylthionine chloride, and methylene azure A.
20. Other names for this compound are 3-(dimethylamino)-7-(methylamino)phenothiazin-5-ium chloride, 7-(dimethylamino)-3-(methylimino)-3*H*-phenothiazine hydrochloride, 3-methylamino-7-dimethylaminophenazathionium chloride, trimethyldiaminophenazathionium chloride, trimethylthionine chloride, and methylene azure B.
21. Other names for this compound are Acid Blue 83, Coomassie Brilliant Blue R, Coomassie Brilliant Blue R 250, and *N*-[4-[[4-[(4-ethoxyphenyl)amino]phenyl][4-[ethyl[(3-sulfo-

phenyl)methyl]amino]phenyl]methylene]-2,5-cyclohexadien-1-ylidene]-*N*-ethyl-3-sulfobenzenemethaminium hydroxide, inner salt, monosodium salt.

22. Other names for this compound are Direct Red 28, 3,3′-[(1,1′-biphenyl)-4,4′-diylbis(azo)]bis(4-amino-1-naphthalenesulfonic acid) disodium salt, sodium diphenyldiazo-bis-α-naphthylaminesulfonate, Atlantic Congo red, Atul Congo red, azocard red Congo, benzo Congo red, Brasilamina Congo 4B, C.I. Direct Red 28, cotton red L, diacotton Congo red, Erie Congo 4B, Hispamin Congo 4B, Kayaku Congo red, Mitsui Congo red, Peeramine Congo red, Sugai Congo red, tertrodirect red C, trisulfon Congo red, and Vondacel red CL.
23. Other names for this compound are Acid Blue 90, Brilliant Blue G, Coomassie Brilliant Blue G 250, and *N*-[4-[[4-[(4-ethoxyphenyl)amino]phenyl][4-[ethyl[(3-sulfophenyl)-methyl]amino]-2-methylphenyl]methylene]-3-methyl-2,5-cyclohexadien-1-ylidene]-*N*-ethyl-3-sulfobenzenemethaminium hydroxide, inner salt, monosodium salt.
24. Another name for this compound is 5,9-diaminobenzo[*a*]phenoxazonium acetate.
25. Other names for this compound are Basic Violet 3, Gentian Violet, *N*-[4-[bis[4-(dimethylamino)phenyl]methylene]-2,5-cyclohexadien-1-ylidene]-*N*-methylmethanaminium chloride, hexamethylpararosaniline, hexamethyl-*p*-rosaniline chloride, aniline violet, methylrosaniline chloride, Adergon, Axuris, Badil, Gentiaverm, Meroxylan, Meroxyl, Pyoktanin, Vianin, and Viocid.
26. Other names for this compound are Acid Red 91, 4′,5′-dibromo-2′,7′-dinitrofluorescein, disodium salt, Eosine I Bluish, 4′,5′-dibromo-3′,6′-dihydroxy-2′,7′-dinitrospiro[isobenzofuran-1(3*H*),9′-[9*H*]xanthen]-3-one disodium salt, and hydroxydibromodinitro-*o*-carboxyphenylfluorone sodium.
27. Other names for this compound are Acid Red 51, 2′,4′,5′,7′-tetraiodofluorescein disodium salt, 3′,6′-dihydroxy-2′,4′,5′,7′-tetraiodospiro[isobenzofuran-1(3*H*),9′-[9*H*]xanthen]-3-one disodium salt, erythrosine BS, erythrosine B, FD & C Red No.3, Food Red 14, Aizen erythrosine, Calcocid erythrosine N, Canacert erythrosine BS, 9-(*o*-carboxyphenyl)-6-hydroxy-2,4,5,7-tetraiodo-3-isoxanthone, C.I. Acid Red 51, Cilefa pink B, D & C Red No. 3, dolkwal, Dye FD & C Red No. 3, E 127, EBS, edicol, supra erythrosine A, erythrosine B-FO (biological stain), Hexacert Red No. 3, Hexacol erythrosine BS, LB-rot 1, maple erythrosine, new pink bluish Geigy, 1427 red, 1671 red, tetraiodofluorescein sodium salt, and Usacert Red No. 3.
28. Other names for this compound are homidium bromide, Novidium bromide, Babidium bromide, RD 1572, Dromilac, 2,7-diamino-9-phenylphenanthridine ethobromide, 3,8-diamino-5-ethyl-6-phenylphenanthridinium bromide, 2,7-diamino-10-ethyl-9-phenylphenanthridinium bromide, and 2,7-diamino-9-phenyl-10-ethylphenanthridinium bromide.
29. Other names for this compound are 3-(diethylamino)-7-[[4-(dimethylamino)phenyl]azo]-5-phenylphenazinium chloride and Janus green V.
30. Other names for this compound are Basic Blue 9, 3,7-bis(dimethylamino)phenothiazin-5-ium chloride, methylthionium chloride, 3,7-bis(dimethylamino)phenazathionium chloride, Swiss blue, solvent blue 8, urolene blue, Aizen methylene blue BH, 3,7-bis(dimethylamino)phenothiazin-5-ium chloride, calcozine blue ZF, chromosmon, C.I. Basic Blue 9, D&C blue number I, external blue I, Hidaco methylene blue salt free,

leather pure blue HB, methylene blue A, methylene blue BB, methylene blue chloride, methylene blue D, methylene blue I, methylene blue polychrome, methylenium ceruleum, methylthionine chloride, methylthionium chloride, Mitsui methylene blue, Modr methylenova, Sandocryl blue BRL, Schultz No. 1038, tetramethylthionine chloride, and Yamamoto methylene blue B.

31. Other names for this compound are C.I. Basic Red 5, Basic Red 5, N^8,N^8,3-trimethyl-2,8-phenazinediamine monohydrochloride, 3-amino-7-dimethylamino-2-methylphenazine hydrochloride, toluylene red, neutral red chloride, nuclear fast red, aminodimethylaminotoluaminozine hydrochloride, Kernechtrot, and Michrome no. 226.
32. Another name for this compound is Acid Black 2.
33. Orcein is a mixture of indefinite composition containing α-, β-, and γ-aminoorcein, iso-α-aminoorcein, α-hydroxyorcein, and β- and γ-aminoorceinimine.
34. Another name for this compound is 3,8-diamino-5-[3-(diethylmethylammonio)propyl]-6-phenylphenanthridinium diiodide.
35. Other names for this compound are Acid Red 94, 4,5,6,7-tetrachloro-3′,6′-dihydroxy-2′,4′,5′,7′-tetraiodospiro[isobenzofuran-1(3*H*),9′-[9*H*]xanthen]-3-one disodium salt, 4,5,6,7-tetrachloro-2′,4′,5′,7′-tetraiodofluorescein disodium salt, Rose Bengale B, Food Red Color No. 105, sodium salt, Food Red No. 105,sodium salt, R105 sodium, 9-(3′,4′,5′,6′-tetrachloro-o-carboxyphenyl)-6-hydroxy-2.4,5,7-tetraiodo3-isoxanthone, and 4,5,6,7-tetrachloro-2′,4′,5′,7′-tetraiodofluorescein disodium salt
36. Other names for this compound are 3,7-diamino-2,8-dimethyl-5-phenylphenazinium chloride and Basic Red 2.
37. Other names for this compound are Basic Blue 17, tolonium chloride, 3-amino-7-(dimethylamino)-2-methylphenothiazin-5-ium chloride, 3-amino-7-dimethylamino-2-methylphenazathionium chloride, blutene chloride, dimethyltoluthionine chloride, Klot, Tolazul, blutene, C.I. 925, C.I. Basic Blue 17, F Klot, and Schultz No. 1041
38. Other names for this compound are Direct Blue 14, Niagara Blue 3B, 3,3′-[(3,3′-dimethyl[1,1′-biphenyl]-4,4′diyl)bis(azo)]bis[5-amino-4-hydroxy-2,7-naphthalenedisulfonic acid] tetrasodium salt, 3,3′-[(3,3′-dimethyl-4,4′-biphenylene)bis(azo)]bis(5-amino-4-hydroxy-2,7-naphthalenedisulfonic acid] tetrasodium salt, tetrasodium 3,3′-[(3,3′-dimethyl-4,4′-biphenylene)bis(azo)]bis(5-amino-4-hydroxy-2,7-naphthalenedisulfonate,sodium ditolyl-diazobis-8-amino-1-naphthol-3,6-disulfonate, Benzamine Blue, Diamine Blue, Benzo Blue, Congo Blue, Dianil Blue, and Naphthylamine Blue, C.I. Direct Blue 14, tetrasodium salt, amanil sky blue, amidine blue 4B, azidine blue 3B, azurro diretto 3B, bencidal blue 3B, bleu diamine, blue emb, brasilamina blue 3B, centraline blue 3B, chloramine blue, chlorazol blue 3B, chrome leather blue 3B, congoblau FB, cresotine blue 3B, dianilblau, diazine blue 3B, diphenyl blue 3B, direct blue 14, hispamin blue 3BX, naphtamine blue 2B, naphthylamine blue, NCI-C61289, Orion blue 3B, paramine blue 3B, parkibleu, parkipan, pontamine blue 3BX, pyrazol blue 3B, pyrotropblau, renolblau 3B, TB, trianol direct blue 3B, tripan blue, and trypanblau.
39. Lunn, G. Unpublished results.
40. Other names for this compound are Acid Red 87, Eosin Yellowish, 2′,4′,5′,7′-tetrabromofluorescein disodium salt, 2′,4′,5′,7′-tetrabromo-3′,6′-dihydroxyspiro[isobenzofuran-1(3*H*),9′-[9*H*]xanthen]-3-one disodium salt, bromoeosine, tetrabromofluorescein

sol, bromofluoresceic acid, D & C Red No. 22, Eosine, Aizen eosine GH, bromo acid, bromo B, bromofluorescein, bronze bromo, Certiqual eosine, C.I. Acid Red 87, eosine FA, eosine lake red Y, fenazo eosine XG, Hidacid bromo acid regular, Hidacid dibromo fluorescein, Irgalite bronze red CL, and phloxine red 20-7600. The CAS Registry number is 548-26-5 and the Colour Index number is 45380.

41. The CAS Registry Number is 51811-82-6. Giemsa stain is a blend of eosin Y, methylene blue, and oxidation products of the latter, including the methylene azures. (see Green, F.J. *The Sigma–Aldrich Handbook of Stains, Dyes and Indicators*; Aldrich Chemical Co.: Milwaukee, WI, 1990; p. 388).

42. Reference 2, p. 552.

BORON TRIFLUORIDE AND INORGANIC FLUORIDES

CAUTION! Refer to safety considerations section on page 7 before starting any of these procedures.

Soluble inorganic fluorides are toxic[1] (e.g., the LD_{50} of sodium fluoride (NaF) [7681-49-4] is 180 mg/kg).[2] Boron trifluoride (BF_3) is highly toxic[3] and, when supplied as the etherate ($BF_3.Et_2O$) [109-63-7], it is corrosive and flammable.[4] These compounds are used industrially and in chemistry laboratories. Potassium fluoride (KF) [7789-23-3][5] and NaF[6] are teratogens, ammonium hydrogen difluoride[7] [$(NH_4)HF_2$] [1341-49-7] is corrosive, and sodium hexafluorosilicate (Na_2SiF_6) [16893-85-9][8] is a skin and severe eye irritant. Sodium fluoride, KF, and Na_2SiF_6 are used as pesticides. Tin(II) fluoride (SnF_2) [7783-47-3] is used as a dental caries prophylactic.

Principles of Destruction and Decontamination

The compound $BF_3.Et_2O$[9] is very rapidly hydrolyzed by H_2O to boric acid and inorganic fluoride.[3,10] Addition of a calcium salt will produce insoluble

Table 1 Removal of Fluoride from Solution Using the above Destruction and Decontamination Procedures

	Compound					
	BF_3	KF	NaF	$(NH_4)HF_2$	SnF_2	Na_2SiF_6
Initial concn (mg/mL)	11.54	10	10	10	10	5
Initial concn (ppm F^-)	4632	3276	4524	6667	2425	3032
Final concn (ppm F^-)	5.2	13.5	27	8	3.7	14.5

(solubility 16 mg/L), nontoxic (LD_{50} > 2.5 g/kg),[11] calcium fluoride (CaF_2), which can easily be removed by filtration. In a similar fashion addition of calcium ions to 10 mg/mL solutions of KF, NaF,[12] $(NH_4)HF_2$,[13] and SnF_2[14] or a 5 mg/mL solution of Na_2SiF_6[15] removed fluoride from solution. A variety of different procedures was tried and it was found that the best results were obtained when calcium oxide (CaO) was used.[10] It was found in control experiments that CaF_2 is more soluble at low pH. Calcium oxide may keep the pH high and thus decrease the solubility of CaF_2, and hence the final fluoride concentration. When CaO was used the final fluoride concentration was less than 30 ppm in each case (Table 1). It was found advisable to allow the reaction mixture to settle, and then to carefully decant the clear liquid into the filter funnel. Finally, the residual sludge was filtered. This procedure helped to prevent clogging of the filter paper.

Destruction and Decontamination Procedures[10]

Destruction of Boron Trifluoride Etherate

Stir 100 mL of H_2O and 2.5 g of CaO, and add 1 mL of $BF_3.Et_2O$. Stir the mixture for 18 h, allow it to settle, filter, check for completeness of decontamination, remove any organic liquid that is present, and discard it.

Decontamination of Solutions Containing Fluoride

If necessary, dilute with H_2O so that the concentration of KF, NaF, $(NH_4)HF_2$, or SnF_2 does not exceed 10 mg/mL and the concentration of

Na_2SiF_6 does not exceed 5 mg/mL. For each 10 mL of solution add 0.5 g of CaO, stir for 18 h, allow to settle, filter, check for completeness of decontamination, and discard it.

Analytical Procedures

Mix 9 mL of H_2O, 1 mL of a total ion strength adjustment buffer (TISAB III), and 1 mL of sample and determine the fluoride concentration using a fluoride combination ion-selective electrode (Orion 960900 or similar). Calibrate the electrode by using solutions containing known concentrations of fluoride ions. Create a calibration curve by graphing the reading (mV) against the concentration of fluoride (ppm) on semilog graph paper. Alternatively, meters can be purchased that can be programmed to display the ion concentration directly.

Related Compounds

This procedure should be generally applicable to other inorganic fluorides and other complexes of BF_3. However, it was found that the procedure cannot be used for disodium fluorophosphate because the fluoride is not completely ionized in solution.

References

1. Lewis, R.J., Sr. *Sax's Dangerous Properties of Industrial Materials*, 8th ed.; Van Nostrand-Reinhold: New York, 1992; p. 1717.
2. Smyth, H.F.; Carpenter, C.P.; Weil, C.S.; Pozzani, U.C.; Striegel, J.A.; Nycum, J.S. Range-finding toxicity data: List VII. *Am. Ind. Hyg. Assoc. J.* **1969**, *30*, 470–476.
3. Reference 1, pp. 532–533.
4. Reference 1, p. 533.
5. Reference 1, p. 2868.
6. Reference 1, pp. 3081–3082.
7. Reference 1, p. 227.
8. Reference 1, p. 1479.
9. Other names for this compound are boron fluoride ethyl ether, boron fluoride etherate, ethyl ether-boron trifluoride complex, and boron trifluoride diethyletherate.
10. Lunn, G. Unpublished observations.
11. Reference 1, p. 667.
12. Other names for this compound are Chemifluor, Duraphat, Fluoros, Luride-SF, Villiaumite, Florocid, Flura-Drops, Karidium, Lemoflur, Ossalin, Ossin, Osteofluor, Zymafluor,

Antibulit, Cavi-trol, Credo, disodium difluoride, FDA 0101, Fl-Tabs, Floridine, Flozenges, Fluoral, Fluorident, Fluorigard, Fluorineed, Fluorinse, Fluoritab, Fluor-O-Kote, Flura-Gel, Flurcare, Fungol B, Gel II, Gelution, Gleem, Iradicav, Karidium, Karigel, Kari-Rinse, Lea-Cov, Lemoflur, Luride, Nafeen, NaFPak, NAFRinse, natrium fluoride, NCI-C55221, Nufluor, Pediaflor, Pedident, Pennwhite, Pergantene, Phos-flur, Point Two, Predent, Rafluor, Rescue Squad, Roach Salt, sodium hydrofluoride, sodium monofluoride, So-Flo, Stay-Flo, Studafluor, Super-Dent, T-Fluoride, Thera-Flur-N, trisodium trifluoride, and villiaumite.

13. Other names for this compound are ammonium bifluoride, acid ammonium fluoride, and ammonium hydrogen fluoride.
14. Other names for this compound are stannous fluoride, and tin difluoride, Fluoristan, and tin bifluoride.
15. Other names for this compound are sodium fluosilicate, sodium silicofluoride, Salufer, disodium hexafluorosilicate, Destruxol Applex, disodium silicofluoride, Ens-Zem Weevil Bait, ENT 1,501, Ortho Earwig Bait, Ortho Weevil Bait, Prodan, PSC Co-op Weevil Bait, Safsan, silicon sodium fluoride, sodium fluorosilicate, sodium hexafluorosilicate, and Super Prodan.

BUTYLLITHIUM

> **CAUTION!** Refer to safety considerations section on page 7 before starting any of these procedures.

Butyllithium (*n*-butyl lithium) [109-72-8] is generally supplied as a solution in a hydrocarbon solvent (e.g., pentane, hexanes, or cyclohexane). Solutions will ignite on contact with H_2O and carbon dioxide (CO_2) and solutions of greater than 20% will ignite spontaneously in air.[1,2] This compound should be handled under nitrogen with special equipment.[3] It is used in organic synthesis.

Principle of Destruction

Butyllithium is allowed to react with 1-butanol in a dry hydrocarbon solvent to give lithium butoxide and butane. The lithium butoxide is subsequently hydrolyzed to butanol and lithium hydroxide.[4] 1,10-Phenanthroline is used as an indicator. In the presence of excess lithium a red color is produced.[5] Alternatively, butyllithium can be reacted with a siloxane polymer to give alkylsilyl groups and lithium silanoate groups. The lithium silanoate is subsequently hydrolyzed with H_2O to lithium hydroxide and silanol groups.[6]

Destruction Procedures

Caution! These reactions should be done under nitrogen and provision should be made for venting, through a bubbler, the considerable amounts of gas that may be generated.

1. Prepare a 10% (v/v) solution of 1-butanol in isooctane (2,2,4-trimethylpentane) and add 1,10-phenanthroline (1 mg/mL) as an indicator. Dry this solution over a 4-Å molecular sieve overnight. Stir 15 mL of the 1-butanol–isooctane mixture under nitrogen in an ice bath and cautiously add 5 mL of a 1.55 *M* butyllithium solution in hexanes. After 10 min, or when the reaction appears to have stopped, check the color of the mixture. If it is red, indicating the presence of excess butyllithium, add more of the 1-butanol–isooctane mixture until a yellow color is produced. When the reaction mixture is yellow add 10 mL of H_2O. Stir the mixture overnight, separate the layers, and discard them.[4]

2. Prepare a mixture of equal volumes of poly(dimethylsiloxane) 200 fluid (5 centistokes, Aldrich 31,766-7 or equivalent) and dry tetrahydrofuran (THF) that contains 2.5 mg/mL of 1,10-phenanthroline as an indicator. Stir 4 mL of the poly(dimethylsiloxane)–THF mixture under nitrogen and cautiously add 10 mL of a 1.55 *M* butyllithium solution in hexanes. After 3 h, or when the color of the solution has turned from red to yellow, add 25 mL of H_2O. If the color does not change from red to yellow, add more of the polydimethylsiloxane–THF mixture. Stir the mixture for 1 h, separate the layers, and discard them.[6]

Analytical Procedures for Alkyllithium Reagents

Instead of destroying aged samples of butyllithium of uncertain titer it may be more efficient to retitrate the reagent. Methods of analysis[7] and indicators[5] have been reviewed. Lithium reagents can be titrated with 2-butanol (*sec*-butyl alcohol) using 1,10-phenanthroline,[8] 2,2′-biquinoline,[8] or *N*-phenyl-1-naphthylamine,[9] which form colored complexes when alkyllithiums are present. When one equivalent of 2-butanol is added the reaction mixture becomes colorless. Alternatively, alkyllithiums can be titrated with diphenylacetic acid,[10] 2,5-dimethoxybenzyl alcohol,[11] 1,3-diphenyl-2-propanone tosylhydrazone,[12] or 4-biphenylmethanol.[13] In each case, adding one equivalent of alkyllithium produces the colorless anion and adding a slight excess produces the colored dianion.

Related Compounds

These techniques should be generally applicable to other alkyllithium reagents. The original reference[6] states that the poly(dimethylsiloxane) technique can also be used for methyllithium and *tert*-butyllithium.

References

1. Lewis, R.J., Sr. *Sax's Dangerous Properties of Industrial Materials*; 8th ed.; Van Nostrand-Reinhold: New York, 1992; p. 623.
2. Bretherick, L. *Bretherick's Handbook of Reactive Chemical Hazards*; 4th ed.; Butterworths: London, 1990; p. 475.
3. Lane, C.F.; Kramer, G.W. Handling air-sensitive reagents. *Aldrichimica Acta* **1977**, *10*, 11–16.
4. Lunn, G. Unpublished observations.
5. Indicators for organolithium assay. *Aldrichimica Acta* **1988**, *21*, 14.
6. Suzuki, T. Alkyllithium Reagents. *Chem. Eng. News* **1991**, *February 11*, 2.
7. Crompton, T.R. *Comprehensive Organometallic Analysis*; Plenum Press: New York, 1987; pp. 181–193.
8. Watson, S.C.; Eastham, J.F. Colored indicators for simple direct titration of magnesium and lithium reagents. *J. Organometal. Chem.* **1967**, *9*, 165–168.
9. Bergbreiter, D.E.; Pendergrass, E. Analysis of organomagnesium and organolithium reagents using *N*-phenyl-1-naphthylamine. *J. Org. Chem.* **1981**, *46*, 219–220.
10. Kofron, W.G.; Baclawski, L.M. A convenient method for estimation of alkyllithium concentrations. *J. Org. Chem.* **1976**, *41*, 1879–1880.
11. Winkle, M.R.; Lansinger, J.M.; Ronald, R.C. 2,5-Dimethoxybenzyl alcohol: a convenient self-indicating standard for the determination of organolithium reagents. *J. Chem. Soc. Chem. Commun.* **1980**, 87–88.
12. Lipton, M.F.; Sorensen, C.M.; Sadler, A.C.; Shapiro, R.H. A convenient method for the accurate estimation of alkyllithium reagents. *J. Organometal. Chem.* **1980**, *186*, 155–158.
13. Juaristi, E.; Martínez-Richa, A.; García-Rivera, A.; Cruz-Sánchez, J.S. Use of 4-biphenylmethanol, 4-biphenylacetic acid, and 4-biphenylcarboxylic acid/triphenylmethane as indicators in the titration of lithium alkyls. Study of the dianion of 4-biphenylmethanol. *J. Org. Chem.* **1983**, *48*, 2603–2606.

CALCIUM CARBIDE

CAUTION! Refer to safety considerations section on page 7 before starting any of these procedures.

Calcium carbide (acetylenogen, calcium acetylide, or CaC_2) [75-20-7] is used in the laboratory to generate acetylene. Calcium carbide reacts with small quantities of H_2O to generate acetylene and when this is done in an uncontrolled fashion explosive acetylene–air mixtures may be formed. Decomposition under controlled conditions generates acetylene, which is vented into the fume hood, and calcium salts. With solutions of silver and copper salts the corresponding, explosive, silver and copper acetylides are formed. Calcium carbide reacts vigorously with methanol after an induction period; it is incompatible with a variety of reagents including hydrogen chloride, iron(III) chloride, iron(III) oxide, lead difluoride, magnesium, and sodium peroxide.[1]

Destruction Procedures

1. Slowly add, in small portions, 5 g of CaC_2 to 250 mL of H_2O stirred in a flask in a fume hood. When the reaction has ceased, neutralize the aqueous solution, and discard it.[2]

2. Place CaC_2 (50 g) in a 2-L flask and suspend it by stirring in 600 mL of toluene or cyclohexane. Surround the flask with an ice bath and pass nitrogen into the flask. Exhaust the acetylene that is generated through a plastic tube into the back of the fume hood. Add hydrochloric acid (6 *M*, 300 mL) dropwise from a dropping funnel over about 5 h. Stir the mixture for another hour, then neutralize the aqueous layer, separate the layers, and discard them.[3]

References

1. Bretherick, L. *Bretherick's Handbook of Reactive Chemical Hazards*; 4th ed.; Butterworths: London, 1990; pp. 203–205.
2. Lunn, G. Unpublished results.
3. National Research Council, Committee on Hazardous Substances in the Laboratory. *Prudent Practices for Disposal of Chemicals from Laboratories;* National Academy Press: Washington, DC, 1983; p. 96.

CARBAMIC ACID ESTERS

CAUTION! Refer to safety considerations section on page 7 before starting any of these procedures.

Four carbamic acid esters were investigated: methyl carbamate [598-55-0] (MC, $CH_3OC(O)NH_2$);[1] ethyl carbamate,[2] which is more commonly called urethane [51-79-6] [UT, $CH_3CH_2OC(O)NH_2$]; *N*-methylurethane [105-40-8] [MUT, $CH_3CH_2OC(O)NHCH_3$]; and *N*-ethylurethane [623-78-9] [EUT, $CH_3CH_2OC(O)NHCH_2CH_3$]. Methyl carbamate (mp 56–58°C) and urethane (mp 48.5–50°C) are solids and *N*-methylurethane (bp 170°C) and *N*-ethylurethane (bp 176°C) are liquids. However, liquified methyl carbamate (bp 176–177°C) and urethane (bp 182–184°C) have fairly low boiling points, so all these compounds should be regarded as volatile and should only be handled inside a chemical fume hood. Urethane,[3] *N*-methylurethane,[4] and *N*-ethylurethane[4] are carcinogenic in experimental animals. Evidence for the carcinogenicity of methyl carbamate is inconclusive.[5] Urethane is a teratogen, causes depression of bone marrow and nausea, and can affect the brain and central nervous system.[6] Urethane is used industrially as an intermediate and it also appears to form naturally by fermentation in some alcoholic beverages. Methyl carbamate is used industrially in the textile and pharmaceutical industries. All of these compounds are soluble in H_2O and ethanol.

Principles of Destruction

These compounds are all hydrolyzed using 5 *M* sodium hydroxide (NaOH) solution,[7] although the reaction times vary. Methyl carbamate is hydrolyzed to methanol and carbamic acid and UT is hydrolyzed to ethanol and carbamic acid. *N*-Methylurethane is hydrolyzed to ethanol and *N*-methylcarbamic acid and EUT is hydrolyzed to ethanol and *N*-ethylcarbamic acid. Carbamic acid decomposes to carbon dioxide (CO_2) and ammonia, *N*-methylcarbamic acid decomposes to CO_2 and methylamine, and *N*-ethylcarbamic acid decomposes to CO_2 and ethylamine. In all cases destruction is greater than 99% and good accountances are obtained for the products (Table 1).

Destruction Procedures

Destruction of N-*Methylurethane, Methyl Carbamate, and Urethane*

Add 50 mg of the compound to 10 mL of 5 *M* NaOH solution and stir at room temperature for 24 h (MC, UT) or 48 h (MUT). Check the reaction mixture for completeness of destruction, neutralize, and discard it.
Note: This procedure is **not** suitable for *N*-ethylurethane.

Destruction of N-*Methylurethane,* N-*Ethylurethane, Methyl Carbamate, and Urethane*

Add 50 mg of the compound to 10 mL of 5 *M* NaOH solution and reflux for 4 h. Cool, check the reaction mixture for completeness of destruction, neutralize, and discard it.

Analytical Procedures

The carbamic acid esters were determined by gas chromatography (GC) using a 1.8-m × 2-mm i.d. packed column filled with 5% Carbowax 20 M on 80/100 Chromosorb W HP.[7] The column was fitted with a precolumn and it was found helpful to change it regularly. The oven temperature was 120°C (MUT and EUT) or 140°C (MC and UT), the injection temperature

Table 1 Summary of Reaction Conditions for the Hydrolysis of RNHC(O)OR′

Compound	Conditions	Time (h)	RNH_2 (%)	R′OH (%)	Amount Remaining (%)
MC	Room temperature	24		98	<0.61
	Reflux	4		75	<0.61
UT	Room temperature	24		103	<0.15
	Reflux	4		51	<0.16
MUT	Room temperature	48	60	92	<0.075
	Reflux	4	39	69	<0.075
EUT	Reflux	4	43	74	<0.15

was 200°C, and the flame ionization detector operated at 300°C. Injection of reaction mixtures containing 5 *M* NaOH solution onto the hot GC column degraded any residual carbamate and gave unreliable results. To get around this, 2 mL of the reaction mixture were acidified, before analysis, with 1 mL of concentrated hydrochloric acid (**Caution!** Exothermic) and this mixture was then neutralized by adding solid sodium bicarbonate. Injection of this solution onto the column gave reliable results. The absence of any carbamate could be confirmed by spiking the reaction mixture with a small amount of a dilute solution of the carbamate in question. The products of the reaction were determined using the same conditions using a column packed with 10% Carbowax 20 M + 2% KOH in 80/100 Chromosorb W AW with an oven temperature of 150°C.

The GC conditions given above are only a guide; the exact conditions would have to be determined experimentally. Using 5-μL injections our detection limits were about 30 μg/mL (MC), 7 μg/mL (UT), and 4 μg/mL (MUT and EUT).

Mutagenicity Assays

The mutagenicity assays were carried out as described on page 4 using tester strains TA98, TA100, TA1530, and TA1535. The final reaction mixtures (tested at a level corresponding to 0.5 mg of undegraded material per plate) were not mutagenic. The only pure compound that was mutagenic was MC [tested in dimethyl sulfoxide (DMSO) solution], which was mutagenic to TA98 with activation. Ammonium carbamate, which is related to a possible intermediate in the degradation reactions, was not mutagenic.

Related Compounds

The procedure should be generally applicable to the destruction of carbamic acid esters, but it should be carefully checked to ensure that the compounds are completely degraded. The resistance to hydrolysis appears to increase as the degree of substitution increases. More highly substituted carbamic esters may require prolonged refluxing.

References

1. Other names for this compound are urethylane, Bendiocarb, and NCI-C55594.
2. Other names for this compound are urethan, A 11032, Estane 5703, ethylurethan, Leucethane, Leucothane, NSC-746, pracarbamin, pracarbamine, and U-compound.
3. International Agency for Research on Cancer. *IARC Monographs on the Evaluation of Carcinogenic Risk of Chemicals to Man.* Volume 7, *Some Anti-thyroid and Related Substances, Nitrofurans and Industrial Chemicals*; International Agency for Research on Cancer: Lyon, 1974; pp. 111–140.
4. Larsen, C.D. Pulmonary-tumor induction with alkylated urethans. *J. Natl. Cancer Inst.* **1948**, *9*, 35–37.
5. International Agency for Research on Cancer. *IARC Monographs on the Evaluation of Carcinogenic Risk of Chemicals to Man.* Volume 12, *Some Carbamates, Thiocarbamates and Carbazides*; International Agency for Research on Cancer: Lyon, 1976; pp. 151–159.
6. Lewis, R.J., Sr. *Sax's Dangerous Properties of Industrial Materials*, 8th ed.; Van Nostrand-Reinhold: New York, 1992; pp. 3471–3472.
7. Lunn, G.; Sansone, E. B. Validated methods for degrading hazardous chemicals: Some alkylating agents and other compounds. *J. Chem. Educ.* **1990**, *67*, A249–A251.

CHLOROMETHYLSILANES AND SILICON TETRACHLORIDE

CAUTION! Refer to safety considerations section on page 7 before starting any of these procedures.

The chloromethylsilanes, chlorotrimethylsilane (or trimethylsilyl chloride) [75-77-4] [$(CH_3)_3SiCl$],[1] dichlorodimethylsilane [$(CH_3)_2SiCl_2$] [75-78-5],[2] and methyltrichlorosilane (CH_3SiCl_3) [75-79-6][3] are flammable, volatile, toxic, corrosive liquids used in organic chemistry. Silicon tetrachloride ($SiCl_4$) [10026-04-7][4] is a volatile, toxic, corrosive liquid[5] used in the preparation of pure silicon, for producing smokescreens, and in inorganic chemistry.

Destruction Procedure[6]

Hydrolyze by cautiously adding 5 mL of the compound to 100 mL of vigorously stirred H_2O in a flask. The reaction produces hydrochloric acid and polymeric silicon-containing material. Remove any insoluble material and discard it with the solid or liquid waste. Neutralize the aqueous layer and discard it.

References

1. Lewis, R.J., Sr. *Sax's Dangerous Properties of Industrial Materials*, 8th ed.; Van Nostrand-Reinhold: New York, 1992; p. 886.
2. Reference 1, p. 1140.
3. Reference 1, p. 2408.
4. Other names for this compound are silicon(IV) chloride and Extrema.
5. Reference 1, pp. 3042–3043.
6. Patnode, W.; Wilcock, D.F. Methylpolysiloxanes. *J. Am. Chem. Soc.* **1946**, *68*, 358–363.

N-CHLOROSUCCINIMIDE

CAUTION! Refer to safety considerations section on page 7 before starting any of these procedures.

N-Chlorosuccinimide (NCS, I) [128-09-6][1] is an oxidizing and a chlorinating agent. It is used in organic chemistry. It is a white crystalline solid (mp 150–151 °C). It may be a carcinogen.[2] *N*-Chlorosuccinimide reacts violently with aliphatic alcohols, benzylamine, and hydrazine hydrate[3] and it has been known to smoulder on storage.[2,4]

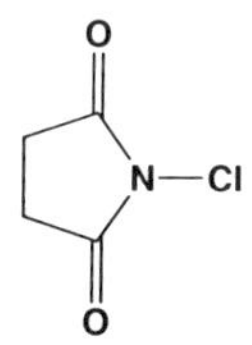

O

N—Cl

O

I

Destruction Procedure[5]

Add 5 g of the compound to 100 mL of 10% (w/v) sodium metabisulfite solution and stir the mixture at room temperature. Test for completeness of destruction by adding a few drops of the reaction mixture to an equal volume of 10% (w/v) potassium iodide solution, acidifying with 1 *M* hydrochloric acid, and adding a drop of starch as an indicator. A deep blue color indicates the presence of excess oxidant. If destruction is complete, discard the mixture. If destruction is not complete, add more sodium metabisulfite solution until a negative test is obtained.

Related Compounds

This technique should be generally applicable to other *N*-chloro compounds.

References

1. Other names for this compound are 1-chloro-2,5-pyrrolidinedione, succinchlorimide, succinochlorimide, and *N*-chlorsuccinimide.
2. Lewis, R.J., Sr. *Sax's Dangerous Properties of Industrial Materials*, 8th ed.; Van Nostrand-Reinhold: New York, 1992; p. 3150.
3. Bretherick, L. *Bretherick's Handbook of Reactive Chemical Hazards*; 4th ed.; Butterworths: London, 1990; pp. 1622–1623.
4. Reference 3, p. 424.
5. Lunn, G. Unpublished observations.

CHLOROSULFONIC ACID

CAUTION! Refer to safety considerations section on page 7 before starting any of these procedures.

Chlorosulfonic acid ($ClSO_2OH$) [7790-94-5][1] is a volatile (bp 151–152 °C), corrosive liquid that is used in organic chemistry as a chlorosulfonating and condensing agent. It is used for preparing sulfate esters, sulfones, and saccharin. It reacts violently with H_2O but can be hydrolyzed in a controlled fashion by adding it to crushed ice. It is corrosive, causes severe acid burns, and is very irritating to the eyes, lungs, and mucous membranes.[2] It reacts violently with a wide variety of compounds.[2,3]

Destruction Procedure[4]

Cautiously add 5 mL of chlorosulfonic acid to 100 g of crushed ice. Stir the reaction mixture until it reaches room temperature and the reaction is over, neutralize, and discard it.

Related Compounds

This procedure is specific for chlorosulfonic acid. Acid halides, sulfonyl halides, and acid anhydrides can generally be degraded by adding them to 2.5 *M* sodium hydroxide solution (see the monograph on Acid Halides and Anhydrides).

References

1. Other names for this compound are chlorosulfuric acid, monochlorosulfuric acid, sulfonic acid monochloride, and sulfuric chlorohydrin.
2. Lewis, R.J., Sr. *Sax's Dangerous Properties of Industrial Materials*, 8th ed.; Van Nostrand-Reinhold: New York, 1992; pp. 876–877.
3. Bretherick, L. *Bretherick's Handbook of Reactive Chemical Hazards*; 4th ed.; Butterworths: London, 1990; p. 951.
4. Lunn, G. Unpublished observations.

CHROMIUM(VI)

CAUTION! Refer to safety considerations section on page 7 before starting any of these procedures.

Chromium(VI) is an oxidizer and a carcinogen in humans and experimental animals.[1-3] It is widely used in organic synthesis and is one of the hazardous constituents of chromic acid. Because compounds containing chromium(VI) are powerful oxidizers they can react violently with a variety of organic and inorganic compounds.[4] Chromic acid and its salts are poisonous and corrosive to the skin and mucous membranes forming ulcers that are slow to heal.[4] Less hazardous alternatives to chromic acid are available and they should be used whenever possible. EOSULF, an ethylenediaminetetraacetic acid (EDTA)–organosulfonate-based detergent, has been tested and found to be just as effective as chromic acid for cleaning glassware.[5]

The compounds for which this procedure has been validated are chromium trioxide [1333-82-0] [CrO_3, chromium(VI) oxide, chromic anhydride], sodium dichromate [10588-01-9] ($Na_2Cr_2O_7.2H_2O$, sodium bichromate), potassium dichromate [7778-50-9] ($K_2Cr_2O_7$, potassium bichromate), ammonium dichromate [7789-09-5] [$(NH_4)_2Cr_2O_7$, ammonium bichromate], chromic acid (a solution of chromium(VI) in concentrated sulfuric acid (H_2SO_4)), and the commercially available solution Chromerge, both in the concentrated and diluted forms.

Principles of Destruction

Chromium(VI) is reduced to chromium(III), which is not an oxidizer, using sodium metabisulfite and the chromium(III) is precipitated by basification with magnesium hydroxide [$Mg(OH)_2$]. The International Agency for Research on Cancer (IARC) has reported that there is inadequate evidence that chromium(III) is carcinogenic.[3] However, chromium(III) should not be discharged to the environment[6] because it may become reoxidized to chromium(VI) in soil[7–10] or in water treatment plants.[6] Sodium or potassium hydroxide give a gelatinous precipitate, which is hard to filter. Precipitates that are easier to filter can be obtained by careful control of the pH,[11] but $Mg(OH)_2$ automatically produces the right pH and the sludgelike precipitate is relatively easy to filter.[12] The clear filtrate is just slightly basic (pH 7.1–9.2) and contains no trace of chromium(VI) (<0.25 ppm) and only trace amounts of chromium(III).

Destruction Procedures

Disposal of Bulk Quantities of Chromium(VI)-Containing Compounds (Sodium Dichromate, Potassium Dichromate, Ammonium Dichromate, Chromium Trioxide, and Chromerge Concentrate)

Stir the chromium compound (5 g) in 100 mL of 0.5 *M* H_2SO_4. When it has completely dissolved add 10 g of sodium metabisulfite. Stir the mixture for 1 h and allow to cool, then check for the presence of chromium(VI). Mix a few drops of the reaction mixture with a few drops of 100 mg/mL potassium iodide (KI) solution. A dark color indicates that chromium(VI) is still present. (If necessary, the color can be made more apparent by adding a drop of starch solution.) If chromium(VI) is still present, add sodium metabisulfite until a negative test is obtained. Add $Mg(OH)_2$ (6 g) to the reaction mixture and stir the mixture for 1 h, then allow it to stand overnight. Decant the mixture into a suction filter apparatus so that the clear liquid is filtered first, then the green precipitate is sucked dry. If the filtrate is yellow, this may indicate the presence of chromium(VI). Check using the KI test described above (acidify first with a little dilute H_2SO_4).

If chromium(VI) is present in the filtrate, acidify with H_2SO_4 and repeat the process. The filtrate contains no trace of chromium(VI) and only traces of chromium(III). The sludge does not contain chromium(VI) but does contain large amounts of chromium(III). The sludge is, however, no longer an oxidizer. Dispose of appropriately.

If the reaction is performed on a larger scale than that described above, considerable heat is generated, particularly when the sodium metabisulfite is added, and it may be necessary to let a longer time elapse between stages to allow for cooling.

Disposal of Solutions Containing Chromium(VI) (For Example, New or Used Chromic Acid or Chromerge Solutions)

Carefully add the chromium solution (10 mL), with stirring, to 60 mL of H_2O and stir the mixture for at least 1 h until cool. Add sodium metabisulfite solution (100 mg/mL, 10 mL) and stir the mixture for a few minutes, then check for the presence of chromium(VI). Mix a few drops of the reaction mixture with a few drops of 100 mg/mL KI solution. A dark color indicates that chromium(VI) is still present. (If necessary the color can be made more apparent by adding a drop of starch solution.) If chromium(VI) is still present, add sodium metabisulfite until a negative test is obtained. Add $Mg(OH)_2$ (12 g) to the reaction mixture and stir the mixture for 1 h, then allow it to stand overnight. Decant the mixture into a suction filter apparatus so that the clear liquid is filtered first, then the green precipitate is sucked dry. If the filtrate is yellow, this may indicate the presence of chromium(VI). Check using the KI test described above (acidify first with a little dilute H_2SO_4). If chromium(VI) is present, acidify the filtrate with H_2SO_4 and repeat the process. The filtrate contains no trace of chromium(VI) and only traces of chromium(III). The sludge does not contain chromium(VI) but does contain large amounts of chromium(III). The sludge is, however, no longer an oxidizer. Dispose of appropriately.

If the reaction is performed on a larger scale than that described above, considerable heat is generated, particularly when the solution is initially diluted and when the $Mg(OH)_2$ is added, and it may be necessary to do these procedures more slowly and let a longer time elapse between stages to allow for cooling.

Analytical Procedures

Total chromium can be determined by flame atomic absorption spectroscopy using a hollow cathode lamp at 357.9 nm.

Chromium(VI) may be determined by using a colorimetric procedure.[13] Dissolve *sym*-diphenylcarbazide (0.20 g) in 100 mL of ethanol to prepare a reagent solution. Add 200 μL of 3 *M* H_2SO_4 to 3 mL of the reaction mixture and check to make sure that this mixture is acidic. Add 100 μL of the reagent solution and shake the mixture for a few seconds. Let it stand for 10 min, then determine the violet color at 540 nm against a suitable blank. About 0.25 ppm chromium(VI) produces a just visible violet color; high concentrations of chromium(VI) produce a very intense violet color that fades rapidly. If high concentrations of chromium(VI) are present, dilute the sample mixture and repeat the procedure. The response is linear to 4 ppm. The method is quite insensitive to chromium(III), but the method can be adapted to measuring total chromium by oxidizing all the chromium present to chromium(VI).[13]

Mutagenicity Assays

To test the filtrates mutagenicity assays were carried out as described on page 4 using tester strains TA98, TA100, TA1530, and TA1535. To each plate 100 μL of filtrate was applied. No mutagenic activity was found.

Related Compounds

This procedure is specific for chromium(VI) and should not be used for any other heavy metal (see also monographs on Mercury and Heavy Metals).

References

1. International Agency for Research on Cancer. *IARC Monographs on the Evaluation of Carcinogenic Risk of Chemicals to Man*. Volume 2, *Some Inorganic and Organometallic Compounds*; International Agency for Research on Cancer: Lyon, 1972; pp. 100–125.
2. International Agency for Research on Cancer. *IARC Monographs on the Evaluation of the Carcinogenic Risk of Chemicals to Humans*. Volume 23, *Some Metals and Metallic Compounds;* International Agency for Research on Cancer: Lyon, 1980; pp. 205–323.
3. International Agency for Research on Cancer. *IARC Monographs on the Evaluation of the Carcinogenic Risk of Chemicals to Humans, Supplement No. 7, Overall Evaluations of Carcinogenicity: An Updating of* IARC Monographs *Volumes 1 to 42*; International Agency for Research on Cancer: Lyon, 1987; pp. 165–168.
4. Lewis, R.J., Sr. *Sax's Dangerous Properties of Industrial Materials*, 8th ed.; Van Nostrand-Reinhold: New York, 1992; pp. 898–900.

5. Manske, P.L.; Stimpfel, T.M.; Gershey, E.L. A less hazardous chromic acid substitute for cleaning glassware. *J. Chem. Educ.* **1990**, *67*, A280–A282.
6. Environmental Protection Agency, Chromium(III) compounds; Toxic chemical release reporting; Community right-to-know. *Fed. Reg.* **1991**, *56*, 58859–58862.
7. Bartlett, R.; James, B. Behavior of chromium in soils: III. Oxidation. *J. Environ. Qual.* **1979**, *8*, 31–38.
8. James, B.R.; Bartlett, R.J. Behavior of chromium in soils. VI. Interactions between oxidation–reduction and organic complexation. *J. Environ. Qual.* **1983**, *12*, 173–176.
9. Bartlett, R.J. Chromium oxidation in soils and water: Measurements and mechanisms. In *Proceedings, Chromium Symposium 1986: An Update*; Serrone, D.M., Ed., Industrial Health Foundation, Inc.: Pittsburgh, PA, 1986; pp. 310–330.
10. Bartlett, R.J.; James, B.R. Mobility and bioavailability of chromium in soils. In *Advances in Environmental Science and Technology, Vol. 20*; Wiley: New York, 1988; pp. 267–304.
11. Armour, M-.A.; Browne, L.M.; Weir, G.L. Tested disposal methods for chemical wastes from academic laboratories. *J. Chem. Educ.* **1985**, *62*, A93–A95.
12. Lunn, G.; Sansone, E.B. A laboratory procedure for the reduction of Chromium(VI) to Chromium(III). *J. Chem. Educ.* **1989**, *66*, 443–445.
13. *Standard Methods for the Examination of Water and Wastewater*, 16th ed.; American Public Health Association: Washington DC, 1985; pp. 201–204.

CISPLATIN

CAUTION! Refer to safety considerations section on page 7 before starting any of these procedures.

The degradation of a number of antineoplastic drugs, including cisplatin, was investigated by the International Agency for Research on Cancer (IARC).[1] Cisplatin [*cis*-$Pt(NH_3)_2Cl_2$] [15663-27-1][2] is a solid (mp 270°C) slightly soluble in H_2O but insoluble in most organic solvents, except *N,N*-dimethylformamide (DMF). Cisplatin is mutagenic and carcinogenic in animals[3–5] and it is a teratogen and has effects on the bone marrow and kidney.[6] Cisplatin is used as an antineoplastic drug.

Principles of Destruction

Cisplatin can be destroyed by reduction to elemental platinum with zinc (~99% destruction) or by forming an inactive complex with sodium diethyldithiocarbamate. In the latter case the extent of destruction is unknown but the reaction mixtures formed were not mutagenic. Oxidation of cisplatin with potassium permanganate was found to give mutagenic residues.[1]

Destruction Procedures

Destruction of Bulk Quantities of Cisplatin

1. Dissolve the cisplatin in 2 *M* sulfuric acid (H_2SO_4) solution so that its concentration does not exceed 0.6 mg/mL. Stir the reaction mixture and add 3 g of zinc powder, in portions to avoid frothing, for each 100 mL of solution. Stir the reaction mixture overnight, check for completeness of destruction, neutralize, and discard it.

2. Dissolve the cisplatin in H_2O and for every 100 mg of cisplatin add 30 mL of a 0.68 *M* solution of sodium diethyldithiocarbamate in 0.1 *M* sodium hydroxide (NaOH). Add 30 mL of saturated aqueous sodium nitrate ($NaNO_3$) solution (a yellow precipitate may form), check for completeness of destruction, and discard the mixture.

Destruction of Cisplatin in Aqueous Solutions and Injectable Pharmaceutical Preparations of 5% Dextrose or 0.9% Saline

1. Dilute the solution with H_2O so that the concentration of the drug does not exceed 0.6 mg/mL and add concentrated H_2SO_4, with stirring, until a 2 *M* solution is obtained. After cooling, stir the reaction mixture and add 3 g of zinc powder, in portions to avoid frothing, for each 100 mL of solution. Stir the reaction mixture overnight, check for completeness of destruction, neutralize, and discard it.

2. For every 100 mg of cisplatin add 30 mL of a 0.68 *M* solution of sodium diethyldithiocarbamate in 0.1 *M* NaOH. Then add 30 mL of saturated aqueous $NaNO_3$ solution (a yellow precipitate may form), check for completeness of destruction, and discard the mixture.

Destruction of Cisplatin in Organic Solvents Miscible with Water

Add at least an equal volume of 4 *M* H_2SO_4, more if required, so that the concentration of the drug does not exceed 0.6 mg/mL. Stir the reaction mixture and add 3 g of zinc powder, in portions to avoid frothing, for each 100 mL of solution. Stir the reaction mixture overnight, neutralize, check for completeness of destruction, and discard it.

Destruction of Cisplatin in Urine[7]

For each 3 mL of urine containing cisplatin add 1 mL of a 10% solution of sodium diethyldithiocarbamate in 0.1 *M* NaOH and 2 mL of saturated aqueous $NaNO_3$ solution. After 10 min check for completeness of destruction and discard the mixture.

Decontamination of Glassware Contaminated with Cisplatin

1. Rinse the glassware at least four times with enough H_2O to completely wet the glass. Combine the rinses and dilute with H_2O, if necessary, so that the concentration of cisplatin does not exceed 0.6 mg/mL. Add concentrated H_2SO_4, with stirring, until a 2 *M* solution is obtained. After cooling, stir the reaction mixture and add 3 g of zinc powder, in portions to avoid frothing, for each 100 mL of solution. Stir the reaction mixture overnight, check for completeness of destruction, neutralize, and discard it.

2. Immerse the glassware in a mixture of equal volumes of sodium diethyldithiocarbamate solution (0.68 *M* in 0.1 *M* NaOH) and $NaNO_3$ solution (saturated aqueous). Check for completeness of destruction and discard it.

Decontamination of Spills of Cisplatin

Allow any organic solvent to evaporate and remove as much of the spill as possible by high efficiency particulate air (HEPA) vacuuming (not sweeping), then rinse the area with H_2O. Take up the rinse with absorbent material and allow the rinse and any absorbent material used to react with a mixture of equal volumes of sodium diethyldithiocarbamate solution (0.68 *M* in 0.1 *M* NaOH) and $NaNO_3$ solution (saturated aqueous). Check for completeness of decontamination by using a wipe moistened with H_2O. Analyze the wipe for the presence of the drug.

Analytical Procedures

Cisplatin presents an analytical problem. Solutions can be analyzed by atomic absorption using an acetylene–air flame and a platinum lamp at 260 nm, but this only shows the presence or absence of platinum. It does not show if the platinum exists as cisplatin or if it has been complexed in an inactive form. Since cisplatin is highly mutagenic, the absence of significant mutagenicity (defined as a number of revertants that is more than twice the background) is a reasonable test for the absence of cisplatin. A 5-μg/mL aqueous solution of cisplatin produced a significant mutagenic response in TA100 with or without activation. If atomic absorption is not available, mix 9 mL of the solution to be tested with 1 mL of 10% sodium diethyldithiocarbamate in 0.1 *M* NaOH solution and 1 mL of saturated aqueous $NaNO_3$ solution.[1] Shake the reaction mixture and allow to react for 1 h. Add 1 mL of H_2O-saturated chloroform and shake. Centrifuge this mixture (1200 × *g*) for 5 min, mix on a vortex mixer, and centrifuge for 10 min more. Analyze the organic layer by high-performance liquid chromatography (HPLC) using a 300 × 3.6-mm i.d. column of μ Bondapak CN with heptane : isopropanol (82:18) flowing at 2 mL/min as the mobile phase and a UV detector set at 254 nm. This technique will not distinguish between platinum in cisplatin and platinum bound in an inactive form.

Other methods of analyzing for cisplatin have been reviewed.[8] Electrochemical detection,[9,10] differential pulse amperometric detection,[11] and post-column reaction detection[12] have all been recommended for use with HPLC.

Mutagenicity Assays

In the IARC study[1] tester strains TA98, TA100, and TA1535 of *Salmonella typhimurium* were used with and without metabolic activation. The reaction mixtures were not mutagenic. When cisplatin in urine was treated, tester strains TA98, TA100, UTH8413, and UTH8414 were used without metabolic activation. No mutagenic activity was observed.[7]

Related Compounds

These procedures may be applicable to other platinum containing compounds, but any new application should be thoroughly validated both for complete destruction of the compound and for the production of nonmutagenic reaction mixtures. It has recently been reported[13] that inactivation by complexation with sodium diethyldithiocarbamate can also be used to treat 5 mg/mL saline solutions of *cis*-dichloro-*trans*-dihydroxybis(isopropylamine)platinum(IV) [62928-11-4]).[14]

References

1. Castegnaro, M.; Adams, J.; Armour, M-.A.; Barek, J.; Benvenuto, J.; Confalonieri, C.; Goff, U.; Ludeman, S.; Reed, D.; Sansone, E. B.; Telling, G., Eds., *Laboratory Decontamination and Destruction of Carcinogens in Laboratory Wastes: Some Antineoplastic Agents*; International Agency for Research on Cancer: Lyon, 1985 (IARC Scientific Publications No. 73).

2. Other names for this compound are *cis*-platinous diammine dichloride, *cis*-diamminedichloroplatinum, *cis*-dichlorodiammine platinum(II), *cis*-platinum(II) diaminedichloride, *cis*-DDP, CACP, CPDC, DDP, Cisplatyl, Neoplatin, Platinex, Platinol, Peyrone's chloride, NSC-119875, Briplatin, Cismaplat, Citoplatino, Lederplatin, Platamine, Platiblastin, Platinoxan, Platistin, Platosin, Randa, (SP-4-2)-diamminedichloroplatinum, *cis*-platinum II, CDDP, CPDD, and NCI-C55776.
3. International Agency for Research on Cancer. *IARC Monographs on the Evaluation of the Carcinogenic Risk of Chemicals to Humans, Supplement No. 4, Chemicals, Industrial Processes and Industries Associated with Cancer in Humans. IARC Monographs, Volumes 1 to 29*; International Agency for Research on Cancer: Lyon, 1982; pp. 93–94.
4. International Agency for Research on Cancer. *IARC Monographs on the Evaluation of the Carcinogenic Risk of Chemicals to Humans.* Volume 26, *Some Antineoplastic and Immunosuppressive Agents*; International Agency for Research on Cancer: Lyon, 1981; pp. 151–164.
5. International Agency for Research on Cancer. *IARC Monographs on the Evaluation of the Carcinogenic Risk of Chemicals to Humans, Supplement No. 7, Overall Evaluations of Carcinogenicity: An Updating of* IARC Monographs *Volumes 1 to 42*; International Agency for Research on Cancer: Lyon, 1987; pp. 170–171.
6. Lewis, R.J., Sr. *Sax's Dangerous Properties of Industrial Materials*, 8th ed.; Van Nostrand-Reinhold: New York, 1992; p. 2824.
7. Monteith, D.K.; Connor, T.H.; Benvenuto, J.A.; Fairchild, E.J.; Theiss, J.C. Stability and inactivation of mutagenic drugs and their metabolites in the urine of patients administered antineoplastic therapy. *Environ. Mol. Mutagenesis* **1987**, *10*, 341–356.
8. Riley, C.M.; Sternson, L.A. Recent advances in the clinical analysis of cisplatin. *Pharm. Int.* **1984**, *5*, 15–19.
9. Parsons, P.J.; LeRoy, A.F. Determination of *cis*-diamminedichloroplatinum(II) in human plasma using ion-pair chromatography with electrochemical detection. *J. Chromatogr.* **1986**, *378*, 395–408.
10. Parsons, P.J.; Morrison, P.F.; LeRoy, A.F. Determination of platinum-containing drugs in human plasma by liquid chromatography with reductive electrochemical detection. *J. Chromatogr.* **1987**, *385*, 323–335.
11. Elferink, F.; van der Vijgh, W.J.F.; Pinedo, H.M. Analysis of antitumour [1,1-bis(aminomethyl)cyclohexane]platinum(II) complexes derived from spiroplatin by high-performance liquid chromatography with differential pulse amperometric detection. *J. Chromatogr.* **1985**, *320*, 379–392.
12. Marsh, K.C.; Sternson, L.A.; Repta, A.J. Post-column reaction detector for platinum(II) antineoplastic agents. *Anal. Chem.* **1984**, *56*, 491–497.
13. Benvenuto, J.A.; Connor, T.H.; Monteith, D.K.; Laidlaw, J.L.; Adams, S.C.; Matney, T.S.; Theiss, J.C. Degradation and inactivation of antitumor drugs. *J. Pharm. Sci.* **1993**, *82*, 988–991.
14. Other names for this compound are iproplatin, CHIP, JM-28, and NSD 256 927.

CITRININ

CAUTION! Refer to safety considerations section on page 7 before starting any of these procedures.

Citrinin (**I**) [518-75-2][1] is a fungal metabolite from *Aspergillus niveus* and also from *Penicillium citrinum*. This compound is a teratogen and a severe skin irritant.[2] Citrinin may be carcinogenic in experimental animals[3] but it has not been found to be mutagenic.[4–6] In a recent collaborative study organized by the International Agency for Research on Cancer (IARC) the safe disposal of citrinin was investigated.[7]

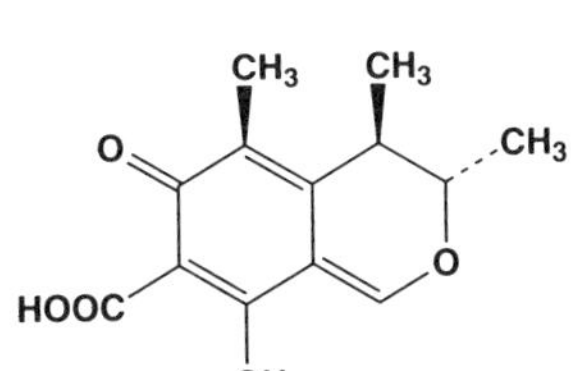

CH_3 CH_3 CH_3 O HOOC O OH

I

Citrinin is a yellow crystalline solid (mp 170–173°C). This compound is practically insoluble in H_2O but soluble in organic solvents. The solid compound may become electrostatically charged and cling to glassware or protective clothing.

Principle of Destruction

Citrinin may be degraded using dilute sodium hypochlorite (NaOCl) solution and by using potassium permanganate in sodium hydroxide solution ($KMnO_4$ in NaOH). Degradation efficiency was greater than 99.5% using NaOCl and greater than 99.9% using $KMnO_4$.

Destruction Procedures[7]

Destruction of Bulk Quantities of Citrinin

1. Prepare a dilute solution of NaOCl by adding 100 mL of commercial 5.25% NaOCl solution (Clorox bleach) to 200 mL of H_2O. Use fresh NaOCl solution (see assay procedure below). Dissolve each 1 mg of citrinin in 2 mL of methanol. For each 2 mL of solution add 50 mL of the dilute NaOCl solution. Sonicate to improve solubilization, allow to react for at least 30 min, check for completeness of destruction, and discard the reaction mixture.

2. Prepare a 0.3 *M* solution of $KMnO_4$ in 2 *M* NaOH solution by stirring the mixture for at least 30 min but no more than 2 h. Dissolve 2 mg of citrinin in 5 mL of acetonitrile and add 10 mL of $KMnO_4$ in NaOH. Stir for at least 3 h. The color should be either green or purple. If it is not, add more $KMnO_4$ in NaOH until the green or purple color persists for at least 1 h. For each 10 mL of $KMnO_4$ in NaOH add 0.8 g of sodium metabisulfite (more if necessary for complete decolorization), dilute with an equal volume of H_2O, filter to remove the manganese salts,[8] check for completeness of destruction, and discard the solid and filtrate appropriately.

Destruction of Citrinin in Aqueous Solution

1. Prepare a dilute solution of NaOCl by adding 100 mL of commercial 5.25% NaOCl solution (Clorox bleach) to 200 mL of H_2O. Use fresh NaOCl solution (see assay procedure below). If necessary adjust the pH of the citrinin solution to neutral or alkaline. For each 1 mg of citrinin present add 2 mL of methanol. For each 1 mg of citrinin add 50 mL of the dilute NaOCl solution. Sonicate to improve solubilization, allow to react for at

least 30 min, check for completeness of destruction, and discard the reaction mixture.

2. Dilute with H_2O, if necessary, so that the concentration of citrinin does not exceed 200 μg/mL. Add sufficient NaOH, with stirring, to make the concentration 2 *M*, then add sufficient solid $KMnO_4$ to make the concentration 0.3 *M*. Stir for at least 3 h. The color should be either green or purple. If it is not, add more $KMnO_4$ in NaOH until the green or purple color persists for at least 1 h. For each 10 mL of $KMnO_4$ in NaOH add 0.8 g of sodium metabisulfite (more if necessary for complete decolorization), dilute with an equal volume of H_2O, filter to remove the manganese salts,[8] check for completeness of destruction, and discard the solid and filtrate appropriately.

Destruction of Citrinin in Volatile Organic Solvents

1. Prepare a dilute solution of NaOCl by adding 100 mL of commercial 5.25% NaOCl solution (Clorox bleach) to 200 mL of H_2O. Use fresh NaOCl solution (see assay procedure below). Remove the solvent under reduced pressure using a rotary evaporator. Add enough methanol to wet the glass, adding at least 2 mL of methanol for each 1 mg of citrinin present. For each 2 mL of solution add 50 mL of the dilute NaOCl solution. Sonicate to improve solubilization, allow to react for at least 30 min, check for completeness of destruction, and discard the reaction mixture.

2. Prepare a 0.3 *M* solution of $KMnO_4$ in 2 *M* NaOH solution by stirring the mixture for at least 30 min but no more than 2 h. Remove the organic solvent under reduced pressure using a rotary evaporator. For each 2 mg of citrinin present add 5 mL of acetonitrile and swirl until dissolved. Next add 10 mL of the solution of $KMnO_4$ in NaOH. Stir for at least 3 h. The color should be either green or purple. If it is not, add more $KMnO_4$ in NaOH until the green or purple color persists for at least 1 h. For each 10 mL of $KMnO_4$ in NaOH add 0.8 g of sodium metabisulfite (more if necessary for complete decolorization), dilute with an equal volume of H_2O, filter to remove the manganese salts,[8] check for completeness of destruction, and discard the solid and filtrate appropriately.

Destruction of Citrinin in Dimethyl Sulfoxide or N,N-Dimethylformamide

Prepare a dilute solution of NaOCl by adding 100 mL of commercial 5.25% NaOCl solution (Clorox bleach) to 200 mL of H_2O. Use fresh NaOCl solution (see assay procedure below). Dilute the dimethyl sulfoxide (DMSO) or *N,N*-dimethylformamide (DMF) solution with two volumes of H_2O and extract three times with equal volumes of dichloromethane, pool the extracts, and dry them over anhydrous sodium sulfate. Remove the sodium sulfate by filtration and wash it with one volume of dichloromethane. Evaporate to dryness and make sure that all the dichloromethane is removed under reduced pressure using a rotary evaporator. Add enough methanol to wet the glass adding at least 2 mL of methanol for each 1 mg of citrinin present. For each 2 mL of solution add 50 mL of the dilute NaOCl solution. Sonicate to improve solubilization, allow to react for at least 30 min, check for completeness of destruction, and discard the reaction mixture.

Decontamination of Glassware

1. Prepare a dilute solution of NaOCl by adding 100 mL of commercial 5.25% NaOCl solution (Clorox bleach) to 200 mL of H_2O. Use fresh NaOCl solution (see assay procedure below). Add enough methanol to wet the glassware and immerse it in the dilute NaOCl solution for at least 30 min, check for completeness of destruction, and discard the decontaminating solution.

2. Prepare a 0.3 *M* solution of $KMnO_4$ in 2 *M* NaOH solution by stirring the mixture for at least 30 min but no more than 2 h. Rinse the glassware five times with small portions of dichloromethane. Combine the rinses and evaporate the dichloromethane under reduced pressure using a rotary evaporator. Dissolve 2 mg of citrinin in 5 mL of acetonitrile and add 10 mL of $KMnO_4$ in NaOH. Stir for at least 3 h. The color should be either green or purple. If it is not, add more $KMnO_4$ in NaOH until the green or purple color persists for at least 1 h. For each 10 mL of $KMnO_4$ in NaOH add 0.8 g of sodium metabisulfite (more if necessary for complete decolorization), dilute with an equal volume of H_2O, filter to remove the manganese salts,[8] check for completeness of destruction, and discard the solid and filtrate appropriately.

Decontamination of Protective Clothing

Prepare a dilute solution of NaOCl by adding 100 mL of commercial 5.25% NaOCl solution (Clorox bleach) to 200 mL of H_2O. Use fresh NaOCl solution (see assay procedure below). Add enough methanol to wet the protective clothing and immerse it in the dilute NaOCl solution for at least

30 min, check for completeness of destruction, and discard the decontaminating solution.

Decontamination of Spills

1. Prepare a dilute solution of NaOCl by adding 100 mL of commercial 5.25% NaOCl solution (Clorox bleach) to 200 mL of H_2O. Use fresh NaOCl solution (see assay procedure below). Collect spills of liquid with a dry cloth and spills of solid with a tissue wetted with sodium bicarbonate solution (5% w/v). Wipe the area with a cloth wetted with sodium bicarbonate solution (5% w/v). Immerse all cloths in the dilute NaOCl solution, allow to react for at least 30 min, check for completeness of decontamination, and discard the decontaminating solution. Cover the spill area with the dilute NaOCl solution. After at least 30 min absorb the liquid with cloths and discard it. Check the surface for completeness of decontamination by using a wipe moistened with methanol and analyzing the wipe for the presence of citrinin.

2. Prepare a 0.3 *M* solution of $KMnO_4$ in 2 *M* NaOH solution by stirring the mixture for at least 30 min but no more than 2 h. Collect spills of liquid with a dry tissue and spills of solid with a tissue wetted with dichloromethane. Immerse all tissues in the $KMnO_4$ in NaOH solution. Allow to react for at least 3 h. The color should be either green or purple. If it is not, add more $KMnO_4$ in NaOH until the green or purple color persists for at least 1 h. For each 10 mL of $KMnO_4$ in NaOH add 0.8 g of sodium metabisulfite (more if necessary for complete decolorization), dilute with an equal volume of H_2O, filter to remove manganese salts,[8] check for completeness of destruction, and discard the solid and filtrate appropriately. Cover the spill area with an excess of the $KMnO_4$ in NaOH solution and allow to react for 3 h. Collect the solution on a tissue and immerse the tissue in 2 *M* sodium metabisulfite solution. If the pH of this solution is acidic, make it alkaline with NaOH. Rinse the spill area with a 2 *M* solution of sodium metabisulfite. Check the surface for completeness of decontamination by using a wipe moistened with methanol and analyzing the wipe for the presence of citrinin.

Decontamination of Thin-Layer Chromatography Plates

Prepare a dilute solution of NaOCl by adding 100 mL of commercial 5.25% NaOCl solution (Clorox bleach) to 200 mL of H_2O. Use fresh NaOCl solution (see assay procedure below). Spray the plate with the dilute NaOCl solution and allow to react for at least 30 min. Check for completeness of destruction by scraping the plate and eluting any remaining citrinin with a suitable solvent.

Analytical Procedures

1. For reaction mixtures from the NaOCl procedures acidify an aliquot of the final reaction mixture to pH 3–4 with concentrated hydrochloric acid (HCl) and pass nitrogen through the mixture for at least 1 min to remove chlorine. Analyze by reverse phase high-performance liquid chromatography (HPLC) using acetonitrile:0.25 *M* aqueous phosphoric acid 75:25 flowing at 1.5 mL/min and a UV detector set at 254 nm or a spectrofluorometric detector using 330 nm for excitation and 480 nm for emission.[7] An alternative mobile phase is methanol:0.02 *M* potassium dihydrogen phosphate (pH 4.7) 50:50.[9]

2. For reaction mixtures obtained using the $KMnO_4$ procedures, acidify an aliquot to pH 2–3 using concentrated HCl. Extract this mixture three times with an equal volume of dichloromethane, pool the extracts, and dry them over anhydrous sodium sulfate. Remove the sodium sulfate by filtration, evaporate to dryness and take up the residue in 0.5 mL of methanol:water 75:25. Analyze by HPLC as above.

Mutagenicity Assays

The residues from these degradation procedures were tested using tester strains TA97, TA98, TA100, and TA102 of *Salmonella typhimurium* with and without metabolic activation. No mutagenic activity was found.[7] Citrinin has not been found to be mutagenic.[4–6]

Related Compounds

The above techniques were investigated for citrinin but they may also be applicable to some other mycotoxins. However, these techniques should be thoroughly investigated before being applied to other compounds. See also monographs on Aflatoxins, Ochratoxin A, Patulin, and Sterigmatocystin.

Assay of Sodium Hypochlorite Solution

Sodium hypochlorite solutions tend to deteriorate with time, so these solutions should be periodically checked for the amount of active chlorine they contain. Pipette 10 mL of the NaOCl solution into a 100-mL volumetric flask and fill it to the mark with distilled H_2O. Pipette 10 mL of this solution into a conical flask containing 50 mL of distilled H_2O, 1 g of potassium iodide, and 12.5 mL of 2 *M* acetic acid. Titrate this solution against a 0.1 *N* sodium thiosulfate solution using starch as an indicator. Each 1 mL of the sodium thiosulfate solution corresponds to 3.545 mg of active chlorine. Commercially available NaOCl solution (Clorox bleach) contains 5.25% NaOCl and should contain about 45–50 g of active chlorine per liter.

References

1. Other names for this compound are (3*R-trans*)-4,6-dihydro-8-hydroxy-3,4,5-trimethyl-6-oxo-3*H*-2-benzopyran-7-carboxylic acid, (3*R*,4*S*)-4,6-dihydro-8-hydroxy-3,4,5-trimethyl-6-oxo-3*H*-2-benzopyran-7-carboxylic acid, antimycin, and S-52.
2. Lewis, R.J., Sr. *Sax's Dangerous Properties of Industrial Materials*; 8th ed.; Van Nostrand-Reinhold: New York, 1992; p. 917.
3. International Agency for Research on Cancer. *IARC Monographs on the Evaluation of the Carcinogenic Risk of Chemicals to Humans*, Volume *40*, *Some Naturally Occurring and Synthetic Food Components, Furocoumarins and Ultraviolet Radiation*; International Agency for Research on Cancer: Lyon, 1986; pp. 67–82.
4. Hayes, A.W. *Mycotoxin Teratogenicity and Mutagenicity*; CRC Press: Boca Raton, FL, 1981.
5. Bendele, A.M.; Neal, S.B.; Oberley, T.J.; Thompson, C.Z.; Bewsey, B.J.; Hill, L.E.; Rexroat, M.A.; Carlton, W.W.; Probst, G.S. Evaluation of ochratoxin A for mutagenicity in a battery of bacterial and mammalian cell assays. *Fd. Chem. Toxicol.* **1985**, *23*, 911–918.
6. Würgler, F.E.; Friedrich, U.; Schlatter, J. Lack of mutagenicity of ochratoxin A and B, citrinin, patulin and cnestine in *Salmonella typhimurium* TA102. *Mutat. Res.* **1991**, *261*, 209–216.
7. Castegnaro, M.; Barek, J.; Frémy, J-.M.; Lafontaine, M.; Miraglia, M.; Sansone, E.B.; Telling, G.M., Eds. *Laboratory Decontamination and Destruction of Carcinogens in Laboratory Wastes: Some Mycotoxins*; International Agency for Research on Cancer: Lyon, 1991 (IARC Scientific Publications No. 113).
8. Lunn, G.; Sansone, E.B.; De Méo, M.; Laget, M.; Castegnaro, M. Potassium permanganate can be used for degrading hazardous compounds. *Am. Ind. Hyg. Assoc. J.* **1994**, *55*, 167–171.
9. Lunn, G. Unpublished observations.

COMPLEX METAL HYDRIDES

CAUTION! Refer to safety considerations section on page 7 before starting any of these procedures.

Complex metal hydrides are widely used in organic synthesis. These hydrides are generally air and H_2O sensitive and may be pyrophoric. Vigorous reaction with H_2O may ignite the flammable hydrocarbon solvents generally used with these reagents. On reaction with H_2O, flammable hydrogen is released.[1]

Compound Name	Reference	Formula	Registry Number
Calcium hydride	2	CaH_2	[7789-78-8]
Borane.THF complex	3	$BH_3 \cdot C_4H_8O$	[14044-65-6]
Lithium aluminum hydride	4,5	$LiAlH_4$	[16853-85-3]
Lithium hydride	6	LiH	[7580-67-8]
Potassium hydride	7	KH	[7693-26-7]
Sodium borohydride (sodium tetrahydroborate)	8	$NaBH_4$	[16940-66-2]
Sodium cyanoborohydride (sodium cyanotrihydridoborate)		$NaBH_3CN$	[25895-60-7]
Sodium hydride	9	NaH	[7646-69-7]

All of these reagents can react violently with a variety of organic and inorganic compounds. Sodium borohydride is toxic by ingestion,[8] borane.THF complex may explode on storage,[3] and lithium hydride[6] is irritating to the skin and eyes. However, all these compounds should be handled as if they were strong irritants and toxic. Lithium aluminum hydride may ignite when ground.[4] Fires involving $LiAlH_4$ should be smothered with dry sand.[10] As an added hazard sodium cyanoborohydride generates one equivalent of cyanide when it is used as a reducing agent.[11]

Principles of Destruction

In general, the hydride is allowed to react slowly with H_2O, an alcohol, or ethyl acetate under controlled conditions. Although sodium and potassium hydride react readily with butanol, lithium hydride does not and so it is necessary to use H_2O to destroy this compound. Sodium borohydride is more stable and it is necessary to use acetic acid to cause decomposition. Because sodium cyanoborohydride generates an equivalent of cyanide as it reacts,[11] sodium hypochlorite (NaOCl) solution (Clorox bleach) is used. The hydride is oxidized to hydrogen chloride and the cyanide is oxidized to the much less toxic cyanate. A typical reduction procedure with $NaBH_3CN$ involves the use of methanol as a reaction solvent.[12] The reaction takes place at pH 3 and some cyanide is evolved as the reaction proceeds. The reaction should be done in a hood but the cyanide can be removed from the exhaust gas by bubbling it through NaOCl solution. The methanol is evaporated using a rotary evaporator and the residue is taken up in H_2O and extracted with diethyl ether. The distillate collected in the evaporator also contains cyanide. All of these solutions can be decontaminated with NaOCl solution.[13] Destruction is $>99.7\%$. Methanol is added to solubilize the ether layer. The researcher should be aware that when the reaction mixture is evaporated most of the cyanide disappears, presumably through the exhaust of the aspirator and down the sink. This condition may be important if the reaction is carried out on a large scale.

Destruction Procedures

Lithium Aluminum Hydride

1. Stir the $LiAlH_4$ in a suitable solvent and, for each *n* grams of $LiAlH_4$ present, slowly add *n* mL of H_2O under nitrogen.[14] Use an ice bath and

an efficient bubbler that can deal with the large quantities of gas produced in the course of the reaction. Add *n* mL of 15% sodium hydroxide solution and 3*n* mL of H_2O in succession and stir the mixture vigorously for 20 min. Filter the granular precipitate that forms. Separate the organic and aqueous layers of the filtrate and discard them.

2. Stir the $LiAlH_4$ under nitrogen in a suitable solvent using an ice bath and slowly add 7 mL of 95% ethanol for each gram of $LiAlH_4$. The 95% ethanol reacts less vigorously than H_2O.[15] Use an efficient bubbler that can deal with the large quantities of gas produced in the course of the reaction. When the reaction is complete cautiously add a volume of H_2O equal to the initial reaction volume, separate the organic and aqueous layers, and discard them.

3. Stir the $LiAlH_4$ under nitrogen in a suitable solvent using an ice bath and slowly add 11 mL of ethyl acetate for each gram of $LiAlH_4$. Ethyl acetate reacts less vigorously than H_2O and generates no hydrogen.[16] When the reaction is complete, add a volume of H_2O equal to the initial reaction volume, separate the organic and aqueous layers of the filtrate and discard them.

Sodium Borohydride

Sodium borohydride is relatively stable in H_2O and acid is needed for its decomposition.[15] Dissolve the solid compound in H_2O and dilute aqueous solutions with H_2O, if necessary, so that the concentration does not exceed 3%. For each 100 mL of solution add 1 mL of 10% (v/v) aqueous acetic acid with stirring under nitrogen.[13] Discard when the reaction has ceased.

Borane.THF Complex

Add 30 mL of a 1 *M* solution of borane.tetrahydrofuran (BH_3.THF) in THF to 30 mL of acetone with stirring under nitrogen. After 5 min add 30 mL of H_2O. When the reaction has ceased add 30 mL of H_2O and discard the mixture.[13]

Sodium Hydride and Potassium Hydride

Sodium hydride and potassium hydride are generally supplied as dispersions in mineral oil. For each gram of the hydride dispersion add 25 mL of dry isooctane (2,2,4-trimethylpentane) and stir the mixture under nitrogen. For each gram of hydride dispersion slowly add 10 mL of *n*-butyl alcohol. Ensure that no unreacted material remains on the side of the flask. After 30 min, or when the reaction appears to have stopped, add 25 mL of cold H_2O for each gram of hydride dispersion.[13] Separate the layers and discard them.

Lithium Hydride

Add 1 gram of LiH to 50 mL of H_2O with stirring. When the reaction has finished discard the reaction mixture.[13]

Calcium Hydride

1. Add 1 g of CaH_2 to 25 mL of 95% ethanol, with stirring, under nitrogen. When the reaction has finished add an equal volume of H_2O and discard it.[13]

2. Add 1 g of CaH_2 to 50 g of crushed ice behind a safety shield. When the reaction has ceased discard the reaction mixture.[13]

Sodium Cyanoborohydride

1. Dissolve each gram of solid $NaBH_3CN$ in 10 mL of H_2O.[13] If necessary, dilute reaction mixtures with H_2O so that the concentration of $NaBH_3CN$ does not exceed 10%. Stir the $NaBH_3CN$ solution and cautiously add 200 mL of a 5.25% NaOCl solution (Clorox bleach) for each gram of $NaBH_3CN$ present. Stir the reaction mixture for 3 h, check that the solution is still oxidizing, analyze for complete destruction of cyanide, and discard it. Use fresh NaOCl solution (see below for assay procedure).

2. For each 80 mL of ether extract that may contain cyanide add 200 mL of a 5.25% NaOCl solution (Clorox bleach) and 150 mL of methanol to produce one phase. Stir the reaction mixture for 3 h, check that the solution is still oxidizing, analyze for complete destruction of cyanide, and discard it. Use fresh NaOCl solution (see below for assay procedure).

3. Pass the exhaust gases through 5.25% NaOCl solution (Clorox bleach). After standing for several hours check that the solution is still oxidizing, test for completeness of destruction, and discard it. Use fresh NaOCl solution (see below for assay procedure).

Assay of Sodium Hypochlorite Solution

Sodium hypochlorite solutions tend to deteriorate with time so they should be periodically checked for the amount of active chlorine they contain.

Pipette 10 mL of NaOCl solution into a 100-mL volumetric flask and fill to the mark with distilled H_2O. Pipette 10 mL of this solution into a conical flask containing 50 mL of distilled H_2O, 1 g of potassium iodide (KI), and 12.5 mL of 2 *M* acetic acid. Titrate this solution against 0.1 *N* sodium thiosulfate solution using starch as an indicator. Each 1 mL of the sodium thiosulfate solution corresponds to 3.545 mg of active chlorine. The NaOCl solution used in these degradation reactions should contain 45–50 g of active chlorine per liter.

Analytical Procedures

For oxidizing power. Add a few drops of the reaction mixture to an equal volume of 10% (w/v) KI solution then acidify with a drop of 1 *M* hydrochloric acid (HCl) and add a drop of starch solution as an indicator. The deep blue color of the starch–iodine complex indicates that excess oxidant is present.

For residual cyanide.[17] Prepare the following solutions. The buffer solution is prepared by dissolving 13.6 g of potassium phosphate, monobasic (KH_2PO_4), 0.28 g of sodium phosphate, dibasic (Na_2HPO_4), and 3.0 g of potassium bromide (KBr) in distilled H_2O and making up to 1 L with distilled H_2O. Note that the presence of KBr is necessary for the assay procedure to work correctly. The reagent is prepared by stirring 3.0 g of barbituric acid in 10 mL of H_2O and adding 15 mL of 4-methylpyridine and 3 mL of concentrated HCl while continuing to stir. After cooling the solution is diluted to 50 mL with H_2O. The sodium ascorbate solution is 10 mg/mL in H_2O, the NaCN solution is 100 mg/L in H_2O, and the Chloramine-T solution is 10 mg/mL in H_2O (*not* 100 mg/mL as stated in the paper[17]). The sodium ascorbate and Chloramine-T solutions are prepared fresh daily and the standard NaCN solution is prepared fresh every week. The other solutions appear to be quite stable.

Centrifuge two portions of the reaction mixture, if necessary to remove suspended solids, and add 1 mL of each to 4 mL of buffer. If an orange or yellow color appears, add sodium ascorbate solution dropwise until the mixtures are colorless (but do not add more than 2 mL). Spike one solution with 200 μL of NaCN solution, and add 1 mL of Chloramine-T solution to each solution. Shake the solutions and allow to stand for 1–2 min, then add 1 mL of the reagent and shake the mixtures and allow them to stand for 5 min. A blue color indicates the presence of cyanide. For complete destruction the unspiked solution should be colorless and the spiked solution should be blue. Measure the absorbance at 605 nm using 10-mm cuvettes (after centrifuging again if necessary to remove suspended solids). Appropriate standards and blanks should always be run. The limit of detection is about 3 μg/mL.

Related Compounds

Complex metal hydrides vary greatly in their reactivity and the application of any of the above methods to another hydride should be carefully, and cautiously, investigated before employing it. In general one should start with a less reactive substrate such as butanol and then, if necessary, move on to more reactive substrates such as 95% ethanol and H_2O.

References

1. Lewis, R.J., Sr. *Sax's Dangerous Properties of Industrial Materials*, 8th ed.; Van Nostrand-Reinhold: New York, 1992; pp. 1890–1891.
2. Reference 1, p. 668.
3. Reference 1, p. 529.
4. Reference 1, pp. 2130–2131.
5. Other names for this compound are lithium aluminohydride, lithium aluminum tetrahydride, lithium alanate, aluminum lithium hydride, LAH, Lithal, and lithium tetrahydroaluminate.
6. Reference 1, p. 2128.
7. Reference 1, pp. 2870–2871.
8. Reference 1, pp. 3066–3067.
9. Reference 1, p. 3085.
10. Bretherick, L., Ed., *Hazards in the Chemical laboratory*, 4th ed.; Royal Society of Chemistry: London, 1986; p. 169.
11. Lane, C.F. Sodium cyanoborohydride—A highly selective reducing agent for organic functional groups. *Synthesis* **1975**, 135–146.
12. Borch, R.F.; Bernstein, M.D.; Durst, H.D. The cyanohydridoborate anion as a selective reducing agent. *J. Am. Chem. Soc.* **1971**, *93*, 2897–2904.
13. Lunn, G. Unpublished observations.
14. Micovic, V.M.; Mihailovic, M.L. The reduction of acid amides with lithium aluminum hydride. *J. Org. Chem.* **1953**, *18*, 1190–1200.
15. National Research Council, Committee on Hazardous Substances in the Laboratory.

Prudent Practices for Disposal of Chemicals from Laboratories; National Academy Press: Washington, DC, 1983; p. 85.

16. Fieser, L.F.; Fieser, M. *Reagents for Organic Synthesis*; Wiley: New York, 1967; Vol. 1, p. 584.
17. Lunn, G.; Sansone, E.B. Destruction of cyanogen bromide and inorganic cyanides. *Anal. Biochem.* **1985**, *147*, 245–250.

CYANIDES AND CYANOGEN BROMIDE

CAUTION! Refer to safety considerations section on page 7 before starting any of these procedures.

Inorganic cyanides are acutely toxic compounds, for example, the LD_{50} in the rat is 15 mg/kg for sodium cyanide (NaCN) [143-33-9][1] and 10 mg/kg for potassium cyanide (KCN) [151-50-8].[2] These compounds should be handled with great care and not allowed to come in contact with acid that will generate hydrogen cyanide (HCN) [74-90-8], a volatile, highly toxic, flammable gas,[3] which forms explosive mixtures with air. Acute cyanide poisoning is rapidly fatal. Less acute cases may produce symptoms such as headache, dizziness, nausea, and so on.[3,4] Inorganic cyanides have a number of uses in organic synthesis. Cyanogen bromide (bromine cyanide, CNBr) [506-68-3] is a highly toxic, volatile crystalline solid (mp 49–51°C) with its toxic effects being those of HCN.[5] A concentration of 92 ppm for 10 min has caused a fatality[6] and low concentrations (10 ppm) are irritating to the eyes, nose, and respiratory tract.[5,6] It should be handled with great care only in a properly functioning chemical fume hood. Cyanogen bromide is used in laboratories for the activation of agarose beads in affinity chromatography,[7] in the analysis of protein structure,[8] and for the detection of pyridine compounds.[9]

Principles of Destruction

Inorganic cyanides and CNBr are oxidized by sodium or calcium hypochlorite [NaOCl or $Ca(OCl)_2$],[10,11] in basic solution, to the much less toxic cyanate ion (mouse LD_{50} for sodium cyanate = 260 mg/kg[12]). Further hydrolysis of the cyanate ion is also possible. Some reactions (e.g., reduction with sodium cyanoborohydride) release HCN. This compound can be removed by passing the exhaust gases through NaOCl solution.

Destruction Procedures

Destruction of Bulk Quantities[10]

1. Dissolve NaCN or CNBr in H_2O so that the concentration does not exceed 25 mg/mL for NaCN or 60 mg/mL for CNBr. Mix one volume of this solution with one volume of sodium hydroxide (NaOH) solution (1 *M*) and two volumes of 5.25% NaOCl [i.e., cyanide solution:NaOH:NaOCl (1:1:2)]. Stir the mixture for 3 h, test for completeness of destruction, neutralize, and discard it. Use fresh NaOCl solution (see below for assay procedure).

2. Dissolve NaCN or CNBr in H_2O so that the concentration does not exceed 25 mg/mL for NaCN or 60 mg/mL for CNBr. Mix one volume of this solution with one volume of NaOH solution (1 *M*) and add 60 g of $Ca(OCl)_2$ per liter of basified solution. Stir the mixture for 3 h, test for completeness of destruction, neutralize, and discard it.

Destruction of Sodium Cyanide or Cyanogen Bromide in Solution[10]

1. If necessary dilute the NaCN solution with H_2O so that the concentration does not exceed 25 mg/mL. If necessary dilute aqueous solutions of CNBr with H_2O so that the concentration does not exceed 60 mg/mL. If necessary dilute solutions of CNBr in organic solvents with the same organic solvent so that the concentrations do not exceed 60 mg/mL for acetonitrile, 30 mg/mL for dimethyl sulfoxide (DMSO), *N,N*-dimethylformamide (DMF), 2-methoxyethanol, or 0.1 *M* hydrochloric acid (HCl), 25 mg/mL for ethanol, or 19 mg/mL for *N*-methyl-2-pyrrolidinone. For each volume of solution add one volume of 1 *M* NaOH solution and two volumes of 5.25%

NaOCl solution. Stir the mixture for 3 h, test for completeness of destruction, neutralize, and discard it. Use fresh NaOCl solution (see below for assay procedure).

2. If necessary dilute the NaCN solution with H_2O so that the concentration does not exceed 25 mg/mL. If necessary dilute aqueous solutions of CNBr with H_2O so that the concentration does not exceed 60 mg/mL. If necessary dilute solutions of CNBr in organic solvents with the same organic solvent so that the concentrations do not exceed 60 mg/mL for acetonitrile, 30 mg/mL for DMSO, DMF, 2-methoxyethanol, or 0.1 *M* HCl, 25 mg/mL for ethanol, or 19 mg/mL for *N*-methyl-2-pyrrolidinone. For each volume of solution add one volume of 1 *M* NaOH solution, then add 60 g of $Ca(OCl)_2$ per liter of basified solution. Stir the mixture for 3 h, test for completeness of destruction, neutralize, and discard it.

Destruction of Cyanogen Bromide in 70% Formic Acid[10]

1. If necessary dilute the solution so that the concentration of CNBr does not exceed 60 mg/mL and basify the solution by the **slow** addition of two volumes of 10 *M* potassium hydroxide (KOH) solution (a **very** exothermic process). Cool, then for each volume of solution add one volume of 1 *M* NaOH solution and two volumes of 5.25% NaOCl solution. Stir the mixture for 3 h, test for completeness of destruction, neutralize, and discard it. Use fresh NaOCl solution (see below for assay procedure).

2. If necessary dilute the solution so that the concentration of CNBr does not exceed 60 mg/mL and basify the solution by the **slow** addition of two volumes of 10 *M* KOH solution (a **very** exothermic process). Cool, then for each volume of solution add one volume of 1 *M* NaOH solution, then add 60 g of $Ca(OCl)_2$ per liter of basified solution. Stir the mixture for 3 h, test for completeness of destruction, neutralize, and discard it.

Destruction of Hydrogen Cyanide[11]

Dissolve the HCN in several volumes of ice water and add one molar equivalent of NaOH solution at 0–10°C. (Do **not** add NaOH, NaCN, or any base to liquid HCN; a violent reaction may occur.) Add a 50% excess of 5.25% NaOCl solution (80 mL of solution for each gram of HCN) at 0–10°C with stirring and allow the mixture to warm to room temperature. After standing for several hours test for completeness of destruction, neutralize, and discard it. Use fresh NaOCl solution (see below for assay procedure).

Decontamination of Hydrogen Cyanide from Exhaust Gases[13]

Pass the exhaust gases through 5.25% NaOCl solution. After standing for several hours test for completeness of destruction, neutralize, and discard it. Use fresh NaOCl solution (see below for assay procedure).

Assay of Sodium Hypochlorite Solution

Sodium hypochlorite solutions tend to deteriorate with time so they should be periodically checked for the amount of active chlorine they contain. Pipette 10 mL of NaOCl solution into a 100-mL volumetric flask and fill to the mark with distilled H_2O. Pipette 10 mL of this solution into a conical flask containing 50 mL of distilled H_2O, 1 g of potassium iodide, and 12.5 mL of 2 *M* acetic acid. Titrate this solution against 0.1 *N* sodium thiosulfate solution using starch as an indicator. Each 1 mL of the sodium thiosulfate solution corresponds to 3.545 mg of active chlorine. The NaOCl solution used in these degradation reactions should contain 45–50 g of active chlorine per liter.

Analytical Procedures[10]

Prepare the following solutions. The buffer solution was prepared by dissolving 13.6 g of potassium phosphate, monobasic (KH_2PO_4), 0.28 g of sodium phosphate, dibasic (Na_2HPO_4), and 3.0 g of potassium bromide (KBr) in distilled H_2O and making up to 1 L with distilled H_2O. Note that the presence of KBr is necessary for the assay procedure to work correctly. The reagent was prepared by stirring 3.0 g of barbituric acid in 10 mL of H_2O and adding 15 mL of 4-methylpyridine and 3 mL of concentrated HCl while continuing to stir. After cooling the solution was diluted to 50 mL with H_2O. The sodium ascorbate solution was 10 mg/mL in H_2O, the NaCN solution was 100 mg/L in H_2O, and the Chloramine-T solution was 10 mg/mL in H_2O (*not* 100 mg/mL as stated in the paper[10]). The sodium ascorbate and Chloramine-T solutions were prepared fresh daily and the standard NaCN solution was prepared fresh every week. The other solutions appeared to be quite stable.

Two portions of the reaction mixtures were centrifuged, if necessary to remove suspended solids, and 1 mL of each was added to 4 mL of buffer. If an orange or yellow color appeared, sodium ascorbate solution was added dropwise until the mixtures were colorless (but no more than 2 mL should be added). One solution was spiked with 200 μL of NaCN solution, and 1 mL of Chloramine-T solution was added to each solution. The solutions were shaken and allowed to stand for 1–2 min, then 1 mL of the reagent was added and the mixtures were shaken and allowed to stand for 5 min. A blue color indicates the presence of cyanide. For complete destruction the unspiked solution should be colorless and the spiked solution should be blue. The absorbance was measured at 605 nm using 10-mm cuvettes (after centrifuging again if necessary to remove suspended solids). Appropriate standards and blanks should always be run. The limit of detection was about 3 μg/mL.

Mutagenicity Assays

Since CNBr or cyanide ion have not been reported to be mutagenic no studies were performed.

Related Compounds

These procedures should be applicable to cyanogen chloride [506-77-4], cyanogen iodide [506-78-5], and various inorganic cyanides although we have carried out no tests. Similar procedures can be used for the destruction of sodium cyanoborohydride (see Complex Metal Hydrides monograph). These procedures are not applicable to organic nitriles (see the Organic Nitriles monograph for methods of disposing of these compounds).

References

1. Smyth, H.F.; Carpenter, C.P.; Weil, C.S.; Pozzani, U.C.; Striegel, J.A.; Nycum, J.S. Range-finding toxicity data: List VII. *Am. Ind. Hyg. Assoc. J.* **1969**, *30*, 470–476.
2. Hayes, W.J. The 90-dose LD_{50} and a chronicity factor as measures of toxicity. *Toxicol. Appl. Pharmacol.* **1967**, *11*, 327–335.
3. Lewis, R.J., Sr. *Sax's Dangerous Properties of Industrial Materials*, 8th ed.; Van Nostrand-Reinhold: New York, 1992; pp. 1896–1897.
4. Reference 3, pp. 974–975.
5. Parmeggiani, L., Ed., *Encyclopedia of Occupational Safety and Health*, 3rd ed.; International Labour Office: Geneva, 1983; Vol. 1, pp. 574–577.
6. Prentiss, A.M. *Chemicals in War*; McGraw-Hill: New York, 1937; p. 174.
7. Cuatrecasas, P.; Anfinsen, C.B. Affinity chromatography. In *Methods in Enzymology*; Jakoby, W.B., Ed., Academic Press: New York, 1977; Vol. 22, pp. 345–378.
8. Huang, H.V.; Bond, M.W.; Hunkapiller, M.W.; Hood,L.E. Cleavage at tryptophanyl

residues with dimethyl sulfoxide-HCl and cyanogen bromide. In *Methods in Enzymology*; Hirs, C.H.W., Timasheff, S.N., Eds., Academic Press: New York, 1983; Vol. 91, Part 1, pp. 318–324.

9. Fuentes-Duchemin, J.; Casassas, E. Photometric determination of traces of pyridine by reaction with cyanogen bromide and 4,4′-diaminostilbene-2,2′-disulphonic acid. *Anal. Chim. Acta* **1969**, *44*, 462–466.
10. Lunn, G.; Sansone, E.B. Destruction of cyanogen bromide and inorganic cyanides. *Anal. Biochem.* **1985**, *147*, 245–250.
11. National Research Council, Committee on Hazardous Substances in the Laboratory. *Prudent Practices for Disposal of Chemicals from Laboratories;* National Academy Press: Washington, DC, 1983; pp. 86–87.
12. Cerami, A.; Allen, T.A.; Graziano, J.H.; deFuria, F.G.; Manning, J.M.; Gillette, P.N. Pharmacology of cyanate. I. General effects on experimental animals. *J. Pharmacol. Exp. Ther.* **1973**, *185*, 653–666.
13. Lunn, G. Unpublished results.

CYCLOSERINE

CAUTION! Refer to safety considerations section on page 7 before starting any of these procedures.

Cycloserine (**I**) [68-41-7][1] is an antibacterial, tuberculostatic drug produced by *Streptomyces orchidaceus*. This compound is moderately toxic and can produce a variety of effects in humans.[2] Cycloserine is a solid (mp 155–156°C) and is soluble in H_2O.

I

Principle of Destruction

Cycloserine is oxidized by potassium permanganate in 3 *M* sulfuric acid ($KMnO_4$ in H_2SO_4).[3] The products have not been determined. Destruction is complete and less than 0.4% of the original compound remains.

Destruction Procedure

Take up 100 mg of cycloserine in 20 mL of 3 *M* H_2SO_4 and add 0.96 g of $KMnO_4$ in portions with stirring. Stir the reaction mixture at room temperature for 18 h, then decolorize it with sodium metabisulfite, make strongly basic with 10 *M* potassium hydroxide solution, dilute with H_2O, filter to remove manganese salts,[4] and neutralize the filtrate. Test for completeness of destruction and discard it.

Analytical Procedure

A modification of a previously reported spectrofluorometric method[5] was used. A pH 8.1 buffer was prepared by dissolving 11 g of sodium tetraborate decahydrate in 950 mL of H_2O and adding 42 mL of 1 *M* hydrochloric acid. A 0.06% solution of 1,4-benzoquinone in 95% ethanol was also prepared. Cycloserine was determined by adding 4 mL of buffer to 100 μL of the neutralized reaction mixture followed by 100 μL of the benzoquinone solution. This mixture was heated at 100 °C for 30 min, then allowed to cool for 10 min. Fluorescence was determined by using a spectrophotofluorometer set at an excitation wavelength of 381 nm and an emission wavelength of 502 nm. A relatively high background was observed with blanks made with H_2O instead of the reaction mixture and blanks from reactions in which the cycloserine was omitted. However, the method is quite sensitive and spiking experiments showed that if any cycloserine was present it was there at a concentration of less than 20 μg/mL.

Mutagenicity Assays

The mutagenicity assays were carried out as described on page 4 using tester strains TA98, TA100, TA1530, and TA1535. The final reaction mixture (tested at a level corresponding to 0.5 mg of undegraded material per plate) was not mutagenic. The pure compound was toxic to the cells at levels of 1000 and 500 μg per plate but not mutagenic or toxic at lower levels (250–10 μg per plate).

Related Compounds

Potassium permanganate in 3 *M* H_2SO_4 is a general oxidative procedure and it should be applicable to related compounds. Full validation should, however, be carried out in each case.

References

1. Other names for this compound are D-4-amino-3-isoxazolidinone, D-4-amino-3-isoxazolidone, orientomycin, Cyclomycin, Farmiserine, Miroserina, Novoserin, Closina, Farmiserina, Micoserina, Oxamycin, Oxymycin, Tisomycin, Wasserina, Seromycin, PA-94, 106-7, cyclo-D-serine, E-733-A, I-1431, JN-21, K-300, and RO-1-9213.
2. Lewis, R.J., Sr. *Sax's Dangerous Properties of Industrial Materials*, 8th ed.; Van Nostrand-Reinhold: New York, 1992; pp. 1009–1010.
3. Lunn, G; Sansone, E. B. Validated methods for degrading hazardous chemicals: Some alkylating agents and other compounds. *J. Chem. Educ.* **1990**, *67*, A249–A251.
4. Lunn, G.; Sansone, E.B.; De Méo, M.; Laget, M.; Castegnaro, M. Potassium permanganate can be used for degrading hazardous compounds. *Am. Ind. Hyg. Assoc. J.* **1994**, *55*, 167–171.
5. El-Sayed, L.; Mohamed, Z.H.; Wahbi, A-.A.M. Spectrophotometric and spectrofluorimetric determination of cycloserine with *p*-benzoquinone. *Analyst* **1986**, *111*, 915–917.

DICHLOROMETHOTREXATE, VINCRISTINE, AND VINBLASTINE

> **CAUTION!** Refer to safety considerations section on page 7 before starting any of these procedures.

The degradation of a number of antineoplastic drugs, including dichloromethotrexate(**I**) [528-74-5], vincristine(**II**) [57-22-7(base), 2068-78-2(sulfate)], and vinblastine(**III**) [865-21-4(base), 143-67-9(sulfate)], was investigated by the International Agency for Research on Cancer (IARC).[1]

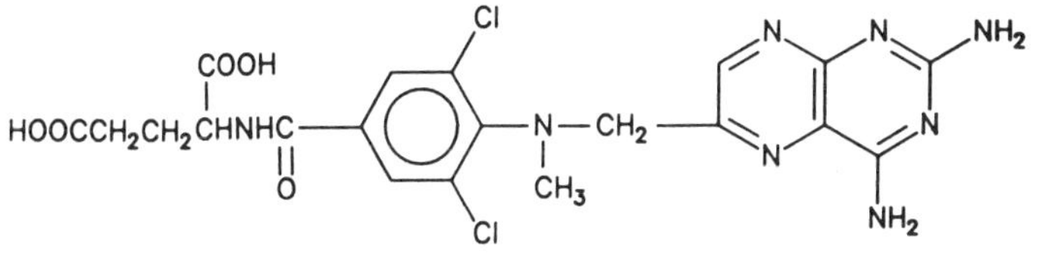

I

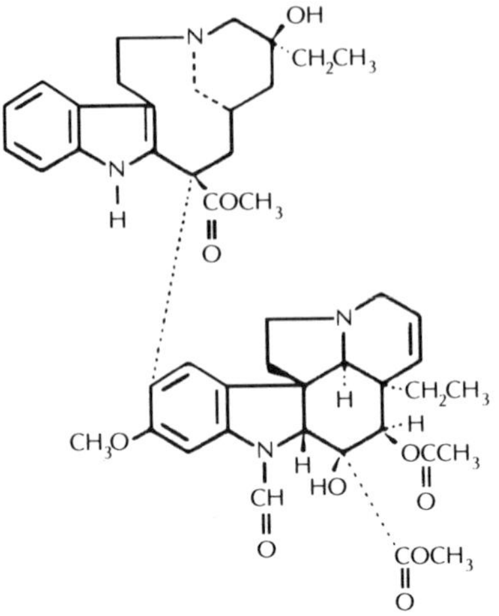

II

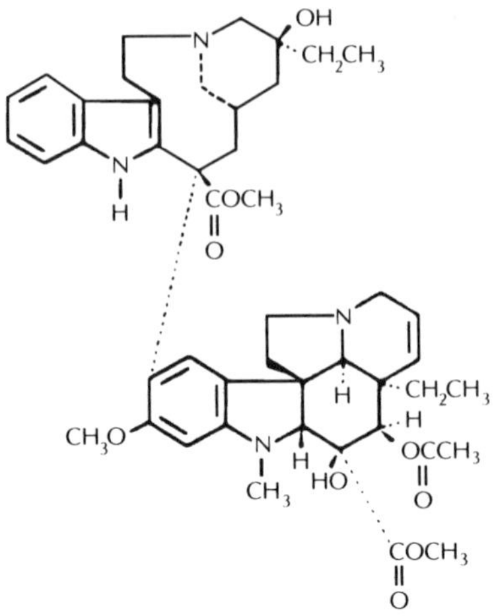

III

Dichloromethotrexate (mp 185–204°C),[2] vincristine sulfate (mp 273–278°C),[3] and vinblastine sulfate (mp 284–285°C)[4] are solids. Dichloromethotrexate is not soluble in H_2O or organic solvents, but it is soluble in dilute acid or base. Vincristine sulfate and vinblastine sulfate are soluble in H_2O, chloroform, and methanol, but not in ether. Vincristine is a teratogen and produces various effects on the body,[5] and vinblastine is a teratogen and has effects on the eye, heart, and bone marrow.[6] Dichloromethotrexate may be teratogenic and carcinogenic.[7] These compounds are all used as antineoplastic drugs.

Principle of Destruction

Dichloromethotrexate, vincristine sulfate, and vinblastine sulfate are destroyed by oxidation with potassium permanganate in sulfuric acid ($KMnO_4$ in H_2SO_4).[1] Destruction is greater than 99%. The products of these reactions have not been determined.

Destruction Procedures

Destruction of Bulk Quantities of Dichloromethotrexate, Vincristine Sulfate, and Vinblastine Sulfate

Dissolve in 3 *M* H_2SO_4 so that the concentration does not exceed 1 mg/mL, then add 0.5 g of $KMnO_4$ for each 10 mL of solution and stir for 2

h. Decolorize with sodium metabisulfite, make strongly basic with 10 *M* potassium hydroxide (KOH) solution (**Caution!** Exothermic), dilute with H_2O, filter to remove manganese compounds,[8] neutralize the filtrate, check for completeness of destruction, and discard it.

Destruction of Aqueous Solutions of Dichloromethotrexate, Vincristine Sulfate, and Vinblastine Sulfate

Dilute with H_2O, if necessary, so that the concentration does not exceed 1 mg/mL, then add enough concentrated H_2SO_4 to obtain a 3 *M* solution and allow it to cool to room temperature. For each 10 mL of solution add 0.5 g of $KMnO_4$ and stir for 2 h. Decolorize with sodium metabisulfite, make strongly basic with 10 *M* KOH (**Caution!** Exothermic), dilute with H_2O, filter to remove manganese compounds,[8] neutralize the filtrate, check for completeness of destruction, and discard it.

Destruction of Pharmaceutical Preparations of Vincristine Sulfate and Vinblastine Sulfate Containing 1 mg of Compound, 1.275 mg of Methyl p-*Hydroxybenzoate, 1.225 mg of Propyl* p-*Hydroxybenzoate, and 100 mg of Mannitol*

Dissolve in 3 *M* H_2SO_4 to obtain a drug concentration of 0.1 mg/mL then, for each 10 mL of solution, add 0.5 g of $KMnO_4$, in small portions to avoid frothing, and stir for 2 h. Decolorize with sodium metabisulfite, make strongly basic with 10 *M* KOH solution (**Caution!** Exothermic), dilute with H_2O, filter to remove manganese compounds,[8] neutralize the filtrate, check for completeness of destruction, and discard it.

Destruction of Injectable Pharmaceutical Preparations of Dichloromethotrexate Containing 2–5% Glucose and 0.45% Saline

Dilute with H_2O so that the concentration does not exceed 2.5 mg/mL, then add enough concentrated H_2SO_4 to obtain a 3 *M* solution and allow it to cool to room temperature. For each 10 mL of solution add 1 g of $KMnO_4$, in small portions to avoid frothing, and stir for 1 h. Decolorize with sodium metabisulfite, make strongly basic with 10 *M* KOH solution (**Caution!** Exothermic), dilute with H_2O, filter to remove manganese compounds,[8] neutralize the filtrate, check for completeness of destruction, and discard it.

Destruction of Solutions of Dichloromethotrexate, Vincristine Sulfate, and Vinblastine Sulfate in Volatile Organic Solvents

Remove the solvent under reduced pressure using a rotary evaporator and take up the residue in 3 *M* H_2SO_4 so that the concentration does not exceed 1 mg/mL. For each 10 mL of solution add 0.5 g of $KMnO_4$ and stir for 2 h. Decolorize with sodium metabisulfite, make strongly basic with 10 *M* KOH solution (**Caution!** Exothermic), dilute with H_2O, filter to remove manganese compounds,[8] neutralize the filtrate, check for completeness of destruction, and discard it.

Destruction of Dimethyl Sulfoxide or N,N-Dimethylformamide Solutions of Dichloromethotrexate, Vincristine Sulfate, and Vinblastine Sulfate

Dilute with H_2O so that the concentration of dimethyl sulfoxide (DMSO) or *N,N*-dimethylformamide (DMF) does not exceed 20% and the concentration of the drug does not exceed 1 mg/mL, then add enough concentrated H_2SO_4 to obtain a 3 *M* solution and allow it to cool to room temperature. For each 10 mL of solution add 1 g of $KMnO_4$, in portions to avoid frothing, and stir for 2 h. Decolorize with sodium metabisulfite, make strongly basic with 10 *M* KOH solution (**Caution!** Exothermic), dilute with H_2O, filter to remove manganese compounds,[8] neutralize the filtrate, check for completeness of destruction, and discard it.

Decontamination of Glassware Contaminated with Dichloromethotrexate, Vincristine Sulfate, and Vinblastine Sulfate

Immerse the glassware in a 0.3 *M* solution of $KMnO_4$ in 3 *M* H_2SO_4 for 2 h, then clean it by immersion in sodium metabisulfite solution. Decolorize the permanganate solution with sodium metabisulfite, make strongly basic with 10 *M* KOH solution (**Caution!** Exothermic), dilute with H_2O, filter to remove manganese compounds,[8] neutralize the filtrate, check for completeness of destruction, and discard it.

Decontamination of Spills of Vincristine Sulfate and Vinblastine Sulfate

Allow any organic solvent to evaporate and remove as much of the spill as possible by high efficiency particulate air (HEPA) vacuuming (not sweeping), then rinse the area with H_2O. Take up the rinse with absorbents and decontaminate them by reaction with a 0.3 *M* solution of $KMnO_4$ in 3 *M* H_2SO_4 for 2 h. If the color fades, add more solution. Check for completeness of decontamination by using a wipe sample moistened with H_2O. Analyze the wipe for the presence of the drug. Decolorize the permanganate solution with sodium metabisulfite, make strongly basic with 10 *M* KOH solution (**Caution!** Exothermic), dilute with H_2O, filter to remove manganese compounds,[8] neutralize the filtrate, check for completeness of destruction, and discard it.

Decontamination of Spills of Dichloromethotrexate

Allow any organic solvent to evaporate and remove as much of the spill as possible by HEPA vacuuming (not sweeping), then rinse the area with 3 *M* H_2SO_4. Take up the rinse with absorbents and allow the rinse and absorbents to react with 0.3 *M* $KMnO_4$ solution in 3 *M* H_2SO_4 for 1 h. If the color fades, add more solution. Check for completeness of decontamination by using a wipe moistened with 0.1 *M* H_2SO_4. Analyze the wipe for the presence of the drug. Decolorize the permanganate solution with sodium metabisulfite, make strongly basic with 10 *M* KOH solution (**Caution!** Exothermic), dilute with H_2O, filter to remove manganese compounds,[8] neutralize the filtrate, check for completeness of destruction, and discard it.

Analytical Procedures

These drugs can be analyzed by high-performance liquid chromatography (HPLC) using a 25-cm reverse phase column and UV detection at 254 nm. The mobile phases that have been recommended are as follows:

Dichloromethotrexate. 5 m*M* Tetrabutylammonium phosphate (adjusted to pH 3.5 with phosphoric acid) : methanol : acetonitrile (66:11:22) at 1.5 mL/min **or** 20 m*M* ammonium formate : methanol (65:35) at 1 mL/min.

Vincristine sulfate. 5 m*M* Tetrabutylammonium phosphate (adjusted to pH 3.5 with phosphoric acid) : acetonitrile : tetrahydrofuran (THF) (54:26:20) at 1.5 mL/min **or** 20 m*M* ammonium formate : methanol (12:88) at 1 mL/min.

Vinblastine sulfate. 5 m*M* Tetrabutylammonium phosphate (adjusted to pH 3.5 with phosphoric acid) : acetonitrile : THF (54:26:20) at 1.5 mL/min **or** 20 m*M* ammonium formate : methanol (10:90) at 1 mL/min.

Mutagenicity Assays

In the IARC study[1] tester strains TA98, TA100, TA1530, and TA1535 of *Salmonella typhimurium* were used with and without metabolic activation (not all strains were used for each drug). Generally, the reaction mixtures were not mutagenic although degradation of pharmaceutical preparations of dichloromethotrexate gave two to three times the background activity with TA1530.

Related Compounds

Potassium permanganate in H_2SO_4 is a general oxidative method and should, in principle, be applicable to many drugs. However, any new application should be thoroughly validated both for complete destruction of the compound and for the production of nonmutagenic reaction mixtures.

References

1. Castegnaro, M.; Adams, J.; Armour, M-.A.; Barek, J.; Benvenuto, J.; Confalonieri, C.; Goff, U.; Ludeman, S.; Reed, D.; Sansone, E. B.; Telling, G., Eds., *Laboratory Decontamination and Destruction of Carcinogens in Laboratory Wastes: Some Antineoplastic Agents*; International Agency for Research on Cancer: Lyon, 1985 (IARC Scientific Publications No. 73).
2. Other names for this compound are 3′,5′-dichloroamethopterin, 4-amino-10-methyl-3′,5′-dichloropteroylglutamic acid, 3′,5′-dichloro-4-amino-4-deoxy-N^{10}-methylpteroglutamic acid, *N*-[3,5-dichloro-4-[[(2,4-diamino-6-pteridinyl)methyl]methylamino]benzoyl]glutamic acid, DCM, NCI-C04875, and NSC-29630.
3. Other names for this compound are 22-oxovincaleukoblastine, leurocristine, VCR, LCR, NCI-C04864, NSC-67574, Oncovin, vincrystine, vinkristin, Kyocristine, Vincosid, Vincrex, Lilly 37231, NSC-67574, and Vincrisul.
4. Other names for this compound are vincaleukoblastine, VLB, nincaluicolflastine, NCI-C04842, NDC 002-1452-01, vincoblastine, Exal, 29060 LE, NSC-49842, Velban, and Velbe.
5. Lewis, R.J., Sr. *Sax's Dangerous Properties of Industrial Materials*, 8th ed.; Van Nostrand-Reinhold: New York, 1992; pp. 2112–2113.
6. Reference 5, pp. 3491–3492.
7. Reference 5, p. 1149.
8. Lunn, G.; Sansone, E.B.; De Méo, M.; Laget, M.; Castegnaro, M. Potassium permanganate can be used for degrading hazardous compounds. *Am. Ind. Hyg. Assoc. J.* **1994**, 55, 167–171.

DIISOPROPYL FLUOROPHOSPHATE

CAUTION! Refer to safety considerations section on page 7 before starting any of these procedures.

Diisopropyl fluorophosphate [55-91-4] [DFP, $[(CH_3)_2CHO]_2P(O)F$][1] is a practically odorless,[2] colorless, volatile (bp 73°C/16 mm Hg[2]) liquid that is widely used as an enzyme inhibitor.[3-9] The density is 1.055 g/mL.[10] This compound is also used to treat glaucoma.[11] Diisopropyl fluorophosphate is highly toxic (LC_{50} 360 μg/L for rats, LD_{50} 500 μg/kg for rabbits) with a toxicity comparable to hydrogen cyanide.[2] This compound is a cholinesterase inhibitor,[11,12] a neurotoxin,[13] and has reproductive effects.[14]

Principle of Destruction

Diisopropyl fluorophosphate can be degraded by adding 1 *M* sodium hydroxide (NaOH) to bulk quantities of DFP and to solutions of DFP in buffer, H_2O, or *N,N*-dimethylformamide (DMF). The product is diisopropylphosphate.[2,15] The pH should be greater than or equal to 12. For solutions of DFP in buffer or H_2O the destruction efficiency was greater than

99.8%, for solutions of DFP in DMF the destruction efficiency was greater than 99.97%, and when bulk quantities of DFP were degraded the destruction efficiency was greater than 99.98%. When bulk quantities are degraded a limitation is the speed at which the oily DFP dissolves in the NaOH. Agitation speeds the dissolution and decreases the decontamination time. Destruction was complete in 2 h, however, even when the reactants were just placed together with minimal agitation. The supplier (Aldrich Chemical Co., Milwaukee, WI) recommends the use of 2% (0.5 *M*) NaOH[16] and 1 *M* NaOH in 50% ethanol has been recommended[3] although validation details are not given. The nerve gases Sarin (GB)[17–19] and Soman (GD),[19] which also possess P—F bonds, are readily hydrolyzed by strong base. The stability of DFP in buffer solution has been determined. Table 1 shows the time required until the addition of buffer solution containing DFP did not degrade chymotrypsin activity by more than 50%. At this point the amount of DFP was less than 0.17% of the initial amount. The stability is pH dependent and decreases markedly as the pH increases. Diisopropyl fluorophosphate is somewhat less stable in pH 7 buffer than in phosphate buffered saline (PBS) or Dulbecco's buffer (both pH 7.2), but this may be due to the higher phosphate concentration in the pH 7 buffer (50 m*M*) than in the other buffers (10 m*M*).

Table 1 Degradation of a 1 m*M* Solution of DFP in Buffer

Buffer	pH	Time Until Addition of Buffer Solution Did Not Degrade Chymotrypsin Activity by > 50% (h)
pH 3	3.0	> 330
pH 5	5.0	> 330
Hanks'	6.4	> 330
pH 7	7.0	168
PBS	7.2	330
Dulbecco's	7.2	330
HEPES	7.5	168
TRIS	8.0	168
pH 9	9.1	18
pH 11	11.0	4

Destruction Procedures

Destruction of Diisopropyl Fluorophosphate in Buffer or Water

To each 1 mL of 10 m*M* DFP in buffer or H_2O add 200 μL of 1 *M* NaOH and check that the reaction mixture is strongly basic (pH ≥12). Allow to stand at room temperature for 18 h, analyze for completeness of destruction, neutralize, and discard it.

Destruction of Diisopropyl Fluorophosphate in N,N-Dimethylformamide

To each 1 mL of 200 m*M* DFP in DMF add 2 mL of 1 *M* NaOH and check that the reaction mixture is strongly basic (pH ≥12). Allow to stand at room temperature for 18 h, analyze for completeness of destruction, neutralize, and discard it.

Destruction of Bulk Quantities of Diisopropyl Fluorophosphate

To each 40 μL of pure DFP add 1 mL of 1 *M* NaOH and check that the reaction mixture is strongly basic (pH ≥12). Stir at room temperature for 1 h, analyze for completeness of destruction, neutralize, and discard it.

Decontamination of Spills or Equipment

To decontaminate spills or equipment add at least 1 mL of 1 *M* NaOH for each 40 μL of DFP that is estimated to be present. Add more NaOH if required to thoroughly wet all contaminated surfaces. Check that the reaction mixture is strongly basic (pH ≥12) and make sure that all the oily DFP has dissolved. Allow to stand at room temperature for at least 2 h and preferably 18 h, analyze the solution for completeness of destruction, neutralize, and discard it. Decontamination is more efficient if it is possible to agitate or stir the reaction mixture, (e.g., when the inside of a flask is being decontaminated). Clean the equipment or the spill area in a conventional fashion.

Buffers

The buffer solutions employed when the destruction of DFP in buffer was investigated were Hanks' Balanced Salts (Hanks') (pH 6.4), Dulbecco's phosphate buffered saline (Dulbecco's) (pH 7.2), and PBS (pH 7.2); which were purchased from Sigma. Hanks' was 0.78 m*M* in phosphate and Dulbecco's and PBS were 10 m*M* in phosphate. The Tris–Borate–EDTA (TBE) buffer contained tris(hydroxymethyl)aminomethane (TRIS, 50 m*M*, 6.1 g/L), boric acid (5.5 m*M*, 0.34 g/L), and ethylenediaminetetraacetic acid (EDTA, 1.7 m*M*, 0.5 g/L) and was adjusted to pH 8.0 with 1 *M* hydrochloric acid (HCl). The pH 3 buffer was a 50 m*M* phthalate buffer (10.2 g of potassium hydrogen phthalate in 1 L of H_2O adjusted to pH 3.0 with 1 *M* HCl) and the pH 5 buffer was a 50 m*M* phthalate buffer (10.2 g of potassium hydrogen phthalate in 1 L of H_2O adjusted to pH 5.0 with 1 *M* NaOH). The pH 7 buffer was a 50 m*M* phosphate buffer (6.8 g of potassium dihydrogen phosphate in 1 L of H_2O adjusted to pH 7.0 with 1 *M* NaOH). The pH 9 buffer was 50 m*M* borax (19 g/L, actual pH 9.1) and the pH 11 buffer was 50 m*M* sodium carbonate (6.2 g/L). The 4-(2-hydroxyethyl)-1-piperazineethanesulfonic acid (HEPES) buffer was 50 m*M* HEPES (11.9 g/L) and 500 m*M* NaCl (29.25 g/L) adjusted to pH 7.5 with 1 *M* NaOH.

Analytical Procedures

The analytical system measures the rate at which *N*-benzoyl-L-tyrosine ethyl ester (BTEE) is hydrolyzed by chymotrypsin using a procedure based on the method of Hummel[20] with the modifications of Rao and Lombardi.[21] The DFP inhibits the activity of chymotrypsin and so the rate of hydrolysis acts as an indicator for the presence or absence of DFP. The hydrolysis of BTEE is measured by determining the increase in absorbance at 256 nm. There are a number of complicating features, however, which make this analysis rather less straightforward than the average colorimetric procedure. The initial absorbance of the BTEE is quite high [typically 2 AU (absorbance units)] and the increase in absorbance is small (typically 0.2 to 0.4 AU). Consequently, a conventional blank cannot be used because small differences in the BTEE solution or in the ultraviolet (UV) cell might have large effects on the initial absorbance, and hence on the final absorbance value. Another complicating factor is that, in order to measure low concentrations of DFP, low concentrations of chymotrypsin must be used. Consequently the reactions take a long time, typically 16–20 h. Because the analytical system uses an enzyme, some variability was seen in the measured rates of hydrolysis and so it was necessary to run a number of controls.

The best procedure is to add a buffered aliquot of the reaction mixture to a solution of chymotrypsin, and then wait for 1 h for any DFP present to inhibit, or partially inhibit, the enzyme. The BTEE solution is then added and the absorbance immediately determined. After standing for 16–20 h the final absorbance of the same solution in the same cell is measured and hence the rate of increase in absorbance [R_{rm}, in milliabsorbance units per hour (mAU/h)] can be calculated. At the same time the spontaneous hydrolysis rate of BTEE is determined by using a blank that contains neither enzyme nor DFP (R_{sp}). The spontaneous hydrolysis of BTEE (typically 6 mAU/h) is significant over the long time course of the reaction at the pH (7.8) employed. A reaction blank that contains enzyme but no DFP is also run to determine the rate of hydrolysis due to both spontaneous and enzymatic hydrolysis (R_{rb}). The reaction blank was processed in exactly the same way as the degradation experiment except that no DFP was added. By subtraction the activity of the enzyme, measured by measuring the rate of enzymatic hydrolysis (typically 12 mAU/h) can be calculated, $A_{rb} = R_{rb} - R_{sp}$. This value is compared with the activity of the enzyme when the reaction mixture being tested for DFP was present, $A_{rm} = R_{rm} - R_{sp}$. The lower limit of detection for DFP was set by the concentration of DFP which would produce 50% inhibition of the activity of the enzyme, that is, $A_{rm} = 0.5 \times A_{rb}$. Note that this does **not** represent a 50% decrease in the initial concentration of DFP because the initial concentration of DFP was many times that needed to completely inhibit the chymotrypsin. Spiking experiments were used to determine the limit of detection.

Prepare the following solutions:

Tris Buffer. Dissolve 12.1 g of TRIS base and 14.7 g of calcium chloride dihydrate in about 900 mL of distilled H_2O. Adjust the pH to 7.8 with 1 *M* HCl and make up to 1 L with distilled H_2O. The buffer is 100 m*M* in Tris and 100 m*M* in calcium.

MOPS Buffer. Dissolve 20.9 g of MOPS in about 900 mL of distilled H_2O. Adjust the pH to 6.5 with 1 *M* NaOH and make up to 1 L with distilled H_2O. The buffer is 100 m*M* in MOPS.

MOPS/Ca^{2+} Buffer. Prepare as above but add 14.7 g of calcium chloride dihydrate. The buffer is 100 m*M* in calcium.

BTEE Solution. Stir 400 μL of Triton X-100 in 200 mL of Tris buffer for about 5 min, then add a solution of 31.34 mg of BTEE in 1 mL of methanol and stir until all the BTEE dissolves. The concentration of BTEE is 0.5 m*M*. The solution should be prepared immediately before use but is usable for several hours.

DFP Solutions. The DFP was stored in the refrigerator. It was supplied in a vial with a septum and portions were removed as required using a Hamilton 100-μL syringe. Prepare a 200 m*M* (37 mg/mL) solution by dissolving 36 μL of DFP in 1 mL of DMF. Prepare a 1 mg/mL solution by adding 100 μL of the 200 m*M* solution to 3.6 mL of DMF. These solutions should be prepared fresh each week. Prepare a 3.1 μg/mL solution of DFP in MOPS by adding 50 μL of the 1 mg/mL solution of DFP in DMF to 16 mL of MOPS (not MOPS/Ca^{2+}) buffer. This solution should be prepared and used immediately. Solutions of DFP in buffer and H_2O that were 10 m*M* were prepared by adding 50 μL of a 200 m*M* solution of DFP in DMF to 950 μL of buffer or H_2O.

Chymotrypsin Solutions. Prepare a 1 mg/mL solution of α-chymotrypsin (Type II from bovine pancreas, Sigma Cat. No. C 4129) using ice-cold 1 m*M* HCl. This solution should be stored in the refrigerator and prepared fresh each week. Prepare the working 0.1 μg/mL solution of chymotrypsin by diluting an aliquot of the 1 mg/mL solution 1:10000 times with ice cold 1 m*M* HCl. This solution should be kept on ice and prepared fresh each day.

Procedure

Perform the destruction procedure as detailed above. At the same time prepare a reaction mixture blank as follows. Add 50 μL of DMF to 950 μL of buffer, then add 200 μL of 1 *M* NaOH. Allow this mixture to stand at room temperature for 18 h, then proceed. Add 100 μL of the reaction mixture to be tested to 1 mL of MOPS/Ca^{2+} buffer. [Buffers that contain phosphate (particularly pH 7 buffer, Dulbecco's buffer, and PBS) may produce a precipitate of calcium phosphate that will interfere with the spectrophotometric determination. If this is the case, add 0.5 mL of the reaction mixture to 5 mL of MOPS/Ca^{2+} buffer, allow to stand for several hours, then centrifuge. Use 1.1 mL of the supernatant.] To each sample add either 100 μL of MOPS buffer or 100 μL of 3.1 μg/mL DFP in MOPS (for spiked reactions). Finally, add 100 μL of the 0.1 μg/mL chymotrypsin solution. The blank for measuring the spontaneous hydrolysis rate consists of 100 μL of 0.2 *M* NaOH, 1 mL of MOPS/Ca^{2+} buffer, 100 μL of MOPS buffer, and 100 μL of 1 m*M* HCl.

Allow these solutions to stand at room temperature for 1 h then add 3 mL of the BTEE solution, place this mixture in a quartz UV cell, and immediately determine the absorbance against an air blank. After 16–20 h again determine the absorbance and calculate the rate of hydrolysis in milli-absorbance units per hour (mAU/h). Table 2 lists a typical set of analytical procedures.

Table 2 Typical Set of Analytical Procedures

Analytical Procedure	Sample	Spike[a]	Enzyme or 1 m*M* HCl
Reaction mixture	Reaction mixture	MOPS	Enzyme
Spiked reaction mixture	Reaction mixture	DFP	Enzyme
Reaction blank	Reaction mixture blank	MOPS	Enzyme
Spontaneous hydrolysis blank	0.2 *M* NaOH	MOPS	HCl

[a] Spiked with MOPS buffer or 3.1 μg/mL DFP in MOPS.

When neat DFP is degraded with NaOH solution the procedure is the same except that only 40 μL of reaction mixture is added to 1 mL of MOPS/Ca^{2+} buffer in order to keep the pH within an acceptable range. This procedure reduced the sensitivity of the analytical technique somewhat but the high initial concentration of DFP more than compensated for this. Sodium hydroxide (1 *M*) was used as a reaction mixture blank.

With the concentrations specified above the rate of hydrolysis (and hence the rate of increase in absorbance) should be linear. However, this should be checked periodically by measuring the absorbance every hour and graphing the readings. The result should be a straight line. If the concentration of the enzyme is too high, the graph will rise sharply, then level off as all the BTEE is consumed. If the concentration of enzyme is too low, the graph will differ little from that for spontaneous hydrolysis.

The absorbance of the analytical samples was determined at 256 nm using quartz UV cells in a Perkin Elmer Lambda 2 UV/vis spectrometer fitted with a 13 cell changer. The cells were cleaned with chromic acid, which was discarded appropriately [see the monograph on Chromium(VI)]. If required the spectrometer could be programmed to record periodically the absorbance of each cell so that the progress of the reaction could be followed graphically. The temperature in the cell compartment was 28°C.

Mutagenicity Assays

The mutagenicity assays were carried out as described on page 4. Tester strains TA98, TA100, TA1530, and TA1535 were used. Before testing, reaction mixtures from the destruction of DFP in buffer, H_2O, or DMF were neutralized by the addition of 100 μL of glacial acetic acid to each milliliter of reaction mixture and reaction mixtures obtained by the degradation of neat DFP were neutralized by the addition of 200 μL of glacial acetic acid to each milliliter of reaction mixture. In general the reaction mixtures were not mutagenic. In some cases mutagenic activity towards strains TA1530 and TA1535 with and without activation was seen when DFP in TBE and Dulbecco's buffers was degraded but the activity was only slightly greater than the level of significance. In addition, the effect was not reproducible and mutagenic activity was sometimes detected in control reactions in which DFP was omitted. Thus we ascribe the activity to some kind of artifact and not to residual DFP. The DFP itself was mutagenic to strains TA100, TA1530, and TA1535 with and without activation. A dose-response effect was seen.

Related Compounds

In principle the procedure should work for all compounds having a P—F bond that can be hydrolyzed. However, many of these compounds are highly toxic (e.g., the nerve gases) and should only be handled in facilities offering high degrees of containment. The decontamination of the chemical warfare agents *O*-ethyl *S*-2-(diisopropylamino)ethyl methylphosphonothiolate (VX), 2,2′-dichlorodiethyl sulfide (mustard gas, mustard, S mustard, sulfur mustard, H, HD), 2-propyl methylphosphonofluoridate (Sarin, GB), and 3,3-dimethyl-2-butyl methylphosphonofluoridate (Soman, GD) have been reviewed.[19]

References

1. Other names for this compound are DIFP, diisopropyl phosphorofluoridate, Isoflurophate, phosphorofluoridic acid bis(1-methylethyl) ester, phosphorofluoridic acid diisopropyl ester, isopropyl fluophosphate, diisopropyl fluorophosphonate, diisopropylphosphorofluoridate, fluostigmine, isofluorphate, Diflupyl, Dyflos, Floropryl, Fluorpryl, diflurophate, diisopropoxyphosphoryl fluoride, diisopropyl-fluorophosphoric acid ester, diisopropyl phosphofluoridate, diisopropyl phosphoryl fluoride, fluophosphoric acid diisopropyl ester, fluorodiisopropylphosphate, isofluorophate, isoflurophate, isopropyl fluophosphate, neoglaucit, PF-3, phosphorofluoridic acid, diisopropyl ester, T-1703, and TL 466.
2. Saunders, B.C.; Stacey, G.J. Esters containing phosphorus. Part IV. Diisopropyl fluorophosphate. *J. Chem. Soc.* **1948**, 695–699.

3. Cohen, J.A.; Oosterbaan, R.A.; Berends, F. Organophosphorus compounds. In *Methods in Enzymology*, Vol. 11; Hirs, C.H.W., Ed., Academic Press: New York, 1967; pp. 686–702.
4. Caughey, G.H.; Viro, N.F.; Lazarus, S.C.; Nadel, J.A. Purification and characterization of dog mastocytoma chymase: Identification of an octapeptide conserved in chymotryptic leukocyte proteinases. *Biochim. Biophys. Acta* **1988**, *952*, 142–149.
5. Park, J.H.; Lee, Y.S.; Chung, C.H.; Goldberg, A.L. Purification and characterization of protease Re, a cytoplasmic endoprotease in *Escherichia coli*. *J. Bacteriol.* **1988**, *170*, 921–926.
6. Zanglis, A.; Lianos, E.A. Platelet activating factor biosynthesis and degradation in rat glomeruli. *J. Lab. Clin. Med.* **1987**, *110*, 330–337.
7. Fried, V.A.; Smith, H.T.; Hildebrandt, E.; Weiner, K. Ubiquitin has intrinsic proteolytic activity: Implications for cellular regulation. *Proc. Natl. Acad. Sci. USA* **1987**, *84*, 3685–3689.
8. Schwartz, L.B.; Bradford, T.R. Regulation of tryptase from human lung mast cells by heparin. Stabilization of the active tetramer. *J. Biol. Chem.* **1986**, *261*, 7372–7379.
9. Momand, J.; Clarke, S. Rapid degradation of D- and L-succinimide-containing peptides by a post-proline endopeptidase from human erythrocytes. *Biochemistry* **1987**, *26*, 7798–7805.
10. *Aldrich Catalog/Handbook of Fine Chemicals 1992–1993*; Aldrich Chemical Co.: Milwaukee, WI, 1992; p. 471.
11. *Physicians' Desk Reference*, 47th ed.; Medical Economics Data: Montvale, NJ, 1993; pp. 1524–1525.
12. Wilson, B.W.; Walker, C.R. Regulation of newly synthesized acetylcholinesterase in muscle cultures treated with diisopropylfluorophosphate. *Proc. Natl. Acad. Sci. USA* **1974**, *71*, 3194–3198.
13. Gordon, C.J.; MacPhail, R.C. Strain comparisons of DFP neurotoxicity in rats. *J. Toxicol. Environ. Health* **1993**, *38*, 257–271.
14. Lewis, R.J., Sr. *Sax's Dangerous Properties of Industrial Materials*, 8th ed.; Van Nostrand-Reinhold: New York, 1992; p. 2058.
15. Ryan, J.A.; McGaughran, W.R.; Lindemann, C.J.; Zacchei, A.G. Separation and spectral properties of diisopropylphosphate, the major decomposition product of isoflurophate. *J. Pharm. Sci.* **1979**, *68*, 1194–1195.
16. *Diisopropyl Fluorophosphate, Technical Information Bulletin Number AL-122*; Aldrich Chemical Co.: Milwaukee, WI, 1986.
17. Epstein, J. Rate of decomposition of GB in seawater. *Science* **1970**, *170*, 1396–1398.
18. Epstein, J. Properties of GB in water. *J. Am. Water Works Assoc.* **1974**, *66*, 31–37.
19. Yang, Y.-C.; Baker, J.A.; Ward, J.R. Decontamination of chemical warfare agents. *Chem. Rev.* **1992**, *92*, 1729–1743.
20. Hummel, B.C.W. A modified spectrophotometric determination of chymotrypsin, trypsin, and thrombin. *Can. J. Biochem. Physiol.* **1959**, *37*, 1393–1399.
21. Rao, K.N.; Lombardi, B. Substrate solubilization for the Hummel α-chymotrypsin assay. *Anal. Biochem.* **1975**, *65*, 548–551.

DIMETHYL SULFATE AND RELATED COMPOUNDS

> **CAUTION!** Refer to safety considerations section on page 7 before starting any of these procedures.

Dimethyl sulfate [DMS, sulfuric acid dimethyl ester, methyl sulfate, $(CH_3)_2SO_4$] [77-78-1] is a clear, oily, high-boiling (bp 188°C) liquid. This compound is quite volatile, has no characteristic odor,[1] is highly toxic,[2] and causes severe burns and injury to the lungs, kidneys, and liver. Dimethyl sulfate causes cancer in laboratory animals and may be a human carcinogen.[3-8] This compound is slightly soluble in H_2O (2.8%) and is used as an alkylating agent industrially and in the laboratory.

Diethyl sulfate [DES, sulfuric acid diethyl ester, ethyl sulfate, $(C_2H_5)_2SO_4$] [64-67-5] is also a volatile liquid (bp 209°C) with a peppermint odor. This compound is almost insoluble in H_2O and is used as an alkylating agent industrially and in the laboratory. Diethyl sulfate causes cancer in experimental animals and may be a human carcinogen.[9-10] Diethyl sulfate is a teratogen and severe skin irritant.[11]

Methyl methanesulfonate [MMS, methyl mesylate, methanesulfonic acid methyl ester, $CH_3SO_2OCH_3$] [66-27-3] is a volatile liquid (bp 203°C), soluble to the extent of about 1:5 in H_2O. Methyl methanesulfonate causes cancer in experimental animals.[12]

Ethyl methanesulfonate (EMS, methanesulfonic acid ethyl ester, ethyl mesylate, $CH_3SO_2OC_2H_5$) [62-50-0] is a volatile liquid (bp 213°C), which is at least somewhat soluble in H_2O. Ethyl methanesulfonate causes cancer in experimental animals.[13]

Butadiene diepoxide (**I**, BDE, erythritol anhydride, diepoxybutane, butadiene dioxide, Bioxiran, 2,2′-bioxirane, dianhydrothreitol, dianhydroerythritol, 1,1′-biethylene oxide, bp 56–58°C at 25 mm Hg) [298-18-0] is miscible with H_2O. This compound is used industrially, particularly in the polymer industry, and causes cancer in experimental animals.[14]

1,3-Propane sultone (**II**, PS, 1,2-oxathiolane 2,2-dioxide, 3-hydroxy-1-propanesulfonic acid sultone, mp 31–33°C, bp 180°C at 30 mm Hg) [1120-71-4] is somewhat soluble in H_2O. This compound is used industrially, particularly in the detergent industry, and causes cancer in experimental animals.[15]

I 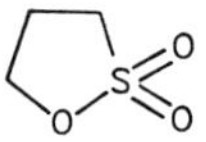 II

Although the toxicological properties of these compounds are not well known, by analogy with DMS they should be regarded as capable of causing lung injury and burns as well as being carcinogens. All of these compounds are mutagenic.

Principles of Destruction

Dimethyl sulfate is hydrolyzed by dilute base [sodium hydroxide (NaOH) solution (1 or 5 *M*), sodium carbonate (Na_2CO_3) solution (1 *M*), or ammonium hydroxide (NH_4OH) solution (1.5 *M*)] to methanol and methyl hydrogen sulfate.[16] Subsequent hydrolysis of methyl hydrogen sulfate to methanol and sulfuric acid is slow. We found, as others have,[17] that methyl hydrogen sulfate was nonmutagenic. Methyl hydrogen sulfate is a very poor alkylating agent.[18] When hydrolyzed using NH_4OH, the products are methylamine, dimethylamine, and trimethylamine. Hydrolysis destroyed DMS, a mutagenic compound, without producing other mutagenic species. The toxicity of methyl hydrogen sulfate is not well established, so appropriate steps should be taken to protect workers handling this material. In a similar fashion, DES, which is also mutagenic, can be hydrolyzed by the above reagents although the process is slower.[19] The products are presumably the analogous ethyl compounds. Ethanol is produced when the hydrolyzing agent is NaOH. Refluxing with alcoholic potassium hydroxide solution has also been reported to degrade dialkyl sulfates but validation details were not provided.[20] Methyl and ethyl methanesulfonates can be hydrolyzed with either 1 or 5 *M* NaOH solution. The products are methanol or ethanol and, presumably, methanesulfonic acid. Methyl methanesulfonate and EMS are mutagenic, but methanesulfonic acid is not.[19] Butadiene diepoxide and 1,3-propane sultone are hydrolyzed with either 1 or 5 *M* NaOH solution.[19] Theoretically, 1,3-propane sultone could reform on acidification, although we could find no evidence for this. It is probably prudent, however, not to acidify the reaction mixtures when the reaction is complete. When NaOH was used the degradation efficiency was greater than 99% in each case (Table 1). The compounds DMS, DES, MMS, and EMS can also be degraded by using a 1 *M* solution of sodium thiosulfate ($Na_2S_2O_3$).[21] The degradation efficiency was greater than 99.5%. The products of the reaction have not been determined.

Destruction Procedures

Destruction of Bulk Quantities of Dimethyl Sulfate and Diethyl Sulfate[16,19]

Note: The following reaction times gave good results in our tests. However, the reaction time may be affected by such factors as the size and shape of the flask and the rate of stirring. If two phases are apparent, this is an indication that the reaction is not complete. Stirring should be continued until the reaction mixture is homogeneous.

To accomplish destruction 10 mL of DMS or DES was added at once to a flask containing 500 mL of rapidly stirred 1 *M* NaOH solution, 1 *M* Na_2CO_3 solution, or 1.5 *M* NH_4OH solution. No DMS could be detected 15 min after the last of the DMS went into solution and no DES could be detected 3 h after the last of the DES went into solution. There was no apparent evolution of gas; the maximum temperature rise observed was

5°C. At the end of the reaction the mixture was neutralized, checked for completeness of destruction, and discarded.

This procedure may also be adapted for the destruction of larger quantities. Thus 100 mL of DMS was added to 1 L of **5** *M* NaOH solution[22] and the reaction mixture was stirred. No DMS could be detected 15 min after the last of the DMS went into solution. Similar results were obtained for DES but dissolution was much slower. No DES could be detected in solution 24 h after the addition of the DES to the base. The maximum temperature rise seen was 11°C. At the end of the reaction the mixture was neutralized, checked for completeness of destruction, and discarded.

Destruction of Bulk Quantities of Methyl Methanesulfonate, Ethyl Methanesulfonate, Butadiene Diepoxide, and 1,3-Propane Sultone[19]

1. To accomplish destruction 1 mL of the compound was added to 50 mL of 1 *M* NaOH solution and the reaction mixture was stirred for 1 h (PS), 6 h (MMS), 20 h (BDE), or 48 h (EMS). The reaction mixture was neutralized, checked for completeness of destruction, and discarded. These times may vary depending on such factors as the flask shape and the stirring rate. If the reaction mixture is not homogeneous, stirring should be continued until it is. The maximum temperature rise observed was 3°C (for PS).

2. To accomplish destruction 1 mL of the compound was added to 10 mL of 5 *M* NaOH solution and the reaction mixture was stirred for 1 h (PS), 2 h (MMS), 22 h (BDE), or 24 h (EMS). The reaction mixture was neutralized, checked for completeness of destruction, and discarded. These times may vary depending on such factors as the shape of the flask and the stirring rate. If the reaction mixture is not homogeneous, stirring should be continued until it is. The maximum temperature rise observed was 17°C (for MMS).

Destruction of Dimethyl Sulfate in Organic Solvents[16]

To accomplish destruction, 1 mL of a solution of DMS in methanol, ethanol, dimethylsulfoxide (DMSO), acetone, or *N,N*-dimethylformamide (DMF) (1 mL of DMS in 10 mL of solvent) was shaken with 4 mL of 1 *M* NaOH solution, 1 *M* Na_2CO_3 solution, or 1.5 *M* NH_4OH solution until the mixture was homogeneous. After 15 min no DMS could be detected when the solvent was methanol, ethanol, DMSO, or DMF ($<0.045\%$). After 1 h, no DMS could be detected when the solvent was acetone

(<0.045%). The reaction mixture was neutralized, checked for completeness of destruction, and discarded.

For solvents not miscible with H_2O (toluene, *p*-xylene, benzene, 1-pentanol, ethyl acetate, chloroform, and carbon tetrachloride) 1 mL of a solution of DMS (1 mL of DMS in 10 mL of solvent) was added to 4 mL of 1 *M* NaOH, 1 *M* Na_2CO_3, or 1.5 *M* NH_4OH and the heterogeneous mixture was rapidly stirred. After 24 h no DMS could be detected in the organic layer (<0.045%).

Acetonitrile solutions can also be decontaminated using the methods described. Although the solution should be homogeneous, it was found that two layers were present in some instanccs. After 3 h the solutions were all homogeneous and no DMS could be detected (<0.045%). We recommend stirring for 24 h to ensure complete destruction. When the reaction was complete the reaction mixture was neutralized and the layers were separated, checked for completeness of destruction, and discarded.

Destruction of Dimethyl Sulfate, Diethyl Sulfate, Methyl Methanesulfonate, and Ethyl Methanesulfonate[21]

Bulk quantities of DMS, DES, MMS, or EMS were dissolved in acetone and for each 1 mL of this solution 49 mL of 1 *M* sodium thiosulfate solution was added. The mixture was stirred for 2 h and discarded.

Spills of Dimethyl Sulfate[23]

The spill was covered with a 1:1:1 mixture of sodium carbonate or calcium carbonate, bentonite, and sand. The mixture was scooped up and added to a 10% NaOH solution. For each milliliter of DMS 10 mL of NaOH solution was used. This mixture was stirred for 24 h, checked for completeness of destruction, and discarded.

Analytical Procedures[7,24]

Note: If reaction mixtures that contain 5 *M* NaOH solution are to be analyzed, an aliquot of the reaction mixture should be diluted with four volumes of H_2O and 100 μL of this solution analyzed.

A 100 μL aliquot of the solution to be analyzed was added to 1 mL of a solution of 2 mL of acetic acid in 98 mL of 2-methoxyethanol. This mixture was swirled and 1 mL of a solution of 5 g of 4-(4-nitrobenzyl)pyridine (4-NBP) in 100 mL of 2-methoxyethanol was added. The solution was heated at 100°C for 10 min, then cooled in ice for 5 min. Piperidine (0.5 mL) and 2-methoxyethanol (2 mL) were added, and the violet color was determined at 560 nm using 10-mm disposable plastic cuvettes in a Gilford 240 UV/vis spectrophotometer.

To check the efficacy of the analytical procedure, a small quantity of DMS can be added to the solution to be analyzed after the acetic acid-2-methoxyethanol has been added but before the 4-NBP is added. A positive response will indicate that the analytical technique is satisfactory. Dichloromethane interfered with the determination, giving false positives. Although we have no reason to believe that DMS was not destroyed in dichloromethane solutions (and no mutagenic responses were obtained in these cases), we were not able to verify that complete destruction occurred. On the other hand, the false positives (obtained also in blank experiments) were small, with the largest value being less than 0.03% after 2 days. This result is equivalent to a destruction efficiency of greater than 99.97%. We recommend carefully checking each layer of each reaction for completeness of destruction.

Using the analytical procedure described above, the limits of detection were DMS 10 mg/L, DES 27 mg/L, MMS 21 mg/L, EMS 275 mg/L, BDE 90 mg/L, and PS 66 mg/L, but these could easily be reduced by increasing the volume of reaction mixture tested. For example, using 150 μL of solution containing EMS the limit of detection could be lowered to about 180 mg/L. Increasing the heating time to 60 min may also increase sensitivity.[25] We observed that a just noticeable violet color corresponded to a concentration that was about twice the detection limit given above, so the method could be used to rapidly screen a number of samples. The warming and cooling cycle may be inconvenient to carry out in many laboratories. For the analysis of DMS we found that it could be omitted provided that the solution was allowed to stand for 4 h before the piperidine and 2-methoxyethanol were added. The appropriate blanks and positive controls should always be run. When diethylamine was substituted for piperidine no noticeable decrease in sensitivity was observed.

The products from these degradation reactions were determined by gas chromatography (GC) using a Hewlett Packard HP5880 gas chromatograph equipped with a 1.8-m × 2-mm i.d. column packed with 10% Carbowax 20 M + 2% KOH on 80/100 Chromosorb W AW with flame ionization detection. The oven temperature was 150°C, the injection temperature was 200°C, the detector temperature was 300°C, and the carrier gas was nitrogen flowing at 30 mL/min. The approximate retention times were 0.5 min for methanol and 0.6 min for ethanol. The GC conditions given are only a guide and the exact conditions would have to be determined experimentally.

Mutagenicity Assays

The mutagenicity assays were carried out as described on page 4. For DMS tester strain TA100 was used and for DES, MMS, EMS, BDE, and PS tester strains TA98, TA100, TA1530, and TA1535 were used. For each plate 100 L of solution (corresponding to 2.6 mg of undegraded DMS, 2.3 mg of DES, 2.5 mg of MMS, 2.3 mg of EMS, 2.0 mg of BDE, and 2.5 mg of PS) was used.[19]

Solutions of DMS in various solvents were degraded using 1 *M* NaOH solution, 1 *M* Na_2CO_3 solution, or 1.5 *M* NH_4OH solution and the reaction mixtures were tested for mutagenicity. No mutagenic response was observed when solutions of DMS in methanol, H_2O, acetone, and DMSO were tested after 2 h of reaction, when the organic layer of solutions of DMS in dichloromethane and benzene were tested after 24 h of reaction, or when the organic layer of a solution of DMS in toluene was tested after 3 days of reaction.

The reaction mixture was cytotoxic (and therefore a determination of mutagenicity could not be made) when solutions of DMS in methanol or acetone were degraded using 1 *M* NaOH and when solutions of DMS in 1-pentanol were degraded using 1 *M* NaOH, 1 *M* Na_2CO_3, or 1.5 *M* NH_4OH.

A solution of DMS in ethanol was allowed to stand and gave a strong mutagenic response that slowly decreased with time, presumably as the DMS degraded. A similar decrease in activity was observed when a solution of DMS in ethanol was measured colorimetrically as described above.

No mutagenic activity was observed when DES, MMS, EMS, BDE, or PS were degraded with NaOH as described using the reaction times specified above with the exception that the degradation of BDE with 1 *M* NaOH solution gave reaction mixtures that were just mutagenic to TA98 without activation (59 revertants observed, control value 27). Degradation of BDE with 5 *M* NaOH did not give reaction mixtures that were significantly mutagenic.

Tester strains TA97, TA98, TA100, and TA102 were used when DMS, DES, MMS, and EMS were degraded using sodium thiosulfate solution.[21] No mutagenic activity was found.

Related Compounds

These destruction procedures and analytical methods should be applicable to other dialkyl sulfates, alkyl methanesulfonates, and related compounds although we have not verified this. Problems might arise when large alkyl groups are present because the compound may not be miscible with H_2O. In addition, large alkyl groups may slow hydrolysis to such an extent that complete destruction may not be obtained. In the work described above the hydrolysis slowed as the size of the alkyl group increased. β-Propiolactone, which is a cyclic ester, was completely degraded using these methods, although there did seem to be some tendency to reform β-propiolactone on acidification and mutagenic reaction mixtures were produced so these procedures cannot be recommended for this compound. See the β-Propiolactone monograph for details of a destruction procedure for this compound.

Table 1 Summary of Degradation Conditions for DMS, DES, MMS, EMS, BDE, and PS with Sodium Hydroxide

Compound	Molarity of NaOH (*M*)	Reaction Time[a]	Residue (%)	Maximum Temperature Rise (°C)	Products (%)
DMS	1	15 min*	<0.06	6	CH_3OH (68)
	5	15 min*	<0.06	11	CH_3OH (71)
DES	1	3 h*	<0.11	1	C_2H_5OH (47)
	5	24 h	<0.11	1	C_2H_5OH (50)
MMS	1	6 h	<0.09	1	CH_3OH (89)
	5	2 h	<0.09	17	CH_3OH (65)
EMS	1	48 h	<0.9	1	C_2H_5OH (88)
	5	24 h	<0.9	1	C_2H_5OH (68)
BDE	1	20 h	<0.4	1	
	5	22 h	<0.4	13	
PS	1	1 h	<0.24	3	
	5	1 h	<0.24	6	

For full details refer to the destruction procedures given above.

[a] Reaction time is normally measured from initial mixing. Reaction times marked with an asterisk were measured from the time the compound had completely dissolved in the NaOH solution.

References

1. E.I. du Pont de Nemours & Co. *Dimethyl Sulfate, Properties, Uses, Storage and Handling*; Du Pont: Wilmington, DE, 1981, *Dimethyl Sulfate, Product Safety Bulletin*; Du Pont: Wilmington, DE, 1980, and *Dimethyl Sulfate, Material Safety Data Sheet*; Du Pont: Wilmington, DE, 1980.
2. Lewis, R.J., Sr. *Sax's Dangerous Properties of Industrial Materials*, 8th ed.; Van Nostrand-Reinhold: New York, 1992; pp. 1416–1417.
3. International Agency for Research on Cancer. *IARC Monographs on the Evaluation of the Carcinogenic Risk of Chemicals to Humans, Supplement No. 4, Chemicals, Industrial Processes and Industries Associated with Cancer in Humans. IARC Monographs, Volumes 1 to 29*; International Agency for Research on Cancer: Lyon, 1982; pp. 119–120.
4. International Agency for Research on Cancer. *IARC Monographs on the Evaluation of Carcinogenic Risk of Chemicals to Man.* Volume 4, *Some Aromatic Amines, Hydrazine and Related Substances,* N-*Nitroso Compounds and Miscellaneous Alkylating Agents*; International Agency for Research on Cancer: Lyon, 1974; pp. 271–276.
5. Althouse, R.; Huff, J.; Tomatis, L.; Wilbourn, J. An evaluation of chemicals and industrial processes associated with cancer in humans based on human and animal data: IARC monographs volumes 1 to 20. *Cancer Res.* **1980**, *40*, 1–12.
6. Druckrey, H.; Preussmann, R.; Nashed, N.; Ivankovic, S. Carcinogene alkylierende Substanzen. I. Dimethylsulfat carcinogene Wirkung an Ratten und wahrscheinliche ursache von Berufskrebs. *Z. Krebsforsch.* **1966**, *68*, 103–111.
7. Druckrey, H.; Kruse, H.; Preussmann, R.; Ivankovic, S.; Landschütz, C. Cancerogene alkylierende Substanzen. III. Alkyl-halogenide, -sulfate, -sulfonate und ringgespannte Heterocyclen. *Z. Krebsforsch.* **1970**, *74*, 241–270.
8. International Agency for Research on Cancer. *IARC Monographs on the Evaluation of the Carcinogenic Risk of Chemicals to Humans, Supplement No. 7, Overall Evaluations of Carcinogenicity: An Updating of* IARC Monographs *Volumes 1 to 42*; International Agency for Research on Cancer: Lyon, 1987; pp. 200–201.
9. Reference 4, pp. 277–281.
10. Reference 8, p. 198.
11. Reference 2, pp. 1240–1241.
12. International Agency for Research on Cancer. *IARC Monographs on the Evaluation of Carcinogenic Risk of Chemicals to Man.* Volume 7, *Some Anti-thyroid and Related Substances, Nitrofurans and Industrial Chemicals*; International Agency for Research on Cancer: Lyon, 1974; pp. 253–260.
13. Reference 12, pp. 245–251.
14. International Agency for Research on Cancer. *IARC Monographs on the Evaluation of Carcinogenic Risk of Chemicals to Man.* Volume 11, *Cadmium, Nickel, Some Epoxides, Miscellaneous Industrial Chemicals and General Considerations on Volatile Anaesthetics*; International Agency for Research on Cancer: Lyon, 1976; pp. 115–123.
15. Reference 4, pp. 253–258.
16. Lunn, G.; Sansone, E.B. Validation of techniques for the destruction of dimethyl sulfate. *Am. Ind. Hyg. Assoc. J.* **1985**, *46*, 111–114.

17. Tan, E.-L; Brimer, P.A.; Schenley, R.L.; Hsie, A.W. Mutagenicity and cytotoxicity of dimethyl and monomethyl sulfates in the CHO/HGPRT system. *J. Toxicol. Environ. Health* **1983**, *11*, 373–380.

18. McCormack, W.B.; Lawes, B.C. Sulfuric and Sulfurous Esters. In *Kirk–Othmer Encyclopedia of Chemical Technology*, 3rd ed.; Wiley: New York, 1983; Vol. 22, pp. 233–254.

19. Lunn, G.; Sansone, E. B. Validated methods for degrading hazardous chemicals: Some alkylating agents and other compounds. *J. Chem. Educ.* **1990**, *67*, A249–A251.

20. National Research Council, Committee on Hazardous Substances in the Laboratory. *Prudent Practices for Disposal of Chemicals from Laboratories;* National Academy Press: Washington, DC, 1983; p. 61.

21. de Méo, M.; Laget, M.; Castegnaro, M.; Duménil, G. Evaluation of methods for destruction of some alkylating agents. *Am. Ind. Hyg. Assoc. J.* **1990**, *51*, 505–509.

22. Personal communication from B.I.Tobias, Chemical Safety Officer, NCI-FCRDC, April 16, 1985.

23. Armour, M-.A.; Browne, L.M.; McKenzie, P.A.; Renecker, D.M.; Bacovsky, R.A., Eds., *Potentially Carcinogenic Chemicals, Information and Disposal Guide*; University of Alberta: Edmonton, Alberta, 1986; p. 60.

24. Epstein, J.; Rosenthal, R.W.; Ess, R.J. Use of γ-(4-nitrobenzyl)pyridine as an analytical reagent for ethylenimines and alkylating agents. *Anal. Chem.* **1955**, *27*, 1435–1439.

25. Preussmann, R.; Schneider, H.; Epple, F. Untersuchungen zum Nachweis alkylierender Agentien. II. Der Nachweis verschiedener Klassen alkylierender Agentien mit einer Modifikation der Farbreaktion mit 4-(4-Nitrobenzyl)-pyridin (NBP). *Arzneimittel-Forsch.* **1969**, *19*, 1059–1073.

DOXORUBICIN AND DAUNORUBICIN

CAUTION! Refer to safety considerations section on page 7 before starting any of these procedures.

The degradation of a number of antineoplastic drugs, including doxorubicin and daunorubicin, was investigated by the International Agency for Research on Cancer (IARC).[1] Doxorubicin (**I**) (mp 204–205°C) [23214–92-8][2] and daunorubicin (**II**) (mp 188–189°C) [20830-81-3][3] are solids and they are soluble in H_2O and alcohols but not in organic solvents.

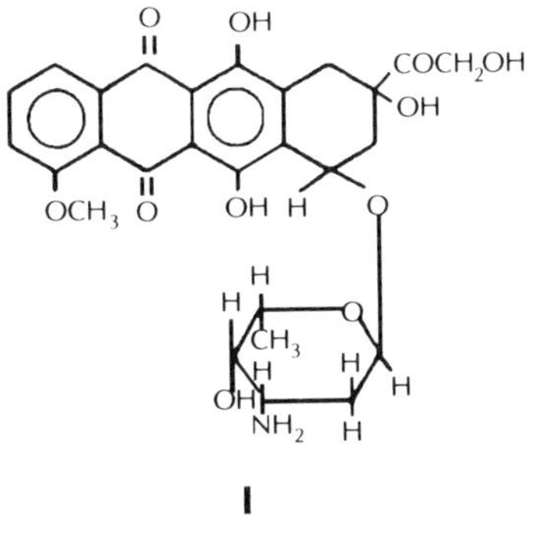

I 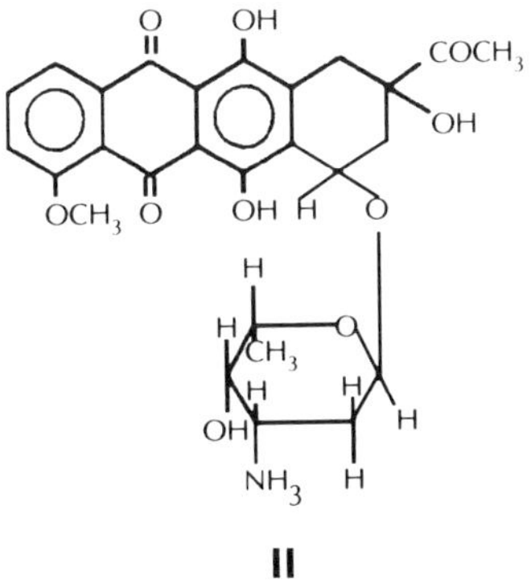 **II**

Doxorubicin[4–6] and daunorubicin[7] are mutagenic and carcinogenic in animals. Doxorubicin is a teratogen and has effects on the heart[8] and daunorubicin is a teratogen.[9] These compounds are employed as antineoplastic drugs.

Principles of Destruction

Doxorubicin and daunorubicin are destroyed by oxidation with potassium permanganate in sulfuric acid ($KMnO_4$ in H_2SO_4).[1] Destruction is greater than 99%. The products of these reactions are unknown. Oxidation of doxorubicin and daunorubicin with sodium hypochlorite (NaOCl) gave mutagenic products although chemical degradation was complete.[1] However, doxorubicin in urine can be completely destroyed by NaOCl followed by sodium thiosulfate without producing mutagenic residues.[10]

Destruction Procedures

Destruction of Bulk Quantities of Doxorubicin and Daunorubicin

Dissolve in 3 *M* H_2SO_4 so that the concentration does not exceed 3 mg/mL, then add 1 g of $KMnO_4$ for every 10 mL of solution, and stir for 2 h. Decolorize with sodium metabisulfite, make strongly basic with 10 *M* potassium hydroxide (KOH) solution (**Caution!** Exothermic), dilute with H_2O, filter to remove manganese compounds,[11] neutralize the filtrate, check for completeness of destruction, and discard it.

Destruction of Aqueous Solutions of Doxorubicin and Daunorubicin

Dilute with H_2O, if necessary, so that the concentration does not exceed 3 mg/mL, then add enough concentrated H_2SO_4 to obtain a 3 *M* solution, and allow it to cool to room temperature. For each 10 mL of solution add 1 g of $KMnO_4$ and stir for 2 h. Decolorize with sodium metabisulfite, make strongly basic with 10 *M* KOH solution (**Caution!** Exothermic), dilute with H_2O, filter to remove manganese compounds,[11] neutralize the filtrate, check for completeness of destruction, and discard it.

Destruction of Solid Pharmaceutical Preparations of Doxorubicin and Daunorubicin

Dissolve in H_2O so that the concentration does not exceed 3 mg/mL, then add enough concentrated H_2SO_4 to obtain a 3 *M* solution, and allow it to cool to room temperature. For each 10 mL of solution add 2 g of $KMnO_4$, in small portions to avoid frothing, and stir for 2 h. Decolorize with sodium metabisulfite, make strongly basic with 10 *M* KOH solution (**Caution!** Exothermic), dilute with H_2O, filter to remove manganese compounds,[11] neutralize the filtrate, check for completeness of destruction, and discard it.

Destruction of Liquid Pharmaceutical Preparations of Doxorubicin and Daunorubicin

Dilute with H_2O, if necessary, so that the concentration does not exceed 3 mg/mL, then add enough concentrated H_2SO_4 to obtain a 3 *M* solution, and allow it to cool to room temperature. For each 10 mL of solution add 2 g of $KMnO_4$, in small portions to avoid frothing, and stir for 2 h. Decolorize with sodium metabisulfite, make strongly basic with 10 *M* KOH solution (**Caution!** Exothermic), dilute with H_2O, filter to remove manganese compounds,[11] neutralize the filtrate, check for completeness of destruction, and discard it.

Destruction of Solutions of Doxorubicin and Daunorubicin in Volatile Organic Solvents

Remove the solvent under reduced pressure using a rotary evaporator and take up the residue in 3 *M* H_2SO_4 so that the concentration does not exceed 3 mg/mL. For each 10 mL of solution add 1 g of $KMnO_4$ and stir for 2 h. Decolorize with sodium metabisulfite, make strongly basic with 10 *M* KOH solution (**Caution!** Exothermic), dilute with H_2O, filter to remove manganese compounds,[11] neutralize the filtrate, check for completeness of destruction, and discard it.

Destruction of Dimethyl Sulfoxide Solutions of Doxorubicin and Daunorubicin

Dilute with H_2O so that the concentration of dimethyl sulfoxide (DMSO) does not exceed 20% and the concentration of the drug does not exceed 3 mg/mL, then add enough concentrated H_2SO_4 to obtain a 3 *M* solution and allow it to cool to room temperature. For each 10 mL of solution add 2 g of $KMnO_4$ and stir for 2 h. Decolorize with sodium metabisulfite, make strongly basic with 10 *M* KOH solution (**Caution!** Exothermic), dilute with H_2O, filter to remove manganese compounds,[11] neutralize the filtrate, check for completeness of destruction, and discard it.

Destruction of Doxorubicin in Urine[10]

For each 4 mL of urine add 1 mL of 5.25% NaOCl solution followed by 100 mg of sodium thiosulfate. Destruction is rapid. Destroy the excess NaOCl by adding 100 mg of sodium bisulfite. Neutralize the reaction mixture, check for completeness of destruction, and discard it. Use fresh NaOCl solution (see below for assay procedure).

Decontamination of Glassware Contaminated with Doxorubicin or Daunorubicin

Immerse the glassware in a 0.3 *M* solution of $KMnO_4$ in 3 *M* H_2SO_4 for 2 h, then clean by immersion in sodium metabisulfite solution. Decolorize the $KMnO_4$ solution with sodium metabisulfite, make strongly basic with 10 *M* KOH solution (**Caution!** Exothermic), dilute with H_2O, filter to remove manganese compounds,[11] neutralize the filtrate, check for completeness of destruction, and discard it.

Decontamination of Spills of Doxorubicin or Daunorubicin

Allow any organic solvent to evaporate then cover the area with a 0.3 *M* solution of $KMnO_4$ in 3 *M* H_2SO_4 for 2 h. If the color fades, add more solution. Decolorize the area with sodium metabisulfite solution and neutralize by the addition of solid sodium carbonate.

Assay of Sodium Hypochlorite Solution

Sodium hypochlorite solutions tend to deteriorate with time so they should be periodically checked for the amount of active chlorine they contain. Pipette 10 mL of NaOCl solution into a 100-mL volumetric flask and fill to the mark with distilled H_2O. Pipette 10 mL of this solution into a conical flask containing 50 mL of distilled H_2O, 1 g of potassium iodide, and 12.5 mL of 2 *M* acetic acid. Titrate this solution against 0.1 *N* sodium thiosulfate solution using starch as an indicator. Each 1 mL of the sodium thiosulfate solution corresponds to 3.545 mg of active chlorine. The NaOCl solution used in these degradation reactions should contain 45–50 g of active chlorine per liter.

Analytical Procedures

These drugs can be analyzed by high-performance liquid chromatography (HPLC) using a 25-cm reverse phase column and UV detection at 254 nm.

The mobile phase was 0.01 *M* potassium phosphate, monobasic (KH_2PO_4) in 0.02 *M* phosphoric acid (H_3PO_4) : acetonitrile (45:55) flowing at 1.5 mL/min. Fluorescence detection with excitation at 470 nm and emission at 565 nm gives more sensitivity, but UV detection at 254 nm is usually satisfactory.

Mutagenicity Assays

In the IARC study[1] tester strains TA98, TA100, and TA102 of *Salmonella typhimurium* were used with and without metabolic activation. Generally the reaction mixtures were not mutagenic although degradation of doxorubicin with $KMnO_4$ gave twice the background activity with TA102.

Related Compounds

Potassium permanganate in H_2SO_4 is a general oxidative method and should, in principle, be applicable to many drugs. However, any new application should be thoroughly validated both for complete destruction of the compound and for the production of nonmutagenic reaction mixtures.

References

1. Castegnaro, M.; Adams, J.; Armour, M-.A.; Barek, J.; Benvenuto, J.; Confalonieri, C.; Goff, U.; Ludeman, S.; Reed, D.; Sansone, E. B.; Telling, G., Eds., *Laboratory Decontamination and Destruction of Carcinogens in Laboratory Wastes: Some Antineoplastic Agents*; International Agency for Research on Cancer: Lyon, 1985 (IARC Scientific Publications No. 73).
2. Other names for this compound are 10-[(3-amino-2,3,6-trideoxy-α-L-lyxo-hexopyranosyl)oxy]-7,8,9,10-tetrahydro-6,8,11-trihydroxy-8-(hydroxyacetyl)-1-methoxy-5,12-naphthacenedione, Adriablastima, adriamycin, 14-hydroxydaunomycin, 14-hydroxydaunorubicine, ADM, adriamycin semiquinone, DX, F.I 106, KW-125, NCI-C01514, NSC-123127, Adriacin, and Adriblastina.
3. Other names for this compound are 8-acetyl-10-[(3-amino-2,3,6-trideoxy-α-L-lyxo-hexopyranosyl)oxy]-7,8,9,10-tetrahydro-6,8,11-trihydroxy-1-methoxy-5,12-naphthacenedione, daunomycin, leukaemomycin C, rubidomycin, Cerubidin, acetyladriamycin, daunamycin, DM, F16339, NCI-C04693, NSC-82151, RP 13057, 13,057 R.P., rubidomycine, rubidomycin C, rubidomycin C1, Streptomyces peucetius, Daunoblastina, and Ondena.
4. International Agency for Research on Cancer. *IARC Monographs on the Evaluation of Carcinogenic Risk of Chemicals to Man.* Volume 10, *Some Naturally Occurring Substances*; International Agency for Research on Cancer: Lyon, 1975; pp. 43–49.
5. International Agency for Research on Cancer. *IARC Monographs on the Evaluation of the Carcinogenic Risk of Chemicals to Humans, Supplement No. 7, Overall Evaluations*

of Carcinogenicity: An Updating of IARC Monographs *Volumes 1 to 42*; International Agency for Research on Cancer: Lyon, 1987; pp. 82–83.

6. International Agency for Research on Cancer. *IARC Monographs on the Evaluation of the Carcinogenic Risk of Chemicals to Humans, Supplement No. 4, Chemicals, Industrial Processes and Industries Associated with Cancer in Humans. IARC Monographs, Volumes 1 to 29*; International Agency for Research on Cancer: Lyon, 1982; pp. 29–31.
7. Reference 4, pp. 145–152.
8. Lewis, R.J., Sr., *Sax's Dangerous Properties of Industrial Materials*, 8th ed.; Van Nostrand-Reinhold: New York, 1992; pp. 84–85.
9. Reference 8, p. 1021.
10. Monteith, D.K.; Connor, T.H.; Benvenuto, J.A.; Fairchild, E.J.; Theiss, J.C. Stability and inactivation of mutagenic drugs and their metabolites in the urine of patients administered antineoplastic therapy. *Environ. Mol. Mutagen.* **1987**, *10*, 341–356.
11. Lunn, G.; Sansone, E.B.; De Méo, M.; Laget, M.; Castegnaro, M. Potassium permanganate can be used for degrading hazardous compounds. *Am. Ind. Hyg. Assoc. J.* **1994**, *55*, 167–171.

DRUGS CONTAINING HYDRAZINE AND TRIAZENE GROUPS

CAUTION! Refer to safety considerations section on page 7 before starting any of these procedures.

The drugs considered in this monograph all have N—N bonds either as part of a triazene group (Dacarbazine) or a hydrazine group (others). These compounds are all solids and are all at least somewhat soluble in H_2O. The basic nature of these drugs makes them very soluble in acid solution. Procarbazine is supplied and used as the hydrochloride.

The following table lists the compounds that are considered:

Compound Name	Reference	MP (°C)	Structure	Registry Number
Procarbazine hydrochloride	1	223–226	(**I.HCl**)	[366-70-1] ([671-16-9], base)
Isoniazid	2	171–173	(**II**)	[54-85-3]
Iproniazid phosphate	3	180–182	(**III.**H_3PO_4)	[305-33-9] ([54-92-2], base)
Dacarbazine	4	250–255	(**IV**)	[4342-03-4]

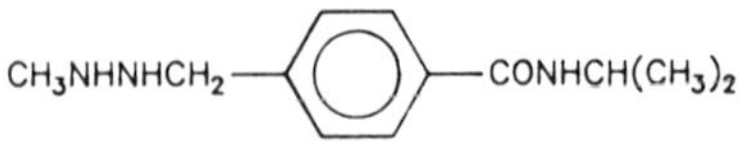

$CH_3NHNHCH_2$—C6H4—$CONHCH(CH_3)_2$

I

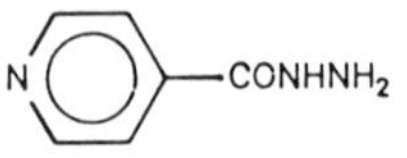

N(C5H4)—$CONHNH_2$

II

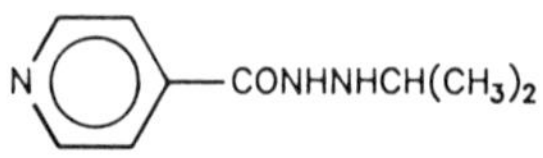

N(C5H4)—$CONHNHCH(CH_3)_2$

III

$CONH_2$, HN, N, $N{=}N{-}N(CH_3)_2$

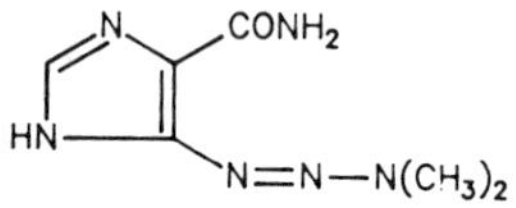

IV

Dacarbazine is mutagenic[5] and also carcinogenic in laboratory animals[6–9] and probably carcinogenic to humans.[7] Procarbazine[10–14] and isoniazid[11,15–17] are carcinogenic in laboratory animals and procarbazine is probably carcinogenic in humans.[12] Procarbazine is a teratogen,[18] isoniazid is a teratogen and produces numerous systemic effects,[19] iproniazid is a teratogen and produces various effects including changes in liver function,[20] and dacarbazine causes nausea and has effects on the blood.[21] Procarbazine and dacarbazine are antineoplastics, isoniazid is a tuberculostatic, and iproniazid is a monoamine oxidase inhibitor.

Principles of Destruction

When these drugs are degraded by reduction with nickel–aluminum (Ni–Al) alloy in potassium hydroxide (KOH) solution, less than 0.03% of the original amount of dacarbazine, less than 0.65% procarbazine, less than 0.2% isoniazid, and less than 0.2% iproniazid remains. The products are 4-aminoimidazole-5-carboxamide and dimethylamine from dacarbazine, *N*-isopropyl-α-amino-*p*-toluamide, *N*-isopropyl-*p*-toluamide, and methylamine from procarbazine, and 4-piperidinecarboxamide from isoniazid and iproniazid. Initially, reduction of the pyridine rings of isoniazid and iproniazid gives the corresponding piperidino hydrazines but reduction for the times specified gives full reduction to 4-piperidinecarboxamide. Destruction of iproniazid also produces isopropylamine. Dacarbazine is generally used in a citric acid solution and this solution will degrade and turn pink upon exposure to light. Such degraded solutions can also be successfully treated with Ni–Al alloy.[5]

Although potassium permanganate in sulfuric acid ($KMnO_4$ in H_2SO_4) oxidation gave complete destruction of dacarbazine in all cases, the reaction mixtures were mutagenic[5] and the carcinogen *N*-nitrosodimethylamine was produced. Increasing the ratio of oxidant to substrate gave nonmutagenic reaction mixtures, but the volumes became too large for a practical destruction procedure. Exposure to sunlight gave only partial degradation of the dacarbazine. Potassium permanganate and calcium hypochlorite (but **not** sodium hypochlorite) can be used to degrade procarbazine. The drug is completely destroyed ($<0.8\%$ remains), the reaction mixtures are nonmutagenic, and nitrosamines are not detected.[22] The products of these reactions are not known.

Destruction Procedures

Destruction of Dacarbazine and Isoniazid

Dissolve bulk quantities in H_2O so that the concentration does not exceed 10 mg/mL. Open capsules to ensure dissolution of the drug. Dilute aqueous solutions and pharmaceutical preparations with H_2O, if necessary, so that the concentration does not exceed 10 mg/mL. Add an equal volume of 1 *M* KOH solution, then add 1 g of Ni–Al alloy for each 20 mL of the basified solution. Perform the reaction in a container that is at least three times as large as the final reaction volume. Add quantities of Ni–Al alloy in excess of 5 g in portions over the course of at least 1 h to avoid frothing. Stir the mixture overnight, then filter it through a pad of Celite®. Neutralize the filtrate, check for completeness of destruction, and discard it. Allow the spent nickel to dry on a metal tray away from flammable solvents for 24 h, then discard it with the solid waste.

Destruction of Iproniazid

Dissolve bulk quantities in H_2O so that the concentration does not exceed 5 mg/mL. Open capsules to ensure dissolution of the drug. Dilute aqueous solutions and pharmaceutical preparations with H_2O, if necessary, so that the concentration does not exceed 5 mg/mL. Add an equal volume of 1 *M* KOH solution, and then add 1 g of Ni–Al alloy for each 20 mL of the basified solution. Perform the reaction in a container that is at least three times as large as the final reaction volume. Add quantities of Ni–Al alloy in excess of 5 g in portions over the course of at least 1 h to avoid frothing. Stir the mixture for 96 h, then filter it through a pad of Celite®. Neutralize the filtrate, check for completeness of destruction, and discard it. Allow the spent nickel to dry on a metal tray away from flammable solvents for 24 h, then discard it with the solid waste.

Destruction of Procarbazine

1. Dissolve bulk quantities in H_2O so that the concentration does not exceed 10 mg/mL. Open capsules to ensure dissolution of the drug. Dilute aqueous solutions and pharmaceutical preparations with H_2O, if necessary, so that the concentration does not exceed 10 mg/mL. Add an equal volume of 1 *M* KOH solution, then add 1 g of Ni–Al alloy for each 20 mL of the basified solution. Perform the reaction in a container that is at least three times as large as the final reaction volume. Add quantities of Ni–Al alloy in excess of 5 g in portions over the course of at least 1 h to avoid frothing. Stir the mixture overnight, then filter it through a pad of Celite®. Neutralize the filtrate, check for completeness of destruction, and discard it. Allow the spent nickel to dry on a metal tray away from flammable solvents for 24 h, then discard it with the solid waste.

2. Take up bulk quantities in H_2O so that the concentration does not exceed 2.5 mg/mL and dilute aqueous solutions, if necessary, so that the concentration does not exceed 2.5 mg/mL. For each 10 mL of solution add 1.5 g of calcium hypochlorite and stir the mixture overnight, add ascorbic acid to destroy excess oxidant, analyze for completeness of destruction, and discard it.

3. Take up bulk quantities in H_2O so that the concentration does not exceed 5 mg/mL and dilute aqueous solutions, if necessary, so that the concentration does not exceed 5 mg/mL. For each 5 mL of solution add 1 mL of concentrated H_2SO_4 and 0.29 g of $KMnO_4$ so that the solution is 3 *M* in H_2SO_4 and 0.3 *M* in $KMnO_4$. Stir the mixture overnight, decolorize with sodium metabisulfite, make strongly basic with 10 *M* KOH solution (**Caution!** Exothermic), dilute with H_2O, filter to remove manganese compounds,[23] neutralize the filtrate, analyze for completeness of destruction, and discard it.

Analytical Procedures

Analysis was by high-performance liquid chromatography (HPLC) using a 250 × 4.6-mm i.d. column of Microsorb C8. The injection volume was

20 μL and UV detection was at 254 nm. The mobile phase was a mixture of methanol and buffer flowing at 1 mL/min. The buffer contained 0.04% ammonium phosphate, monobasic [$(NH_4)H_2PO_4$], and 0.1% triethylamine. The methanol:buffer ratios were 10:90 for isoniazid; 20:80 for dacarbazine and 4-aminoimidazole-5-carboxamide; 50:50 for procarbazine, iproniazid, and *N*-isopropyl-*p*-toluamide; and 80:20 for *N*-isopropyl-α-amino-*p*-toluamide. On our equipment these mobile phase combinations were found to give reasonable retention times (3.4–8.3 min).

Gas chromatography (GC) using a 1.8-m × 2-mm i.d. glass column packed with 10% Carbowax 20 M + 2% KOH on 80/100 Chromosorb W AW was employed to determine the products of these reactions. The injection temperature was 200°C and the flame ionization detector operated at 300°C. The oven temperature was 60°C and approximate retention times were methylamine (0.7 min), dimethylamine (0.7 min), and isopropylamine (0.8 min). At 100°C the retention time of *N*-nitrosodimethylamine was 6.6 min. Using a 1.8-m × 2-mm i.d. glass column packed with 2% Carbowax 20 M + 1% KOH on 80/100 Supelcoport and an oven temperature of 200°C the products of the reductions of isoniazid and iproniazid were detected. The piperidino hydrazines from isoniazid and iproniazid had retention times of 11.8 and 8.2 min, respectively, and 4-piperidinecarboxamide had a retention time of 10.5 min.

Mutagenicity Assays[5]

The mutagenicity assays were carried out as described on page 4 using tester strains TA98, TA100, TA1530, and TA1535. The final reaction mixtures [tested at a level corresponding to 0.5 mg (0.25 mg for iproniazid and the calcium hypochlorite procedures) undegraded material per plate] were not mutagenic. Of the original compounds, only dacarbazine was found to be mutagenic and none of the products detected were found to be mutagenic.

Related Compounds

The Ni–Al alloy procedure should be applicable for related compounds containing hydrazine and triazene groups, but it should be fully validated before routine use. Nickel–aluminum alloy has been shown to reduce 3-methyl-1-*p*-tolyltriazene to its parent amines in good yield although the complete disappearance of the starting material has not been established.[24]

References

1. Other names for this compound are ibenzmethyzin, IBZ, *N*-isopropyl-α-(2-methylhydrazino)-*p*-toluamide, *N*-isopropyl-*p*-[(2-methylhydrazino)methyl]benzamide, *N*-4-isopropylcarbamoylbenzyl-*N'*-methylhydrazine, 2-(*p*-isopropylcarbamoylbenzyl)-1-methylhydrazine, matulane, MBH, 1-methyl-2-[(isopropylcarbamoyl)benzyl]hydrazine, *N*-(1-methylethyl)-4-[(2-methylhydrazino)methyl]benzamide, *p*-(N^1-methylhydrazinomethyl)-*N*-isopropylbenzamide, MIH, Nathulane, natulan, NCI-C01810, NSC-77213, and RO 4-6467. This compound is generally supplied and used as the hydrochloride.
2. Other names for this compound are Amidon, Antimicina, Antituberkulosum, Atcotibine, Azuren, Bacillin, Cedin(Aerosol), Cemidon, Chemiazid, Cotinazin, Defonin, Dibutin, Dinacrin, Ditubin, Eralon, Ertuban, Evalon, Fimaline, Hidranizil, Hidrasonil, Hidrulta, Hycozid, Hydrazid, Hyozid, Hyzyd, Idrazil, INH, Isdonidrin, Isidrina, Ismazide, Isobicina, Isocid, Isocotin, Isolyn, Isonex, Isoniacid, Isonicazide, Isonicid, Isonicotan, isonicotinic acid hydrazide, isonicotinoyl hydrazide, isonicotinoylhydrazine, isonicotinyl hydrazide, isonicotinylhydrazine, Isonide, Isonikazid, Isonilex, Isonin, Isonindon, Isonirit, Isonizide, Isotebezid, Isozide, Isozyd, Laniazid, Mybasan, Neoteben, Neoxin, Neumandin, Nevin, Niadrin, Nicazide, Nicetal, Nicizina, Niconyl, Nicotibina, Nicozide, Nidaton, Nidrazid, Nikozid, Niplen, Nitadon, Niteban, NSC-9659, Nydrazid, Nyscozid, Pelazid, Percin, Pycazide, Pyricidin, 4-pyridinecarboxylic acid hydrazide, Pyrizidin, Raumanon, Retozide, Rimicid, Rimifon, rimitsid, Robisellin, RU-EF-Tb, Sanohydrazina, Sauterazid, Sauterzid, TB-Vis, Tebecid, Tebexin, Teebaconin, Tekazin, Tibazide, Tibinide, Tibivis, Tibizide, Tisin, Tisiodrazida, Tizide, Tubazid, Tubeco, Tubercid, Tubicon, Tubilysin, Tubomel, Tyvid, Tyvid, Unicozyde, USAF CB-2, Vazadrine, Vederon, Zinadon, and Zonazide.
3. Other names for this compound are Euphozid, Fosfazide, IIH, IPN, Iprazid, Iproniazid, Ipronid, Ipronin, isonicotinic acid 2-isopropylhydrazide, 1-isonicotinoyl-2-isopropylhydrazine, 1-isonicotinyl-2-isopropylhydrazine, *N*-isopropyl isonicotinhydrazide, LH, Marsalid, Marsilid, P 887, 4-pyridinecarboxylic acid 2-(1-methylethyl)hydrazide, Rivivol, RO 2-4572, and Yatrozide. This compound is frequently supplied as the phosphate salt.
4. Other names for this compound are 5-(3,3-dimethyl-1-triazenyl)-1*H*-imidazole-4-carboxamide, 4-(3,3-dimethyl-1-triazeno)imidazole-5-carboxamide, 4-(5)-(3,3-dimethyl-1-triazeno)imidazole-5(4)-carboxamide, 5-(3,3-dimethyl-1-triazeno)imidazole-4-carboxamide, DIC, DTIC, DTIC-Dome, Deticene, NSC-45388, Dacatic, and NCI-C04717.
5. Lunn, G.; Sansone, E.B. Reductive destruction of dacarbazine, procarbazine hydrochloride, isoniazid, and iproniazid. *Am. J. Hosp. Pharm.* **1987**, *44*, 2519–2524.
6. Skibba, J.L.; Ertürk, E.; Bryan, G.T. Induction of thymic lymphosarcoma and mammary adenocarcinomas in rats by oral administration of the antitumor agent, 4(5)-(3,3-dimethyl-1-triazeno)imidazole-5(4)-carboxamide. *Cancer* **1970**, *26*, 1000–1005.
7. International Agency for Research on Cancer. *IARC Monographs on the Evaluation of the Carcinogenic Risk of Chemicals to Humans.* Volume 26, *Some Antineoplastic and Immunosuppressive Agents*; International Agency for Research on Cancer: Lyon, 1981; pp. 203–215.
8. International Agency for Research on Cancer. *IARC Monographs on the Evaluation of the Carcinogenic Risk of Chemicals to Humans, Supplement No. 4, Chemicals, Industrial*

Processes and Industries Associated with Cancer in Humans. IARC Monographs, Volumes 1 to 29; International Agency for Research on Cancer: Lyon, 1982; pp. 103–104.

9. International Agency for Research on Cancer. *IARC Monographs on the Evaluation of the Carcinogenic Risk of Chemicals to Humans, Supplement No. 7, Overall Evaluations of Carcinogenicity: An Updating of* IARC Monographs *Volumes 1 to 42*; International Agency for Research on Cancer: Lyon, 1987; pp. 184–185.
10. Kelly, M.G.; O'Gara, R.W.; Yancey, S.T.; Botkin, C. Induction of tumors in rats with procarbazine hydrochloride. *J. Natl. Cancer Inst.* **1968**, *40*, 1027–1051.
11. Kelly, M.G.; O'Gara, R.W.; Yancey, S.T.; Gadekar, K.; Botkin, C.; Oliverio, V.T. Comparative carcinogenicity of *N*-isopropyl-α-(2-methylhydrazino)-*p*-toluamide.HCl (procarbazine hydrochloride), its degradation products, other hydrazines, and isonicotinic acid hydrazide. *J. Natl. Cancer Inst.* **1969**, *42*, 337–344.
12. Reference 7, pp. 311–339.
13. Reference 8, pp. 220–221.
14. Reference 9, pp. 327–328.
15. Reference 9, pp. 227–228.
16. Reference 8, pp. 146–148.
17. International Agency for Research on Cancer. *IARC Monographs on the Evaluation of Carcinogenic Risk of Chemicals to Man.* Volume 4, *Some Aromatic Amines, Hydrazine and Related Substances,* N-*Nitroso Compounds and Miscellaneous Alkylating Agents*; International Agency for Research on Cancer: Lyon, 1974; pp. 159–172.
18. Lewis, R.J., Sr. *Sax's Dangerous Properties of Industrial Materials*, 8th ed.; Van Nostrand-Reinhold: New York, 1992; pp. 2892–2893.
19. Reference 18, p. 2030.
20. Reference 18, pp. 2030–2031.
21. Reference 18, p. 1018.
22. Castegnaro, M.; Brouet, I.; Michelon, J.; Lunn, G.; Sansone, E.B. Oxidative destruction of hydrazines produces *N*-nitrosamines and other mutagenic species. *Am. Ind. Hyg. Assoc. J.* **1986**, *47*, 360–364.
23. Lunn, G.; Sansone, E.B.; De Méo, M.; Laget, M.; Castegnaro, M. Potassium permanganate can be used for degrading hazardous compounds. *Am. Ind. Hyg. Assoc. J.* **1994**, *55*, 167–181.
24. Lunn, G.; Sansone, E.B.; Keefer, L.K. General cleavage of N—N and N—O bonds using nickel/aluminum alloy. *Synthesis* **1985**, 1104–1108.

ETHIDIUM BROMIDE

CAUTION! Refer to safety considerations section on page 7 before starting any of these procedures.

Ethidium bromide (EB) (**I**)[1] is a red solid (mp 260–262°C) [1239-45-8], a potent mutagen,[2] and moderately toxic.[3] This compound is a dye widely used in biomedical laboratories for visualizing nucleic acids. Fluorescent complexes are formed by intercalation and these complexes are readily seen on irradiation with ultraviolet (UV) light.

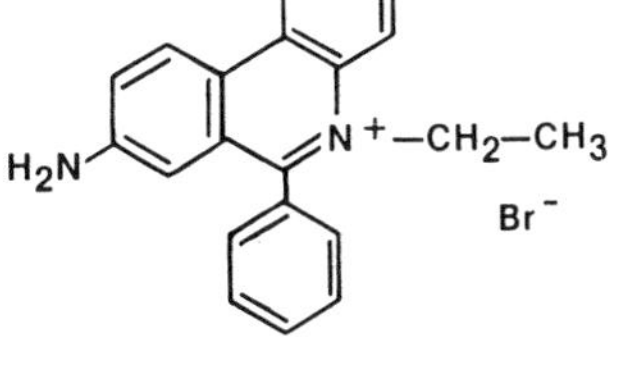

NH_2

H_2N N^+—CH_2—CH_3

Br^-

I

Principles of Destruction and Decontamination

Ethidium bromide in H_2O and buffer solution may be degraded by reaction with sodium nitrite and hypophosphorous acid in aqueous solution.[4] This procedure may also be used to decontaminate equipment.[5] The products have not been determined but appear to consist of compounds in which the amino groups have been removed and replaced, at least partially, with oxygen.[6] Destruction efficiency is greater than 99.87% and the resulting reaction mixtures are nonmutagenic.[7] A modification of this method may be used to degrade EB dissolved in alcohols. The reaction is either one phase (isopropanol saturated with cesium chloride) or two phase (isoamyl alcohol or 1-butanol). Destruction efficiency is greater than 99.75%[8] and, in general, the reaction mixtures are not mutagenic.[9] Ethidium bromide may also be removed from solution by adsorption onto Amberlite XAD-16 resin.[4,10] Removal is greater than 99.95% in most cases. "Blue cotton" (also called Mutasorb) has also been found to remove EB from solution[11] but it is much more expensive and much less efficient than Amberlite XAD-16 so it is not recommended.[4] Potassium permanganate and sodium hypochlorite oxidation and nickel-aluminum alloy reduction produced mutagenic reaction mixtures.[4,7] Potassium permanganate in hydrochloric acid has been reported to give complete destruction and nonmutagenic reaction mixtures,[12] but we have found[7] that this procedure sometimes gives mutagenic reaction mixtures.

The destruction of EB in aqueous or isoamyl alcohol solution with ozone has been described.[13] Destruction efficiency was greater than 99.95% and the final reaction mixtures were nonmutagenic. Ozone is a powerful oxidizing agent and is incompatible with many organic compounds. Further investigation may be required before this procedure can be used to decontaminate isoamyl alcohol solutions on a routine basis.

Procedures for the safe disposal of ethidium bromide have been reviewed.[9] The use of a carbon filter for removing ethidium bromide from aqueous solution has been described.[14] The use of Rezorian A-161, a hydrophobic divinylbenzene polymer resin, has been reported to be useful for decontaminating very dilute (0.5–0.05 μg/mL) solutions of ethidium bromide in a flow-through procedure.[15] A desalting column can be used to separate DNA from ethidium bromide and cesium chloride without the necessity for extraction with organic solvents.[16]

Destruction and Decontamination Procedures

Destruction of Ethidium Bromide in Aqueous Solution

1. Dilute the solution, if necessary, so that the concentration of EB does not exceed 0.5 mg/mL. For each 100 mL of EB in H_2O, 4-morpholine-propanesulfonic acid (MOPS) buffer (see below in Buffer Solutions section), TRIS–borate-ethylenediaminetetraacetic acid (TBE) buffer (see below in Buffer Solutions section), or 1 g/mL cesium chloride (CsCl) solution add 20 mL of 5% hypophosphorous acid solution and 12 mL of 0.5 *M* sodium nitrite solution, stir briefly and allow to stand for 20 h. Neutralize with sodium bicarbonate ($NaHCO_3$), check for completeness of destruction, and discard it. To prepare the hypophosphorous acid solution, add 10 mL of the commercially available 50% solution to 90 mL of H_2O and stir briefly. Prepare both the hypophosphorous acid solution and the sodium nitrite solution fresh each day.

2. **Caution! Ozone is an irritant. This reaction should be carried out in a properly functioning chemical fume hood.** Dilute the solution, if necessary, so that the concentration of EB in H_2O, Tris buffer, MOPS buffer, or CsCl solution does not exceed 0.4 mg/mL. Add hydrogen peroxide (H_2O_2) solution so that the concentration of H_2O_2 in the solution to be decontaminated is 1%. Pass air containing 300–400 ppm of ozone (from an ozone generator) through the solution at a rate of 2 L/min. The red solution will turn light yellow.[13] The destruction process typically takes 1 h. Check the reaction mixture for completeness of destruction and discard it. Degrade residual ozone by making the reaction mixture 1 *M* in sodium hydroxide.[17]

Decontamination of Ethidium Bromide in Aqueous Solution

Dilute the solution, if necessary, so that the concentration of EB does not exceed 0.1 mg/mL. For each 100 mL of EB in H_2O, TBE buffer (see below), MOPS buffer (see below), or CsCl solution add 2.9 g of Amberlite XAD-16 resin, stir for 20 h, then filter the mixture. Place the beads, which now contain the EB, with the hazardous solid waste. Check the liquid for completeness of decontamination and discard it. An alternative procedure for solutions that are more concentrated than 0.1 mg/mL is to increase the relative amount of resin. If this is done, this procedure should be fully validated before employing it on a routine basis.

Decontamination of Equipment Contaminated with Ethidium Bromide

Wash the equipment once with a paper towel soaked in a decontamination solution consisting of 4.2 g of sodium nitrite and 20 mL of hypophosphorous acid (50%) in 300 mL of H_2O. Then wash five times with wet paper towels using a fresh towel each time. Soak all the towels in decontamination solution for 1 h, check for completeness of decontamination, and discard it. Make up the decontamination solution just prior to use.

Glass, stainless steel, Formica, floor tile, and the filters of transilluminators have been successfully decontaminated using this technique.[5] No change in the optical properties of the transilluminator filter could be detected even after a number of decontamination cycles using the decontamination solution.

If the decontamination solution (pH 1.8) is felt to be too corrosive for the surface to be decontaminated, then use six H_2O washes. Again, soak all towels in decontamination solution for 1 h before disposal.

Decontamination of Ethidium Bromide in Isopropanol Saturated with Cesium Chloride

Dilute the solution, if necessary, so that the concentration of EB in the isopropanol saturated with CsCl does not exceed 1 mg/mL. For each volume of EB solution add four volumes of a decontamination solution consisting of 4.2 g of sodium nitrite and 20 mL of hypophosphorous acid (50%) in 300 mL of H_2O and stir the mixture for 20 h, then neutralize with $NaHCO_3$, test for completeness of destruction, and discard it.

Decontamination of Ethidium Bromide in Isoamyl Alcohol and 1-Butanol

Dilute the solution, if necessary, so that the concentration of EB in the alcohol does not exceed 1 mg/mL. For each volume of EB solution add four volumes of a decontamination solution consisting of 4.2 g of sodium nitrite and 20 mL of hypophosphorous acid (50%) in 300 mL of H_2O and stir the two-phase mixture rapidly for 72 h. For each 100 mL of total reaction volume add 2 g of activated charcoal and stir for another 30 min. Filter the reaction mixture, neutralize with $NaHCO_3$, and separate the layers. More alcohol may tend to separate from the aqueous layer on standing. Test the layers for completeness of destruction and discard them. It should be noted that the aqueous layer contains 4.6% of 1-butanol or

2.3% of isoamyl alcohol. Discard the activated charcoal with the solid waste.

This procedure has been tested in three separate experiments for both isoamyl alcohol and 1-butanol. In one experiment one plate (TA1530 with S9 activation) indicated significant mutagenicity. The number of revertants was 2.6 times the control value. All the other plates for this experiment and all the other experiments showed no significant mutagenic activity. For comparison, the numbers of revertants produced by untreated EB solutions were between 39 and 122 times the control value depending on the solvent and the tester strain. The control value is the mean of the number of revertants produced by the cells only and cells plus solvent runs.

Buffer Solutions

The buffer solutions used when the sodium nitrite–hypophosphorous acid method was investigated consisted of a TBE buffer (pH 8.4) containing tris(hydroxymethyl)aminomethane (TRIS) (0.089 *M*), boric acid (0.089 *M*), and ethylenediaminetetraacetic acid (EDTA) (0.002 *M*) and a MOPS buffer (pH 5.3) containing 4-morpholinepropanesulfonic acid (0.04 *M*), sodium acetate (0.0125 *M*), and EDTA (0.00125 *M*).

Analytical Procedures

Although EB is itself fluorescent in solution, the complexes it forms with DNA are much more fluorescent, and so the following procedure is employed. A TBE buffer is prepared containing TRIS (0.089 *M*), boric acid (0.089 *M*), EDTA (0.002 *M*), and sodium chloride (0.1 *M*). A DNA solution is prepared by dissolving 20 μg/mL of calf thymus DNA (Sigma) in the TBE buffer. A 0.1-mL aliquot of the reaction mixture is mixed with 0.9 mL of buffer, then with 1.0 mL of DNA solution. After standing for at least 15 min the fluorescence of the sample is determined using an Aminco-Bowman spectrophotofluorometer (excitation 540 nm, emission 590 nm). Using this procedure the limit of detection of EB in the reaction mixture is about 0.5 μg/mL. If a spectrophotofluorometer is not available a hand-held UV lamp (Mineralight model UVSL-25 from Ultra-Violet Products, Inc., San Gabriel, CA) on the long wave setting can be used instead. Solutions with DNA are prepared as before and visual inspection involving the use of a blank, an EB solution of known concentration, and the unknown solution is used. The limit of detection is about 0.126 μg/

mL.[18] It should be noted that these procedures only determine fluorescent compounds (such as EB) but that the EB is readily changed into nonfluorescent but still toxic compounds. Accordingly, we recommend periodic testing of reaction mixtures for mutagenicity, if possible. This testing becomes even more important if major changes are made in the procedures listed above.

Ethidium bromide can also be determined by thin-layer chromatography (TLC) using silica gel plates eluted with 1-butanol:acetic acid:H_2O (4:1:1).[13]

Mutagenicity Assays

The mutagenicity assays were carried out as described on page 4 using tester strains TA98, TA100, TA1530, TA1535, and TA1538. When isoamyl alcohol or 1-butanol solutions were tested for mutagenicity considerable cell toxicity was seen. To avoid this, isoamyl alcohol solutions were first diluted with three volumes of dimethyl sulfoxide (DMSO) and 1-butanol solutions were first diluted with an equal volume of DMSO. Ethidium bromide itself is mutagenic only to TA98 and TA1538 with activation but, when degradation procedures other than those detailed above were employed, mutagenic activity was frequently found in other strains because the degradation procedures transformed the EB to other compounds that had different mutagenic activities. These results underline the importance of mutagenesis as well as fluorescence testing. Except as mentioned above for one experiment involving the destruction of EB in 1-butanol, all the procedures in this monograph produced nonmutagenic reaction mixtures.

Related Compounds

The procedures involving sodium nitrite–hypophosphorous acid and Amberlite XAD-16 may be used to deal with aqueous or buffer solutions of the related compound propidium iodide [25535-16-4][6] except that 5 g of Amberlite XAD-16 are required to decontaminate 100 mL of propidium iodide solution.[9] The analytical procedure for propidium iodide is the same as for EB except that the excitation wavelength is 350 nm and the emission wavelength is 600 nm.[19] Full details on the use of Amberlite XAD-16 for the decontamination of aqueous solutions of propidium iodide are given in the monograph on Biological Stains. Destruction of propidium iodide in aqueous or phosphate buffered saline solution (500 μg/mL) using the sodium nitrite–hypophosphorous acid method was greater than 99.5% and the final reaction mixtures were not mutagenic. Propidium iodide itself was mutagenic to TA98 (18.6 times background) and TA1538 (23.5 times background) with activation when tested at a level of 1 mg per plate.[6] The sodium nitrite–hypophosphorous acid method was adapted from one recommended for the degradation of aromatic amines[20] and so it may be of use for degrading other aromatic amines (see Aromatic Amines monograph). Amberlite XAD-16 can be used to decontaminate solutions of many other biological stains (see Biological Stains monograph).

References

1. Other names for this compound are homidium bromide, Novidium bromide, Babidium bromide, RD 1572, Dromilac, 2,7-diamino-9-phenylphenanthridine ethobromide, 3,8-diamino-5-ethyl-6-phenylphenanthridinium bromide, 2,7-diamino-10-ethyl-9-phenylphenanthridinium bromide, and 2,7-diamino-9-phenyl-10-ethylphenanthridinium bromide.
2. MacGregor, J.T.; Johnson, I.J. In vitro metabolic activation of ethidium bromide and other phenanthridinium compounds: Mutagenic activity in *Salmonella typhimurium*. *Mutat. Res.* **1977**, *48*, 103–108.
3. Waring, M. Ethidium and propidium. In *Antibiotics*; Corcoran, J.W., Hahn, F.E., Eds., Springer: New York, 1975; Vol. 3, pp. 141–165.
4. Lunn, G.; Sansone, E.B. Ethidium bromide: Destruction and decontamination of solutions. *Anal. Biochem.* **1987**, *162*, 453–458.
5. Lunn, G.; Sansone, E.B. Decontamination of ethidium bromide spills. *Appl. Ind. Hyg.* **1989**, *4*, 234–237.
6. Lunn, G. Unpublished results.
7. Lunn, G.; Sansone, E.B. The use of reductive and oxidative methods to degrade hazardous waste in academic laboratories. In *Waste Disposal in Academic Institutions*; Kaufman, J.A., Ed., Lewis Publishers: Chelsea, MI, 1990; pp. 131–142.
8. Lunn, G.; Sansone, E.B. Degradation of ethidium bromide in alcohols. *BioTechniques* **1990**, *8*, 372–373.
9. Lunn, G.; Sansone, E.B. Review of procedures for the safe handling of ethidium bromide. *Phytochem. Bull.* **1990**, *22*, 21–24.
10. Joshua, H. Quantitative adsorption of ethidium bromide from aqueous solutions by macroreticular resins. *BioTechniques* **1986**, *4*, 207–208.
11. Hayatsu, H.; Oka, T.; Wakata, A.; Ohara, Y.; Hayatsu, T.; Kobayashi, H.; Avimoto, S. Adsorption of mutagens to cotton bearing covalently bound trisulfo-copper-phthalocyanine. *Mutat. Res.* **1983**, *119*, 233–238.
12. Quillardet, P.; Hofnung, M. Ethidium bromide and safety—Readers suggest alternative solutions. *Trends Genet.* **1988**, *4*, 89.

13. Zocher, R.; Billich, A.; Keller, U.; Messner, P. Destruction of ethidium bromide in solution by ozonolysis. *Biol. Chem. Hoppe-Seyler* **1988**, *369*, 1191–1194.
14. Menozzi, F.D.; Michel, A.; Pora, H.; Miller, A.O.A. Absorption method for rapid decontamination of solutions of ethidium bromide and propidium iodide. *Chromatographia* **1990**, *29*, 167–169.
15. Pardue, K.J. Reduce the volume of ethidium bromide waste, using polymer-filled adsorbent cartridges. *Supelco Reporter* **1992**, 9(4), 6–7.
16. Chang, N-.S.; Mattison, J. Rapid separation of DNA from ethidium bromide and cesium chloride in ultracentrifuge gradients by a desalting column. *BioTechniques* **1993**, *14*, 342–346.
17. Budavari, S., Ed., *The Merck Index*, 11th ed.; Merck and Co., Inc.: Rahway, NJ, 1989; p. 1105.
18. Lunn, G.; Sansone, E.B. Decontamination of ethidium bromide spills—authors response. *Appl. Occup. Environ. Hyg.* **1991**, *6*, 644–645.
19. Lunn, G.; Sansone, E.B. Decontamination of aqueous solutions of biological stains. *Biotechnic Histochem.* **1991**, *66*, 307–315.
20. Castegnaro, M.; Barek, J.; Dennis, J.; Ellen, G.; Klibanov, M.; Lafontaine, M.; Mitchum, R.; van Roosmalen, P.; Sansone, E.B.; Sternson, L.A.; Vahl, M., Eds., *Laboratory Decontamination and Destruction of Carcinogens in Laboratory Wastes: Some Aromatic Amines and 4-Nitrobiphenyl*; International Agency for Research on Cancer: Lyon, 1985 (IARC Scientific Publications No. 64).

HALOETHERS

CAUTION! Refer to safety considerations section on page 7 before starting any of these procedures.

Chloromethylmethylether (CMME; $ClCH_2OCH_3$) [107-30-2][1] and bis-(chloromethyl)ether (BCME; $ClCH_2OCH_2Cl$) [542-88-1][2] are both colorless volatile liquids having boiling points of 59 and 104°C, respectively. Both compounds cause cancer in laboratory animals. The compound CMME may be a human carcinogen;[3–5] BCME is a human carcinogen.[4–6] Chloromethylmethylether generally contains some BCME, so conditions employed for the degradation of CMME must also degrade BCME completely. These compounds are used industrially, in organic synthesis, and may also be formed when formaldehyde and hydrogen chloride are mixed.[7]

Principles of Destruction

Both CMME and BCME can be degraded by reaction with aqueous ammonia (NH_3) solution, sodium phenoxide, and sodium methoxide.[7] Degradation efficiency was greater than 99% in all cases. Reaction with sodium phenoxide and sodium methoxide produces ethers and sodium chloride.

Destruction Procedures

Destruction of Bulk Quantities of CMME and BCME

1. Take up bulk quantities in acetone so that the concentration does not exceed 50 mg/mL and add an equal volume of 6% NH_3 solution. Allow the mixture to stand for 3 h, analyze for completeness of destruction, neutralize with 2 *M* sulfuric acid (H_2SO_4), and discard it.

2. Take up bulk quantities in methanol so that the concentration does not exceed 50 mg/mL. For each 1 mL of solution add 3.5 mL of a 15% (w/v) solution of sodium phenoxide in methanol. Allow this mixture to stand for 3 h, analyze for completeness of destruction, and discard it.

3. Take up bulk quantities in methanol so that the concentration does not exceed 50 mg/mL. For each 1 mL of solution add 1.5 mL of an 8–9% (w/v) solution of sodium methoxide in methanol. Allow this mixture to stand for 3 h, analyze for completeness of destruction, and discard it.

Destruction of CMME and BCME in Methanol, Ethanol, Dimethyl Sulfoxide, N,N-*Dimethylformamide, and Acetone*

1. Dilute, if necessary, so that the concentration does not exceed 50 mg/mL. Add an equal volume of 6% NH_3 solution and allow the mixture to stand for 3 h, analyze for completeness of destruction, neutralize with 2 *M* H_2SO_4, and discard it.

2. Dilute, if necessary, so that the concentration does not exceed 50 mg/mL. For each 1 mL of solution add 3.5 mL of a 15% (w/v) solution of sodium phenoxide in methanol. Allow this mixture to stand for 3 h, analyze for completeness of destruction, and discard it.

3. Dilute, if necessary, so that the concentration does not exceed 50 mg/mL. For each 1 mL of solution add 1.5 mL of an 8–9% (w/v) solution of sodium methoxide in methanol. Allow this mixture to stand for 3 h, analyze for completeness of destruction, and discard it. Note that this procedure should not be used if either H_2O or chloroform (**violent reaction!**)[8] is likely to be present.

Destruction of CMME and BCME in Pentane, Hexane, Heptane, and Cyclohexane

1. Dilute, if necessary, so that the concentration does not exceed 50 mg/mL. Add an equal volume of 33% NH_3 solution and shake the mixture

continuously on a mechanical shaker for at least 3 h, analyze for completeness of destruction, neutralize with 2 M H_2SO_4, and discard it.

2. Dilute, if necessary, so that the concentration does not exceed 50 mg/mL. For each 1 mL of solution add 3.5 mL of a 15% (w/v) solution of sodium phenoxide in methanol. Shake this mixture on a mechanical shaker for at least 3 h, analyze for completeness of destruction, and discard it.

3. Dilute, if necessary, so that the concentration does not exceed 50 mg/mL. For each 1 mL of solution add 1.5 mL of an 8–9% (w/v) solution of sodium methoxide in methanol. Shake this mixture on a mechanical shaker for at least 3 h, analyze for completeness of destruction, and discard it. Note that this procedure should not be used if either H_2O or chloroform (**violent reaction!**)[8] is likely to be present.

Decontamination of Glassware and Other Equipment

1. If the equipment is contaminated with CMME or BCME in a solvent that is not miscible with H_2O, rinse with acetone and treat the rinses by one of the methods described above as appropriate for treating solutions of CMME or BCME in acetone. Otherwise, immerse in a 6% NH_3 solution for at least 3 h.

2. Immerse in a 15% (w/v) solution of sodium phenoxide in methanol for at least 3 h.

3. Immerse in an 8–9% (w/v) solution of sodium methoxide in methanol for at least 3 h. Note that this procedure should not be used if either H_2O or chloroform (**violent reaction!**)[8] is likely to be present.

Decontamination of Laboratory Clothing

If the clothing is contaminated with CMME or BCME in a solvent that is not miscible with H_2O, rinse with acetone and treat the rinses by one of the methods described above as appropriate for treating solutions of CMME or BCME in acetone. Otherwise, immerse in a 6% NH_3 solution for at least 3 h.

Decontamination of Spills

Cover the spill area with an absorbent material. Saturate the absorbent material with an excess of 6% NH_3 solution. Isolate the area for at least 3 h, then neutralize with 2 M H_2SO_4 and test for completeness of destruction.

Analytical Procedures

Analysis was by gas chromatography (GC) on a 4.3-m × 2-mm i.d. glass column packed with 10% ethylene glycol adipate (EGA) on 80-100 Chromosorb W AW treated with dimethyldichlorosilane. The oven temperature was 60°C, the injection temperature was 200°C, and the detector temperature was 250°C. An electron capture detector was used. Either direct injection of the sample or injection of a 2-mL head space sample taken with a gas tight syringe may be used. A precolumn should be used and it may be necessary to replace this after each injection.

Mutagenicity Assays

The mutagenicity assays were carried out as described on page 4 using tester strains TA100, TA1530, and TA1535. The final reaction mixtures were not mutagenic; both CMME and BCME were mutagenic.

Related Compounds

These destruction methods might be expected to be applicable to compounds of the general form $ROCH_2Cl$, but no tests have been carried out. Application of these methods **must** be thoroughly validated before being put into routine use.

References

1. Other names for this compound are chloromethoxymethane, methyl chloromethyl ether, monochloromethyl ether, chlorodimethyl ether, dimethylchloroether, and monochlorodimethyl ether.
2. Other names for this compound are chloro(chloromethoxy)methane, *sym*-dichlorodimethyl ether, *sym*-dichloromethylether, dimethyl-1,1′-dichloroether, oxybis(chloromethane), M-chlorex, and Bis-CME.
3. International Agency for Research on Cancer. *IARC Monographs on the Evaluation of Carcinogenic Risk of Chemicals to Man.* Volume 4, *Some Aromatic Amines, Hydrazine and Related Substances,* N-*Nitroso Compounds and Miscellaneous Alkylating Agents*; International Agency for Research on Cancer: Lyon, 1974; pp. 239–245.
4. International Agency for Research on Cancer. *IARC Monographs on the Evaluation of the Carcinogenic Risk of Chemicals to Humans, Supplement No. 4, Chemicals, Industrial Processes and Industries Associated with Cancer in Humans. IARC Monographs, Volumes 1 to 29*; International Agency for Research on Cancer: Lyon, 1982; pp. 64–66.
5. International Agency for Research on Cancer. *IARC Monographs on the Evaluation of the Carcinogenic Risk of Chemicals to Humans, Supplement No. 7, Overall Evaluations*

of Carcinogenicity: An Updating of IARC Monographs *Volumes 1 to 42*; International Agency for Research on Cancer: Lyon, 1987; pp. 131–133.

6. Reference 3, pp. 231–238.
7. Castegnaro, M.; Alvarez, M.; Iovu, M.; Sansone, E. B.; Telling, G.M.; Williams, D.T., Eds., *Laboratory Decontamination and Destruction of Carcinogens in Laboratory Wastes: Some Haloethers*; International Agency for Research on Cancer: Lyon, 1984 (IARC Scientific Publications No. 61).
8. Bretherick, L. *Bretherick's Handbook of Reactive Chemical Hazards*, 4th ed.; Butterworths: London; 1990; p. 133.

HALOGENATED COMPOUNDS

CAUTION! Refer to safety considerations section on page 7 before starting any of these procedures.

Many hazardous compounds, such as pesticides and polychlorinated biphenyls (PCBs), contain halogen atoms. Little work has been done on processes suitable for the chemical degradation of these compounds in the laboratory but some work has been done on model compounds. Validated procedures are available for the following compounds:

Compound Name	Reference	mp or bp (°C)	Registry Number
Iodomethane (methyl iodide)		bp 41–43	[74-88-4]
2-Fluoroethanol		bp 103	[371-62-0]
2-Chloroethanol	1	bp 129	[107-07-3]
2-Bromoethanol	2	bp 56–57/20 mm Hg	[540-51-2]

Compound Name	Reference	mp or bp (°C)	Registry Number
2-Chloroethylamine hydrochloride		mp 143–146	[870-24-6]
2-Bromoethylamine hydrobromide		mp 172–174	[2576-47-8]
2-Chloroacetic acid (chloroethanoic acid)		mp 62–64	[79-11-8]
2,2,2-Trichloroacetic acid	3	mp 54–56	[76-03-9]
1-Chlorobutane (*n*-butyl chloride)		bp 77–78	[109-69-3]
1-Bromobutane (*n*-butyl bromide)		bp 100–104	[109-65-9]
1-Iodobutane (*n*-butyl iodide)		bp 130–131	[542-69-8]
2-Bromobutane (*s*-butyl bromide)		bp 91	[78-76-2]
2-Iodobutane (*s*-butyl iodide)		bp 119–120	[513-48-4]
2-Bromo-2-methylpropane (*tert*-butyl bromide)		bp 72–74	[507-19-7]
2-Iodo-2-methylpropane	4	bp 99–100	[558-17-8]
3-Chloropyridine		bp 148	[626-60-8]
Fluorobenzene		bp 85	[462-06-6]
Chlorobenzene	5	bp 132	[108-90-7]
Bromobenzene (phenyl bromide)		bp 156	[108-86-1]
Iodobenzene (phenyl iodide)		bp 188	[591-50-4]
4-Fluoroaniline		bp 187	[371-40-4]
2-Chloroaniline (2-chlorobenzenamine)		bp 208–210	[95-51-2]
3-Chloroaniline (3-chlorobenzenamine)		bp 230	[108-42-9]
4-Chloroaniline (4-chlorobenzenamine)		mp 68–71	[106-47-8]
4-Fluoronitrobenzene		bp 205	[350-46-9]

Compound Name	Reference	mp or bp (°C)	Registry Number
2-Chloronitrobenzene		mp 33–35	[88-73-3]
3-Chloronitrobenzene		mp 42–44	[121-73-3]
4-Chloronitrobenzene		mp 83–84	[100-00-5]
Benzyl chloride	6	bp 177–181	[100-44-7]
Benzyl bromide	7	bp 198–199	[100-39-0]
α,α-Dichlorotoluene	8	bp 82/10 mm Hg	[98-87-3]
3-Aminobenzotrifluoride (α,α,α-trifluoro-*m*-toluidine)		bp 187	[98-16-8]
1-Bromononane		bp 201	[693-58-3]
1-Chlorodecane		bp 223	[1002-69-3]
1-Bromodecane		bp 238	[112-29-8]

Most of these compounds are volatile liquids and the solids may also have an appreciable vapor pressure. 2-Chloroethylamine and 2-bromoethylamine generally come as their nonvolatile hydrochloride and hydrobromide salts, respectively, but the free bases would be expected to be volatile. Many of these compounds are corrosive to the skin. Iodomethane is a strong narcotic and anesthetic;[9] 2-chloroethanol is a teratogen and may affect the nervous system, liver, spleen, and lungs;[10] 2-chloroethylamine may polymerize explosively;[11] chloroacetic acid is corrosive to the skin, eyes, and mucous membranes;[12] trichloroacetic acid is corrosive and irritating to the skin, eyes, and mucous membranes;[13] 2-bromobutane is narcotic in high concentrations;[14] chlorobenzene is a teratogen and strong narcotic;[15] iodobenzene explodes on heating above 200°C;[16] 3-chloronitrobenzene is a poison and may give rise to cyanosis and blood changes;[17] benzyl chloride is a corrosive irritant to the skin, eyes, and mucous membranes and can decompose explosively under certain circumstances;[18] and benzyl bromide[19] and α,α-dichlorotoluene[20] are lachrymators, intensely irritating to the skin, and cause central nervous system (CNS) depression in large doses. The chloroanilines[21] and chloronitrobenzenes[22] are toxic by inhalation, skin contact, and ingestion. Halogenated compounds may react violently and explosively with alkali metals such as sodium[23] and potassium.[24] With the exception of 2-chloroethanol, 2-bromoethanol, 2-chloroethylamine, 2-bromoethylamine, 2-chloroacetic acid, 2,2,2-trichloroacetic acid, and 3-chloropyridine the compounds are not soluble in H_2O.

Iodomethane is slightly soluble in H_2O. These compounds are all soluble in alcohols and organic solvents.

Iodomethane,[25,26] benzyl chloride,[27–29] α,α-dichlorotoluene,[29,30] 2-bromoethanol,[31] 1-iodobutane,[32] 2-bromobutane,[32] 2-iodobutane,[32] 2-bromo-2-methylpropane,[32] 2-chloronitrobenzene,[33] 4-chloronitrobenzene,[33] trichloroacetic acid,[13] and 4-chloroaniline[34] are carcinogenic in experimental animals.

These compounds are used as intermediates in industry and in organic synthesis. Iodomethane occurs naturally and is used as a chemical intermediate.

Principles of Destruction

The halogenated compounds are reductively dehalogenated with nickel–aluminum (Ni–Al) alloy in dilute base to give the corresponding compound without the halogen (Table 1).[35] When the products are soluble in H_2O the yield of product is good, but when they are not (e.g., toluene from benzyl chloride) they are lost from solution and accountances are not complete. The product from 4-fluoronitrobenzene and the chloronitrobenzenes is aniline (concomitant reduction of the nitro group) and the product from the reduction of 3-chloropyridine is piperidine in 87% yield (concomitant reduction of the initial product pyridine). Because it contains three halogens a higher ratio of Ni–Al alloy:compound was required to obtain complete reduction of 3-aminobenzotrifluoride. In general, where the starting material can be detected by chromatography, less than 1% remains. 1-Bromononane, 1-bromodecane, 1-chlorodecane, and 1-chlorobutane cannot be degraded by this procedure because these compounds are too insoluble in the aqueous methanol solvent used. Although 2-fluoroethanol is soluble in H_2O it was only partially degraded under these conditions.

Iodomethane, benzyl chloride, benzyl bromide, 2-fluoroethanol, 2-chloroethanol, 2- bromoethanol, 2-chloroacetic acid, 1-chlorodecane, 1-bromononane, 1-bromodecane, 1-bromobutane, 1-iodobutane, 2-bromobutane, 2-iodobutane, 2-bromo-2-methylpropane, and 2-iodo-2-methylpropane are completely degraded by refluxing them in 4.5 *M* ethanolic potassium hydroxide (KOH) solution for 2 h and 1-chlorobutane is completely degraded after refluxing for 4 h (Table 2).[35] The products detected are the corresponding ethyl ethers. Of the compounds tested, substantial amounts of 4-fluoroaniline, 2-chloroaniline, 3-chloropyridine, fluorobenzene, chlorobenzene, bromobenzene, iodobenzene, and 3-aminobenzotrifluoride are left; α,α-dichlorotoluene gives incomplete destruction even after refluxing for 4 h; 2-bromoethylamine

Table 1 Degradation of Halogenated Compounds Using Ni–Al Alloy[a]

Compound	Residue (%)	Products	(%)
Iodomethane[b]	<1.1		
2-Chloroethanol	<0.36	EtOH	97
2-Bromoethanol	<0.54	EtOH	93
2-Chloroethylamine	<0.33	$EtNH_2$	100
2-Bromoethylamine		$EtNH_2$	93
2-Chloroacetic acid	<0.6	CH_3CO_2H	101
2,2,2-Trichloroacetic acid		CH_3CO_2H	104
1-Bromobutane[b]	<0.45[c]		
1-Iodobutane[b]	<0.37[c]		
2-Bromobutane[b]	<0.79[c]		
2-Iodobutane[b]	<0.66[c]		
2-Bromo-2-methylpropane[b]	<0.31[c]		
2-Iodo-2-methylpropane[b]	<0.44[c]		
3-Chloropyridine[b]	<0.30[d]	Piperidine	87
Fluorobenzene[b]	<0.22[c]	Benzene	14
Chlorobenzene[b]	<0.14[c]	Benzene	14
Bromobenzene[b]	<0.09[c]	Benzene	17
Iodobenzene[b]	<0.14[c]	Benzene	15
4-Fluoroaniline[b]	<0.50	$PhNH_2$	95
2-Chloroaniline[b]	<0.06[c]	$PhNH_2$	96
3-Chloroaniline[b]	<0.14[c]	$PhNH_2$	103
4-Chloroaniline[b]	<0.06[c]	$PhNH_2$	77
4-Fluoronitrobenzene[b]	<1.0	$PhNH_2$	58; *p*-$CH_3OC_6H_4NH_2$ 39
2-Chloronitrobenzene[b]	<0.20[c]	$PhNH_2$	85
3-Chloronitrobenzene[b]	<0.26[c]	$PhNH_2$	82
4-Chloronitrobenzene[b]	<0.16[c]	$PhNH_2$	104
3-Aminobenzotrifluoride[b]	<0.28[c,e]	*m*-$CH_3C_6H_4NH_2$	52
Benzyl chloride[b]	<0.07[c]	Toluene	29; bibenzyl 7
Benzyl bromide[b]	<0.10[c]	Toluene	27; bibenzyl 6
α,α-Dichlorotoluene[b]	<0.24[c]	Toluene	12; bibenzyl 21

[a] All reactions were carried out as described below.

[b] Reagent was initially dissolved in methanol. Water was the initial solvent for the other reactions.

[c] The final reaction mixture was extracted with ether or dichloromethane and each layer was separately analyzed. The results shown here are those obtained for the organic layer. No halogenated compounds were found in any of the aqueous layers. Control experiments showed that at least 71% of any of the halogenated compounds listed would be extracted into the organic layer.

[d] Pyridine was found as an intermediate but was present at less than 0.3% in the final reaction mixture.

[e] For each 5 g of Ni–Al alloy used 0.1 mL of 3-aminobenzotrifluoride was degraded.

Table 2 Degradation of Halogenated Compounds Using 4.5 *M* KOH in 95% Ethanol[a]

Compound	Residue (%)	Products	(%)
Iodomethane	<0.09		
2-Fluoroethanol	<0.63	$EtOCH_2CH_2OH$	99
2-Chloroethanol	<0.34	$EtOCH_2CH_2OH$	95
2-Bromoethanol	<1.25	$EtOCH_2CH_2OH$	67
2-Chloroacetic acid	<0.63	$EtOCH_2CO_2H$	82
1-Chlorobutane	<0.43[b,c]	BuOEt 48; BuOH	8
1-Bromobutane	<0.20[b]	BuOEt 57; BuOH	5
1-Iodobutane	<0.18[b]	BuOEt 35; BuOH	5
2-Bromobutane	<0.18[b]		
2-Iodobutane	<0.50[b]		
2-Bromo-2-methylpropane	<0.45[b]		
2-Iodo-2-methylpropane	<0.50[b]		
Benzyl chloride	<0.24[b]	$PhCH_2OEt$	85
Benzyl bromide	<0.23[b]	$PhCH_2OEt$	76
1-Bromononane	<0.07[b]	NonOEt	84
1-Chlorodecane	<0.96[b]	DecOEt	97
1-Bromodecane	<0.94[b]	DecOEt	81

[a] All reactions were performed as described below.

[b] The final reaction mixture was extracted with ether or dichloromethane and each layer was separately analyzed. The results shown here are those obtained for the organic layer. No halogenated compounds were found in any of the aqueous layers. Control experiments showed that at least 98% of any of the halogenated compounds listed would be extracted into the organic layer.

[c] The reaction mixture was refluxed for 4 h.

and 2-chloroethylamine give mutagenic residues; and 4-fluoronitrobenzene and 2-, 3-, and 4-chloronitrobenzene are completely degraded but give products that have not yet been completely identified but appear to include azo and azoxy compounds.

Destruction Procedures[35]

With Ni–Al Alloy (not for 2-Fluoroethanol, 1-Bromononane, 1-Bromodecane, 1-Chlorodecane, and 1-Chlorobutane)

Take up 0.5 mL or 0.5 g of the halogenated compound (0.1 mL for 3-aminobenzotrifluoride) in 50 mL of H_2O (chloroacetic acid, trichloroacetic acid, 2-chloroethanol, 2-bromoethanol, 2-chloroethylamine, and 2-bromoethylamine) or methanol (other compounds) and add 50 mL of 2 *M* KOH solution. Stir this mixture and add 5 g of Ni–Al alloy in portions to avoid frothing. Stir the reaction mixture overnight, then filter it through a pad of Celite®. Check the filtrate for completeness of destruction, neutralize, and discard it. Note that the filtrate contains the dehalogenated material. Place the spent nickel on a metal tray and allow it to dry away from flammable solvents for 24 h. Dispose of it with the solid waste.

With Ethanolic Potassium Hydroxide (not for 2-Chloroethylamine, 2-Bromoethylamine, 2,2,2-Trichloroacetic acid, 3-Chloropyridine, Fluorobenzene, Chlorobenzene, Bromobenzene, Iodobenzene, α,α-Dichlorotoluene, 4-Fluoroaniline, 4-Fluoronitrobenzene, the Chloroanilines, the Chloronitrobenzenes, or 3-Aminobenzotrifluoride)

Take up 1 mL of the halogenated compound in 25 mL of 4.5 *M* ethanolic KOH and reflux the mixture with stirring for 2 h (4 h for 1-chlorobutane). Cool the mixture and dilute it with at least 100 mL of H_2O. Separate the layers, if necessary, check for completeness of destruction, neutralize, and discard it. (Prepare the ethanolic KOH by dissolving 79 g of KOH in 315 mL of 95% ethanol. When volatile halides, e.g., iodomethane, are to be degraded, it is important to allow this mixture to cool completely before adding the halide.)

Analytical Procedures

For analysis by gas chromatography (GC) a 1.8-m × 2-mm i.d. packed column was used together with flame ionization detection (Table 3).[35] The injection temperature was 200°C, the detector temperature was 300°C, and the carrier gas was nitrogen flowing at 30 mL/min. It was frequently found advantageous to extract the final reaction mixture with 50 mL of ether or dichloromethane, and then to analyze both the aqueous layer and the extract. In this way trace amounts of compounds that were poorly soluble in H_2O could be detected. The GC conditions shown are only a guide and the exact conditions would have to be determined experimentally.

The limits of detection were in the range of 0.01–0.1 mg/mL. In each case the reaction mixture was analyzed for the presence or absence of the

Table 3 Summary of Gas Chromatography Conditions

Compound	Packing[a]	Temperature (°C)	Approximate Retention Time (min)
Iodomethane	A	60	0.5
2-Fluoroethanol	A	60	6.1
2-Chloroethanol	A	120	1.5
2-Bromoethanol	A	120	7.8
2-Chloroethylamine	B	60	1.9
2-Chloroacetic acid	C	200	2.1
1-Chlorobutane	D	60	5.2
1-Bromobutane[b]	D	100	3.4
1-Iodobutane	E	60	3.9
2-Bromobutane	D	100	2.6
2-Iodobutane	D	100	5.4
2-Bromo-2-methylpropane	D	60	4.1
2-Iodo-2-methylpropane	D	100	3.2
3-Chloropyridine	B	80	2.0
Fluorobenzene	D	60	8.1
Chlorobenzene	B	60	2.0
Bromobenzene	B	60	4.1
Iodobenzene	B	60	11.2
4-Fluoroaniline	B	120	3.8
2-Chloroaniline	B	150	1.8
3-Chloroaniline	B	150	4.2
4-Chloroaniline	B	150	5.3
4-Fluoronitrobenzene	B	150	5.6
2-Chloronitrobenzene	B	150	2.5
3-Chloronitrobenzene	B	130	3.5
4-Chloronitrobenzene	B	130	4.0
3-Aminobenzotrifluoride	B	130	3.2
Benzyl chloride	A	120	2.7
Benzyl bromide	A	150	2.5
α,α-Dichlorotoluene	A	150	2.3
1-Bromononane	B	80	3.7
1-Chlorodecane	B	80	3.9
1-Bromodecane	B	100	1.7

[a] The column packings were A 5% Carbowax 20 M on 80/100 Chromosorb W HP; B 2% Carbowax 20 M + 1% KOH on 80/100 Supelcoport; C 5% FFAP on 80/100 Gas Chrom Q; D 28% Pennwalt 223 + 4% KOH on 80/100 Gas Chrom R; and E 10% Carbowax 20 M + 2% KOH on 80/100 Chromosorb W AW.

[b] When 1-bromobutane was degraded using ethanolic KOH it was found that a small 1-butanol peak interfered with the determination of 1-bromobutane. A column packed with 20% Carbowax 20 M on 80/100 Supelcoport gave sufficient resolution. The oven temperature was 100°C, the carrier gas flowed at 20 mL/min, and the retention time was about 2.5 min.

compound that had been degraded. If the compound was found to be absent, a small quantity of the compound was added to an aliquot of the actual reaction mixture, which was then analyzed again. The presence of an appropriate peak in the spiked reaction mixture confirmed that any halide that survived would be detected. Compounds found to be particularly sensitive to degradation on the column packings that contained base were analyzed as follows. Before analysis, 2 mL of the reaction mixture was acidified with 0.2 mL of concentrated hydrochloric acid and this mixture was neutralized by adding solid sodium bicarbonate until the effervescence of carbon dioxide ceased. This neutralized solution was analyzed using column packing A, which did not contain any base.

2-Bromoethylamine and 2,2,2-trichloroacetic acid could not be determined by GC, but their products could be determined and a good accountance for these products gave some assurance that the compounds were degraded. Products were generally determined using the same chromatographic conditions as used for the parent compounds, although it was frequently necessary to employ a lower oven temperature.

Mutagenicity Assays

The mutagenicity assays were carried out as described on page 4 using tester strains TA98, TA100, TA1530, and TA1535. The final neutralized reaction mixtures were not mutagenic. A 100-μL aliquot of the reaction mixture, which corresponds to a level of 0.5 μL (from Ni–Al alloy reactions) or 0.8 μL (from ethanolic KOH reactions) of undegraded material, was added to each plate. In some cases immiscible layers from the ethanolic KOH reactions were removed and tested directly at 1 mg per plate. None of these layers were mutagenic. Of the pure compounds tested at levels of 1–0.25 mg per plate in dimethyl sulfoxide solution, 2,2,2-trichloroacetic acid, 2-chloroethanol, 2-bromoethanol, 2-chloroethylamine, 2-bromoethylamine, 1-bromobutane, 2-bromobutane, 4-fluoroaniline, 4-fluoronitrobenzene, benzyl chloride, benzyl bromide, and α,α-dichlorotoluene were mutagenic. Iodomethane has been reported to be a weak mutagen at high concentrations.[26] The products that we were able to identify were tested and none of them were found to be mutagenic. When 2-chloroethylamine and 2-bromoethylamine were treated with ethanolic KOH, 2-ethoxyethylamine was obtained, which was mutagenic to TA100, TA1530, and TA1535 with and without activation.

Related Compounds

It has been reported that many compounds can be dehalogenated with Ni–Al alloy.[36] However, although the products were identified, it was not shown that the starting materials were completely degraded. Thus, these procedures cannot be regarded as validated. Based on the results described above, however, it seems likely that Ni–Al alloy should be generally applicable to the reductive dehalogenation of halogenated compounds and should give complete destruction of the starting material. The only exceptions among the compounds studied were 1-chlorobutane, 1-chlorodecane, 1-bromononane, and 1-bromodecane, presumably because they were so insoluble in the aqueous methanol and 2-fluoroethanol for unknown reasons. It should be possible to degrade most halogenated compounds, which have at least some solubility in aqueous methanol. Validation must be performed before any of these compounds are routinely degraded with Ni–Al alloy. Compounds that have been reported by other workers to be reductively dehalogenated, in good yield, to the corresponding compound lacking halogen include: 4-chlorophenol [106-48-9],[37] 2-chlorohydroquinone [615-67-8],[37] 2,4,6-trichlorophenol [88-06-2],[38] 2,4,6-tribromophenol [118-79-6],[39] 5-chloroisophthalic acid [2157-39-3],[40] 4-fluorobenzoic acid [456-22-4],[37] 2-chloro-5-fluorobenzoic acid [2252-50-8],[37] 2-chlorobenzoic acid [118-91-2],[37] 3,4-dichlorobenzoic acid [51-44-5],[37] 4-bromobenzoic acid [586-76-5],[37] 2,4-dichlorobenzoic acid [50-84-0],[37] 2-chlorophenylacetic acid [2444-36-2],[37] 4-bromophenylacetic acid [1878-68-8],[37] 3,4-dichlorophenylacetic acid [5807-30-7],[37] 2,4-dichlorophenylacetic acid [19719-28-9],[37] 4-(4-chloro-3-methylphenyl)butyric acid [91552-42-0],[37] 4-(4-chloro-2-methylphenyl)butyric acid [120347-16-2],[37] and 4-chlorophenoxyacetic acid [122-88-3].[37]

Refluxing in ethanolic KOH solution can also be used to degrade some halides. The procedure failed, however, for a number of compounds, particularly those with unreactive halogen atoms. Thus, it does not appear to have the wide applicability of Ni–Al alloy reduction, but it should be applicable to a number of other halides. Complete validation must be performed before the procedure is used routinely, however. When a higher ratio of organic halide:ethanolic KOH was used the reaction mixtures were frequently found to be mutagenic.[35]

References

1. Other names for this compound are 2-chloroethyl alcohol, ethylene chlorohydrin, glycol chlorohydrin, and NCI-C50135.

2. Other names for this compound are β-bromoethyl alcohol, ethylene bromohydrin, and glycol bromohydrin.
3. Other names for this compound are TCA, trichloroethanoic acid, Aceto-Caustin, Amchem Grass Killer, Konesta, and Varitox.
4. Other names for this compound are *tert*-butyl iodide and trimethyliodomethane.
5. Other names for this compound are phenyl chloride, benzene chloride, MCB, monochlorobenzene, and NCI-C54886.
6. Other names for this compound are tolyl chloride, (chloromethyl)benzene, α-chlorotoluene, ω-chlorotoluene, and NCI-C06360.
7. Other names for this compound are (bromomethyl)benzene, ω-bromotoluene, and α-bromotoluene.
8. Other names for this compound are (dichloromethyl)benzene, benzal chloride, benzyl dichloride, benzylene chloride, and benzylidene chloride.
9. Lewis, R.J., Sr. *Sax's Dangerous Properties of Industrial Materials*, 8th ed.; Van Nostrand-Reinhold: New York, 1992; p. 2333.
10. Reference 9, pp. 1599–1600.
11. Reference 9, p. 801.
12. Reference 9, pp. 761–762.
13. Reference 9, pp. 3346–3347.
14. Reference 9, p. 545.
15. Reference 9, pp. 768–769.
16. Reference 9, p. 1991.
17. Reference 9, pp. 840–841.
18. Reference 9, p. 400.
19. Reference 9, p. 398.
20. Reference 9, p. 348.
21. Bretherick, L., Ed., *Hazards in the Chemical laboratory*, 4th ed.; Royal Society of Chemistry: London, 1986; pp. 238–239.
22. Reference 21, p. 248.
23. Bretherick, L. *Bretherick's Handbook of Reactive Chemical Hazards*, 4th ed.; Butterworths: London, 1990; pp. 1372–1373.
24. Reference 23, pp. 1288–1289.
25. International Agency for Research on Cancer. *IARC Monographs on the Evaluation of the Carcinogenic Risk of Chemicals to Man.* Volume 15, *Some Fumigants,the Herbicides 2,4-D and 2,4,5-T, Chlorinated Dibenzodioxins and Miscellaneous Industrial Chemicals*; International Agency for Research on Cancer: Lyon, 1977; pp. 245–254.
26. International Agency for Research on Cancer. *IARC Monographs on the Evaluation of the Carcinogenic Risk of Chemicals to Humans.* Volume 41, *Some Halogenated Hydrocarbons and Pesticide Exposures*; International Agency for Research on Cancer: Lyon, 1986; pp. 213–227.
27. International Agency for Research on Cancer. *IARC Monographs on the Evaluation of*

Carcinogenic Risk of Chemicals to Man. Volume 11, *Cadmium, Nickel, Some Epoxides, Miscellaneous Industrial Chemicals and General Considerations on Volatile Anaesthetics*; International Agency for Research on Cancer: Lyon, 1976; pp. 217–223.

28. International Agency for Research on Cancer. *IARC Monographs on the Evaluation of the Carcinogenic Risk of Chemicals to Humans.* Volume 29, *Some Industrial Chemicals and Dyestuffs*; International Agency for Research on Cancer: Lyon, 1982; pp. 49–63.
29. International Agency for Research on Cancer. *IARC Monographs on the Evaluation of the Carcinogenic Risk of Chemicals to Humans, Supplement No. 7, Overall Evaluations of Carcinogenicity: An Updating of* IARC Monographs *Volumes 1 to 42*; International Agency for Research on Cancer: Lyon, 1987; pp. 148–149.
30. Reference 28, pp. 65–72.
31. Theiss, J.C.; Shimkin, M.B.; Poirier, L.A. Induction of pulmonary adenomas in strain A mice by substituted organohalides. *Cancer Res.* **1979**, *39*, 391–395.
32. Poirier, L.A.; Stoner, G.D.; Shimkin, M.B. Bioassay of alkyl halides and nucleotide base analogs by pulmonary tumor response in strain A mice. *Cancer Res.* **1975**, *35*, 1411–1415.
33. Weisburger, E.K.; Russfield, A.B.; Homburger, F.; Weisburger, J.H.; Boger, E.; Van Dongen, C.G.; Chu, K.C. Testing of twenty-one environmental aromatic amines or derivatives for long-term toxicity or carcinogenicity. *J. Environ. Path. Toxicol.* **1978**, *2*, 325–356.
34. Reference 9, p. 767.
35. Lunn, G.; Sansone, E.B. Validated methods for degrading hazardous chemicals: Some halogenated compounds. *Am. Ind. Hyg. Assoc. J.* **1991**, *52*, 252–257.
36. Keefer, L.K.; Lunn, G. Nickel–aluminum alloy as a reducing agent. *Chem. Rev.* **1989**, *89*, 459–502.
37. Buu-Hoï, N.P.; Xuong, N.D.; Bac, N.V. Déshalogénation des composés organiques halogénés au moyen des alliages de Raney (nickel ou cobalt), et ses applications à la chimie préparative et structurale. *Bull. Soc. Chim. Fr.* **1963**, 2442–2445.
38. Tashiro, M.; Fukata, G. Studies on selective preparation of aromatic compounds. 12. Selective reductive dehalogenation of some halophenols with zinc powder in basic and acidic media. *J. Org. Chem.* **1977**, *42*, 835–838.
39. Tashiro, M.; Fukata, G. The reductive debromination of bromophenols. *Org. Prep. Proced. Int.* **1976**, *8*, 231–236.
40. Märkl, G. Diensynthesen mit chlorierten α-pyronen. *Chem. Ber.* **1963**, *96*, 1441–1445.

HALOGENS

CAUTION! Refer to safety considerations section on page 7 before starting any of these procedures.

Bromine [7726-95-6] is a dense, dark red liquid and iodine [7553-56-2] is a black solid. Bromine[1,2] and iodine[3,4] are incompatible with various organic and inorganic reagents and these elements are toxic by ingestion and irritating to the eyes and lungs. These halogens are widely used industrially and in organic and inorganic chemistry laboratories.

Principle of Destruction

The halogens are reduced with sodium metabisulfite to the corresponding halide anions.

Destruction Procedure[5]

Add 1 mL of bromine or 5 g of iodine to 100 mL of 10% (w/v) sodium metabisulfite solution. Stir the mixture until the halogen has completely dissolved, test for completeness of reduction, and discard. If reduction is not complete, add more sodium metabisulfite solution.

Analytical Procedure

The filtrate can be tested for residual oxidant by adding a few drops to an equal volume of 10% (w/v) potassium iodide solution. Acidify with 1 *M* hydrochloric acid and add one drop of starch as an indicator. A deep blue color indicates the presence of oxidant and the procedure should be repeated.

References

1. Lewis, R.J., Sr. *Sax's Dangerous Properties of Industrial Materials*, 8th ed.; Van Nostrand-Reinhold: New York, 1992; pp. 539–540.
2. Bretherick, L. *Bretherick's Handbook of Reactive Chemical Hazards*, 4th ed.; Butterworths: London, 1990; pp. 97–103.
3. Reference 1, pp. 1988–1989.
4. Reference 2, pp. 1279–1282.
5. Lunn, G. Unpublished observations.

HEAVY METALS

CAUTION! Refer to safety considerations section on page 7 before starting any of these procedures.

A number of metals are toxic. Although the metals cannot be destroyed they can be removed from dilute aqueous solutions by using ion-exchange resins or by precipitation. In this way the volume of waste that must be disposed of can be greatly reduced.

Solutions of the following compounds were decontaminated:

Compound	Reference	Formula	Registry Number
Cadmium acetate dihydrate		$Cd(CH_3CO_2)_2.2H_2O$	[5743-04-4]
Cobalt(II) sulfate heptahydrate (cobaltous sulfate)		$CoSO_4.7H_2O$	[10026-24-1]
Copper(II) sulfate pentahydrate	1	$CuSO_4.5H_2O$	[7758-99-8]
Iron(II) sulfate heptahydrate	2	$FeSO_4.7H_2O$	[7782-63-0]

Compound	Reference	Formula	Registry Number
Lead(II) acetate trihydrate	3	$Pb(CH_3CO_2)_2.3H_2O$	[6080-56-4]
Manganese(II) sulfate monohydrate		$MnSO_4.H_2O$	[10034-96-5]
Nickel(II) sulfate hexahydrate		$NiSO_4.6H_2O$	[10101-97-0]
Phosphomolybdic acid hydrate (molybdophosphoric acid)		$12MoO_3.H_3PO_4.xH_2O$	[51429-74-4]
Silver nitrate		$AgNO_3$	[7761-88-8]
Thallium(I) nitrate (thallous nitrate)		$TlNO_3$	[10102-45-1]
Tin(II) chloride dihydrate	4	$SnCl_2.2H_2O$	[10025-69-1]
Zinc chloride (butter of zinc)		$ZnCl_2$	[7646-85-7]

All of these compounds are toxic by ingestion. Cadmium,[5] cobalt,[6] and nickel[7] compounds are carcinogenic. Cadmium compounds are teratogenic.[5] Lead[8] and thallium[9] compounds are toxic cumulative poisons, manganese compounds are toxic to the central nervous system (CNS),[10] and molybdenum compounds are highly toxic.[11] Cadmium acetate[12] and lead acetate[13] are carcinogenic and manganese sulfate,[14] nickel sulfate,[15] silver nitrate,[16] and zinc chloride[17] may be carcinogenic. Cadmium acetate,[12] lead acetate,[18] manganese sulfate,[14] and zinc chloride[17] are teratogenic. Silver nitrate[16,19] and tin chloride[20] are incompatible with a variety of reagents.

Principles of Decontamination

Positively charged metal ions can be removed from solution by using Amberlite IR-120(plus) or Dowex 50X8-100, strongly acid gel-type resins with a sulfonic acid functionality. Solutions containing cadmium, cobalt, copper, iron, lead, manganese, nickel, thallium, tin, and zinc containing 1000 ppm of metal were decontaminated in this fashion (Table 1). Solutions of silver nitrate and phosphomolybdic acid that contain negatively charged species can be decontaminated using Amberlite IRA-400(Cl) and Dowex 1X8-50, strongly basic gel-type resins with a quaternary ammonium functionality.[21] The concentrations of these metals were also 1000 ppm. Solutions containing other concentrations can be decontaminated by adjusting the solution volume:weight of resin ratio. On a small scale it is most convenient to stir the resin in the solution to be decontaminated but on a larger scale or for routine use it might be most convenient to pass the solution through a column packed with the resin. Although the volume of waste that must

Table 1 Removal of Metals from 1000 ppm Solution Using Ion Exchange Resins (ppm Remaining)

		Amberlite Resin[a]				Dowex Resin[b]			
Metal	Volume[c]	1 h	2 h	6 h	24 h	1 h	2 h	6 h	24 h
Ag	200	5.98	1.0	<0.25	<0.25	1.6	0.6	0.26	<0.25
Cd	40	<0.1	<0.1	<0.1	<0.1	0.69	0.47	0.86	1.3
Co	40	7.6	6.6	10.2	6.4	8.3	8.6	8.8	10.6
Cu	40	3.7	3.0	2.6	2.8	3.8	3.9	4.5	7.0
Fe	40	1.9	8.1	7.1	8.8	11	11	11	12
Mn	40	9.4	6.0	4.8	4.7	<0.25	<0.25	<0.25	<0.25
Mo	100	481	459	31	<1	52	13	<1	<1
Ni	40	7.0	6.4	5.0	5.2	4.5	4.4	4.7	6.2
Pb	200	15	1.8	<0.45	<0.45	0.45	<0.45	0.45	1.1
Sn	40	<5	<5	<5	<5	<5	<5	<5	<5
Tl	40	13	12	12	12	12	12	12	12
Zn	40	4.6	3.6	3.1	3.0	3.2	3.0	3.7	4.5

[a] Amberlite IRA-400(Cl) for Ag and Mo and Amberlite IR-120(plus) for the other metals.

[b] Dowex 1X8-50 for Ag and Mo and Dowex 50X8-100 for the other metals.

[c] Volume of 1000 ppm solution (mL) that can be decontaminated with 1 g of resin.

be disposed of is greatly reduced by using this technique a small amount of waste (i.e., the resin contaminated with the metal) remains and it should be discarded appropriately. The resin can be regenerated by washing with acid[22] but the concentrated metal-containing solution generated by this technique must also be disposed of appropriately. Metals can also be precipitated from solution using an appropriate anion. Most metals are precipitated by sulfide but this poses toxicity problems. The metals Cd (5.6 ppm left in solution), Co (14.7 ppm), Cu (6.3 ppm), Fe (15.4 ppm), and Mn (1.3 ppm) can be precipitated as the silicate. Although it has been reported that Pb can be precipitated as the silicate,[23] lead silicate is soluble to the extent of 1.74 mg/mL (equivalent to 1275 ppm Pb). Lead can be precipitated as the phosphate (1.27 ppm Pb left in solution). A procedure has been published for recycling silver by recovering silver nitrate from silver chloride residues.[24]

Decontamination Procedures

1. *Solutions containing Cd, Co, Cu, Fe, Mn, Ni, Pb, Sn, Tl, and Zn.* For each 40 mL of solution (200 mL for Pb solutions) containing no more than

1000 ppm of metal add 1 g of Amberlite IR-120(plus) or Dowex 50X8-100 resin. Stir the mixture for 24 h, filter, check the filtrate for completeness of decontamination, and discard it. The speed and efficiency of decontamination will depend on factors such as the size and shape of the flask and the rate of stirring. The beads now contain the metal and should be discarded appropriately.

2. *Solutions containing Ag and Mo.* For each 200 mL of Ag solution and 100 mL of Mo solution containing no more than 1000 ppm of metal add 1 g of Amberlite IRA-400(Cl) or Dowex 1X8-50 resin. Stir the mixture for 24 h, filter, check the filtrate for completeness of decontamination, and discard it. The speed and efficiency of decontamination will depend on factors such as the size and shape of the flask and the rate of stirring. The beads now contain the metal and should be discarded appropriately.

3. *Solutions containing Cd, Co, Cu, Fe, and Mn.* Mix the solution containing no more than 1000 ppm of the metal with an equal volume of 0.5 *M* sodium silicate (Na_2SiO_3) solution. Allow to stand for 1 h, filter, check the filtrate for completeness of decontamination, and discard it.

4. *Solutions containing Pb.* Mix the solution containing no more than 1000 ppm of Pb with an equal volume of 0.5 *M* potassium dihydrogen phosphate (KH_2PO_4) solution. Allow to stand for 1 h, filter, check the filtrate for completeness of decontamination, and discard it.

Analytical Procedures

Atomic absorption spectroscopy can be used to determine the concentrations of the metals in solution (Table 2).

Lead could also be determined by observing the precipitation of lead sulfide. Mix 1 mL of solution to be tested with 1 mL of a 1% w/v solution of Na_2S. A concentration of 10 ppm Pb gave a very light brown precipitate but 6 ppm Pb gave no precipitate. Heavier concentrations of Pb gave much heavier precipitates. Uranium is often found with Pb in waste mixtures and it was found that 1000 ppm U gave no precipitate.

Related Compounds

Ion-exchange resins can be used to decontaminate solutions containing ions of other metals. See also the monographs on Chromium, Osmium Tetroxide, Potassium Permanganate, Mercury, and Uranyl Compounds.

Table 2 Atomic Absorption Parameters

Metal	Wavelength (nm)	Slit Width (nm)	Limit of Detection (ppm)
Ag	328.1	0.7	0.25
Cd	228.8	0.7	0.1
Co	240.7	0.2	1.0
Cu	324.8	0.7	1.0
Fe	248.3	0.2	1.0
Mn	279.5	0.2	0.25
Mo	313.3	0.7	1.0
Ni	232.0	0.2	1.0
Pb	217.0	0.7	0.4
Sn	286.3	0.7	5.0
Tl	276.8	0.7	5.0
Zn	213.9	0.7	0.25

References

1. Other names for this compound are cupric sulfate, bluestone, blue vitriol, Roman vitriol, and Salzburg vitriol.
2. Other names for this compound are ferrous sulfate, copperas, green vitriol, iron vitriol, Feosol, Feospan, Fesofor, Fesotyme, Fero-Gradumet, Fer-in-Sol, Haemofort, Ironate, Irosul, Mol-Iron, Presfersul, and Sulferrous.
3. Other names for this compound are neutral lead acetate, normal lead acetate, sugar of lead, and salt of Saturn.
4. Other names for this compound are stannous chloride, tin dichloride, tin protochloride, and Stannochlor.
5. Lewis, R.J., Sr. *Sax's Dangerous Properties of Industrial Materials*, 8th ed.; Van Nostrand-Reinhold: New York, 1992; pp. 651–652.
6. Reference 5, p. 933.
7. Reference 5, p. 2492.
8. Reference 5, p. 2098.
9. Reference 5, p. 3260.
10. Reference 5, p. 2161.
11. Reference 5, p. 2436.
12. Reference 5, pp. 648–649.
13. International Agency for Research on Cancer. *IARC Monographs on the Evaluation of the Carcinogenic Risk of Chemicals to Humans.* Volume 23, *Some Metals and Metallic Compounds*; International Agency for Research on Cancer: Lyon, 1980; pp. 325–389.

14. Reference 5, p. 2164.
15. Reference 5, pp. 2498–2499.
16. Reference 5, pp. 3049–3050.
17. Reference 5, pp. 3538–3539.
18. Reference 5, pp. 2098–2099.
19. Bretherick, L. *Bretherick's Handbook of Reactive Chemical Hazards*, 4th ed.; Butterworths: London, 1990; pp. 10–13.
20. Reference 5, pp. 3294–3295.
21. Lunn, G. Unpublished results.
22. Wheaton, R.M.; Lefevre, L.J. Ion exchange. In *Kirk-Othmer Encyclopedia of Chemical Technology*, 3rd ed., Vol. 13; Wiley: New York, 1981; pp. 678–705.
23. Armour, M-.A. Tested laboratory disposal methods for small quantities of hazardous chemicals. In *Waste Disposal in Academic Institutions*; Kaufman, J.A., Ed., Lewis Publishers: Chelsea, MI, 1990; pp. 119–130.
24. Thomas, N.C. Recovering silver nitrate from silver chloride residues in about thirty minutes. *J. Chem. Educ.* **1990**, *67*, 794.

HEXAMETHYLPHOSPHORAMIDE

CAUTION! Refer to safety considerations section on page 7 before starting any of these procedures.

Hexamethylphosphoramide (HMPA) ($[(CH_3)_2N]_3P(O)$) [680-31-9][1] is a colorless liquid (bp 230–232°C, mp 7°C). This compound is carcinogenic in experimental animals[2] and is widely used in organic synthesis as an aprotic solvent.

Principle of Destruction

Hexamethylphosphoramide can be hydrolyzed by refluxing in concentrated hydrochloric acid (HCl) to dimethylamine and phosphoric acid.[3,4] The yield of dimethylamine was 88% and less than 0.06% of the HMPA remained in the final reaction mixture.[4]

Destruction Procedure

Take up 5 mL of HMPA in 25 mL of concentrated HCl and reflux for 4 h. Cool, cautiously neutralize by the addition of 1 *M* potassium hydroxide (KOH) solution, and discard the reaction mixture.

Analytical Procedures

For analysis by gas chromatography (GC) a 1.8-m × 2-mm i.d. column packed with 10% Carbowax 20M + 2% KOH on 80/100 Chromosorb W AW was used together with flame ionization detection.[4] The oven temperature was 170°C, the injection temperature was 200°C, the detector temperature was 300°C, and the carrier gas was nitrogen flowing at 30 mL/min. Under these conditions the retention time was 4.9 min and the limit of detection was 0.02 mg/mL. Before analysis 1 mL of the cooled reaction mixture was diluted with 2 mL of H_2O and basified with 2 mL of 10 *M* KOH. The GC column was fitted with a precolumn of the packing material and it was frequently found to be advantageous to replace it as it readily became contaminated. It was particularly important to do this before attempting to determine trace amounts of HMPA. Dimethylamine could also be determined using the same system but with an oven temperature of 100°C (retention time 0.29 min).

Mutagenicity Assays

The mutagenicity assays were carried out as described on page 4 using tester strains TA98, TA100, TA1530, and TA1535. The final neutralized reaction mixture was not mutagenic. Aliquots of 100 μL of the final neutralized reaction mixture, corresponding to a level of 3.43 mg undegraded HMPA, were added to each plate. Hexamethylphosphoramide itself, tested at a level of 1 mg per plate, was just mutagenic to TA1530 with activation.[4]

References

1. Other names for this compound are hexamethylphosphoric triamide, HEMPA, Hexametapol, HMPT, ENT 50882, hexamethylphosphoric acid triamide, hexamethylphosphorotriamide, hexamethylphosphotriamide, HPT, MEMPA, phosphoric tris(dimethylamide), phosphoryl hexamethyltriamide, tri(dimethylamino)phosphineoxide, tris(dimethylamino)phosphine oxide, and tris(dimethylamino)phosphorus oxide.
2. International Agency for Research on Cancer. *IARC Monographs on the Evaluation of the Carcinogenic Risk of Chemicals to Man.* Volume 15, *Some Fumigants, the Herbicides 2,4-D and 2,4,5-T, Chlorinated Dibenzodioxins and Miscellaneous Industrial Chemicals*; International Agency for Research on Cancer: Lyon, 1977; pp. 211–222.
3. National Research Council Committee on Hazardous Substances in the Laboratory. *Prudent Practices for Disposal of Chemicals from Laboratories*; National Academy Press: Washington, DC, 1983; p. 68.
4. Lunn, G. Unpublished observations.

HYDRAZINES

CAUTION! Refer to safety considerations section on page 7 before starting any of these procedures.

Hydrazines constitute a broad class of compounds having two nitrogen atoms linked together by a single bond. These compounds are of the general form $R_1R_2N—NR_3R_4$, where R_1, R_2, R_3, and R_4 may be aryl or alkyl groups or hydrogen. Hydrazine itself,[1-3] 1,1-dimethylhydrazine,[4] 1,2-dimethylhydrazine,[5] methylhydrazine,[6] 1,2-diphenylhydrazine,[7] phenyl hydrazine,[8] procarbazine,[9-13] isoniazid,[10,14-16] and 1,1-dibutylhydrazine[17] are carcinogenic in laboratory animals and procarbazine is probably carcinogenic in humans.[11] Procarbazine,[18] methylhydrazine,[19] and iproniazid[20] are teratogens; isoniazid is a teratogen and produces a number of toxic effects in humans.[21] *p*-Tolylhydrazine has been reported to produce neoplastic effects.[22] 1,1-Dimethylhydrazine,[23,24] 1,2-dimethylhydrazine,[24] 1,1-diethylhydrazine,[23,24] 1,1-dibutylhydrazine,[23,24] 1,1-diisopropylhydrazine,[23] and methylhydrazine[24] have been shown to be mutagenic. Hydrazine presents an explosion hazard under certain conditions, including distillation in air,[25] and can produce skin sensitization, systemic poisoning, and may damage the liver and red blood cells,[26] methylhydrazine is corrosive to skin, eyes, and mucous membranes and may self-ignite in air,[19] 1,1-dimethylhydrazine[27] and 1,2-dimethylhydrazine[28] present a fire hazard when exposed to oxidizers, and phenylhydrazine produces a variety of toxic effects including dermatitis, anemia, and injury to the spleen, liver, bone marrow, and kidneys as well as presenting a fire hazard with oxidizers.[8] Procarbazine is an antineoplastic, isoniazid is a tuberculostatic, and iproniazid is a monoamine oxidase inhibitor. The lower molecular weight hydrazines are generally colorless liquids that are generally soluble in H_2O and organic solvents. Because of their basic nature, most hydrazines are soluble in acid. Strictly speaking those hydrazines with a carbonyl adjacent to one of the nitrogens (e.g., diphenylcarbazide, isoniazid, and iproniazid) are called hydrazides but we will use the generic term hydrazine to embrace both groups here. Methods for the degradation of hydrazine, methylhydrazine, 1,1-dimethylhydrazine, 1,2-dimethylhydrazine dihydrochloride, and procarbazine have been validated by a collaborative study.[29] Some hydrazines whose degradation has been investigated include:

Compound Name	Reference	mp or bp (°C)	Registry Number
Hydrazine (diamide or diamine)		bp 113	[302-01-0]
Methylhydrazine (hydrazomethane or MMH)		bp 87	[60-34-4]
1,1-Dimethylhydrazine	30	bp 63	[57-14-7]
1,2-Dimethylhydrazine dihydrochloride	31	mp 167–169	[306-37-6] (base [540-73-8])
1,1-Diethylhydrazine		bp 96–99	[616-40-0]
1,1-Diisopropyl-hydrazine		bp 41/16 mm Hg	[921-14-2]
1,1-Dibutylhydrazine		bp 87–90/21 mm Hg	[7422-80-2]
N-Aminopyrrolidine hydrochloride		mp 117–119	[63234-71-9]
N-Aminopiperidine (1-piperidinamine)		bp 146/730 mm Hg	[2213-43-6]
N-Aminomorpholine (4-aminomorpholine)		bp 168	[4319-49-7]

Compound Name	Reference	mp or bp (°C)	Registry Number
N,N'-Diaminopiperazine hydrochloride			[40675-64-7] (base [106-59-2])
1-Methyl-1-phenyl-hydrazine		bp 54–55/0.3 mm Hg	[618-40-6]
1,2-Diphenylhydrazine (hydrazobenzene)		mp 123–126	[122-66-7]
Phenylhydrazine (hydrazinobenzene)		bp 238–241	[100-63-0]
p-Tolylhydrazine hydrochloride	32	mp >200	[637-60-5]
1,5-Diphenylcarbazide	33	mp 175–177	[140-22-7]
Procarbazine hydrochloride	34	mp 223–226	[366-70-1] (base [671-16-9])
Isoniazid	35	mp 171–173	[54-85-3]
Iproniazid phosphate	36	mp 180–182	[305-33-9] (base [54-92-2])

Hydrazines find wide use in synthetic organic chemistry, in cancer research laboratories, and in industry (e.g., as rocket fuel). Some hydrazines, for example, procarbazine, isoniazid, and iproniazid, are used as drugs.

Principles of Destruction

Hydrazines may be degraded by reduction with nickel–aluminum (Ni–Al) alloy in potassium hydroxide (KOH) solution[37] or by oxidation with potassium permanganate in sulfuric acid ($KMnO_4$ in H_2SO_4) or sodium hypochlorite (NaOCl).[29] The products of the reductive reactions are the corresponding amines (found in 73–100% yield) with ring reduction also observed in the case of isoniazid and iproniazid;[38] the products of the oxidative reactions are not known. It has been shown, however, that oxidation of hydrazines tends to produce the highly carcinogenic nitrosamines as byproducts, particularly when 1,1-dimethylhydrazine is the substrate, as well as producing mutagenic residues.[39] An additional hazard is the production of nitrosamines from 1,1-substituted hydrazines on exposure to air.[40] In related work we have found that 1,1-dimethylhydrazine is readily converted to *N*-nitrosodimethylamine when exposed to hydrogen peroxide or aqueous solutions of copper salts.[41] The reductive Ni–Al alloy procedure has the added advantage that it will degrade any nitrosamines that have been formed by the exposure of the hydrazine to air.[42] The oxidative methods are only recommended for treating spills and cleaning glassware when the heterogeneous nature of the Ni–Al alloy reductive method renders it inapplicable. All of these methods produced greater than 99% destruction of the hydrazines. Reduction of 1,2-diphenylhydrazine under acidic conditions might give rise to the carcinogen benzidine but it was found that reduction with Ni–Al alloy under basic conditions did not give rise to any detectable amount of benzidine (detection limit = 1% of the theoretical amount).[37]

Destruction Procedures[29,37]

Destruction of Bulk Quantities of Hydrazines

Dissolve the hydrazine in H_2O so that the concentration does not exceed 10 mg/mL. If the hydrazine is not sufficiently soluble in H_2O, use methanol instead. Add an equal volume of KOH solution (1 *M*) and stir the mixture magnetically. For every 100 mL of this solution add 5 g of Ni–Al alloy at such a rate that excessive frothing does not occur. The reaction can be quite exothermic. Do it in a reaction vessel whose volume is at least three times that of the final reaction mixture. Cover the reaction mixture, stir for 24 h, then filter it through a pad of Celite®. Allow the spent nickel to dry on a metal tray for 24 h (away from flammable solvents) and discard it with the solid waste. Check the filtrate for completeness of destruction, neutralize, and discard it.

Destruction of Hydrazines in Aqueous Solution

Dilute the mixture with H_2O, if necessary, so that the concentration does not exceed 10 g/L. Add an equal volume of KOH solution (1 *M*) and stir the mixture magnetically. For every 100 mL of this solution add 5 g of Ni–Al alloy at such a rate that excessive frothing does not occur. The reaction can be quite exothermic. Do it in a reaction vessel whose volume is at least three times that of the final reaction mixture. Cover the reaction mixture, stir for 24 h, then filter it through a pad of Celite®. Allow the spent nickel to dry on a metal tray for 24 h (away from flammable solvents) and discard it with the solid waste. Check the filtrate for completeness of destruction, neutralize, and discard it.

Destruction of Hydrazines in Organic Solvents Not Miscible with Water (For Example, Dichloromethane)

Dilute the solution, if necessary, so that the hydrazine concentration does not exceed 5 g/L. Stir the reaction mixture and add one volume of KOH solution (2 *M*) and three volumes of methanol (i.e., dichloromethane: H_2O: methanol 1:1:3). For every liter of this solution add 100 g of Ni–Al alloy at such a rate that excessive frothing does not occur. The reaction can be quite exothermic. Do it in a reaction vessel whose volume is at least three times that of the final reaction mixture. Cover the reaction mixture, stir for 24 h, then filter it through a pad of Celite®. Allow the spent nickel to dry on a metal tray for 24 h (away from flammable solvents) and discard it with the solid waste. Check the filtrate for completeness of destruction, neutralize, and discard it.

Destruction of Hydrazines in Alcohols

Dilute the solution, if necessary, so that the hydrazine concentration does not exceed 10 g/L. Add an equal volume of KOH solution (1 *M*) and stir the mixture magnetically. For every 100 mL of this solution add 5 g of Ni–Al alloy at such a rate that excessive frothing does not occur. The reaction can be quite exothermic. Do it in a reaction vessel whose volume is at least three times that of the final reaction mixture. Cover the reaction mixture, stir for 24 h, then filter it through a pad of Celite®. Allow the spent nickel to dry on a metal tray for 24 h (away from flammable solvents) and discard it with the solid waste. Check the filtrate for completeness of destruction, neutralize, and discard it.

Destruction of Hydrazines in Dimethyl Sulfoxide

Dilute the mixture, if necessary, with H_2O or methanol so that the hydrazine concentration does not exceed 10 g/L. Add an equal volume of KOH solution (1 *M*) and stir the mixture magnetically. For every 100 mL of this solution add 5 g of Ni–Al alloy at such a rate that excessive frothing does not occur. The reaction can be quite exothermic. Do it in a reaction vessel whose volume is at least three times that of the final reaction mixture. Cover the reaction mixture, stir for 24 h, then filter it through a pad of Celite®. Allow the spent nickel to dry on a metal tray for 24 h (away from flammable solvents) and discard it with the solid waste. Check the filtrate for completeness of destruction, neutralize, and discard it.

Destruction of Hydrazines in Olive Oil and Mineral Oil

For each volume of solution add two volumes of petroleum ether (boiling range 90–100°C) and extract this mixture with one volume of 1 *M* hydrochloric acid (HCl). Extract the organic layer twice more with 1 *M* HCl and combine the extracts. Dilute the extracts, if necessary, so that the concentration of hydrazine does not exceed 10 mg/mL. Add an equal volume of KOH solution (2 *M*) and stir the mixture magnetically. For every 100 mL of this solution add 5 g of Ni–Al alloy at such a rate that excessive frothing does not occur. The reaction can be quite exothermic. Do it in a reaction vessel whose volume is at least three times that of the final reaction mixture. Cover the reaction mixture, stir for 24 h, then filter it through a pad of Celite®. Allow the spent nickel to dry on a metal tray for 24 h (away from flammable solvents) and discard it with the solid waste. Check the filtrate for completeness of destruction, neutralize, and discard it.

Destruction of Hydrazines in Agar Gel

Add the contents of the agar plate (~17 g) to 75 mL of KOH solution (1 *M*) and stir and warm the mixture until the agar dissolves. Note that many hydrazines are liable to volatilize under these conditions. When cool add 2 g of Ni–Al alloy slowly enough so that excessive frothing does not occur. (If >100 mg of hydrazine is present, increase the weight of alloy and volume of KOH solution proportionately.) Stir the mixture for 24 h, then filter it through a pad of Celite®. Allow the spent nickel to dry on a metal tray away from flammable solvents for 24 h, then discard it with the solid waste. Check the filtrate for completeness of destruction, neutralize, and discard it.

Dealing with Spills of Hydrazines

1. Add 3 *M* H_2SO_4 to the spill and remove as much of the spill as possible with an absorbent. Place the absorbent material in an appropriate solvent and decontaminate it. Cover the residue with a 47.4 g/L solution of $KMnO_4$ in 3 *M* H_2SO_4. Leave this mixture overnight, then decolorize with sodium metabisulfite and remove with absorbents. Squeeze out the absorbents, make the solution strongly basic with 10 *M* KOH solution (**Caution!** Exothermic), dilute with H_2O, filter to remove manganese compounds,[43] neutralize the filtrate, check for completeness of destruction, and discard it. If possible, check for completeness of decontamination by taking wipe

samples and analyze these samples. Note that this procedure may damage painted surfaces or Formica.
Note: This procedure frequently generates nitrosamines and results in mutagenic residues. It should only be used with caution and in particular it should not be used to treat spills of 1,1-dimethylhydrazine as the procedure could generate large quantities of *N*-nitrosodimethylamine.[39]

2. Remove as much of the spill as possible with an absorbent, and then cover the residue with 5.25% NaOCl solution and leave overnight. Use fresh NaOCl solution (see below for assay procedure).
Note: This procedure frequently generates nitrosamines and results in mutagenic residues. It should only be used with caution.[39]

Decontamination of Equipment Contaminated with Hydrazines

1. Fill the equipment with a 47.4 g/L solution of $KMnO_4$ in 3 *M* H_2SO_4 (or cover the contaminated surface with this solution). After leaving overnight the solution should still be purple. If it is not, replace with fresh solution, which should stay purple for at least 1 h. At the end of the reaction, decolorize with sodium metabisulfite, make the solution strongly basic with 10 *M* KOH solution (**Caution!** Exothermic), dilute with H_2O, filter to remove manganese compounds,[43] neutralize the filtrate, check for completeness of destruction, and discard it. Finally, clean the equipment in a conventional fashion.
Note: This procedure frequently generates nitrosamines and results in mutagenic residues. It should only be used with caution and in particular it should not be used to treat equipment contaminated with 1,1-dimethylhydrazine as the procedure could generate large quantities of *N*-nitrosodimethylamine.[39]

2. Immerse the equipment in a 5.25% NaOCl solution and leave overnight. Use fresh sodium hypochlorite solution (see below for assay procedures).
Note: This procedure frequently generates nitrosamines and results in mutagenic residues. It should only be used with caution.[39]

Assay of Sodium Hypochlorite Solution

Sodium hypochlorite solutions tend to deteriorate with time so they should be periodically checked for the amount of active chlorine they contain. Pipette 10 mL of NaOCl solution into a 100-mL volumetric flask and fill to the mark with distilled H_2O. Pipette 10 mL of this solution into a conical flask containing 50 mL of distilled H_2O, 1 g of potassium iodide, and 12.5 mL of 2 *M* acetic acid. Titrate this solution against 0.1 *N* sodium thiosulfate solution using starch as an indicator. Each 1 mL of the sodium thiosulfate solution corresponds to 3.545 mg of active chlorine. The NaOCl solution used in these degradation reactions should contain 45–50 g of active chlorine per liter.

Analytical Procedures

Many procedures have been published for the analysis of hydrazines. The more volatile hydrazines are conveniently analyzed by gas chromatography (GC).[37] For analysis by GC a 1.8-m × 2-mm i.d. packed column can be used. For 1,1-dimethylhydrazine (80°C), 1,1-diethylhydrazine (80°C), 1,1-diisopropylhydrazine (80°C), *N*-aminopyrrolidine (80°C), 1,1-dibutylhydrazine (90°C), *N*-aminopiperidine (100°C), and *N*-aminomorpholine (130°C) the packing is 10% Carbowax 20 M + 2% KOH on 80/100 Chromosorb W AW, for *N,N'*-diaminopiperazine (100°C), phenylhydrazine (130°C), and *p*-tolylhydrazine (130°C) the packing is 2% Carbowax 20 M + 1% KOH on 80/100 Supelcoport, for 1,2-dimethylhydrazine (100°C) the packing is 28% Pennwalt 223 + 4% KOH on 80/100 Gas Chrom R, and for 1-methyl-1-phenylhydrazine (110°C) and 1,2-diphenylhydrazine (150°C) the packing is 3% SP 2401-DB on 100/120 Supelcoport. The oven temperatures shown above in parentheses are only a guide; the exact conditions would have to be determined experimentally. The high-performance liquid chromatography (HPLC) conditions for procarbazine and iproniazid are a 250 × 4.6-mm i.d. column of Microsorb C_8 with a mobile phase of methanol: 0.04% ammonium dihydrogen phosphate buffer 50:50 flowing at 1 mL/min. An ultraviolet (UV) detector set at 254 nm was used. For isoniazid the same equipment was used but the methanol : buffer ratio was 10:90.[38] Hydrazine itself can be analyzed using a colorimetric procedure.[39] An aliquot of the reaction mixture (2 mL) is mixed with 2 mL of 1 *M* HCl and, after 1 h, 0.5 g of ascorbic acid is added. After 30 min 2 mL of a determining reagent (1.6 g of *p*-dimethylaminobenzaldehyde in 80 mL of ethanol plus 8 mL of concentrated HCl) is added and the absorbance read at 458 nm. The limit of detection is 0.01 mg/L for hydrazine. Other hydrazines can be determined using this technique but the limits of detection are much higher.[39] The 1 *M* HCl and the ascorbic acid are required to remove any residual oxidizing power due to NaOCl or potassium iodate. If this is not required these reagents may be omitted.

Mutagenicity Assays

A number of hydrazines have been found to be mutagenic. For example, 1,1-dimethylhydrazine,[23,24] 1,2-dimethylhydrazine,[24] 1,1-diethylhydrazine,[23,24] 1,1-dibutylhydrazine,[23,24] 1,1-diisopropylhydrazine,[23] and methylhydrazine[24] have been reported to be mutagenic. The reaction mixtures obtained from the Ni–Al alloy reactions were not tested for mutagenicity but similar reaction mixtures from the degradation of nitrosamines (hydrazines have been shown to be intermediates in this degradation[42]) were found not to be mutagenic.[41]

Related Compounds

The procedures listed above should be generally applicable to hydrazines.

References

1. International Agency for Research on Cancer. *IARC Monographs on the Evaluation of Carcinogenic Risk of Chemicals to Man.* Volume 4, *Some Aromatic Amines, Hydrazine and Related Substances,* N-*Nitroso Compounds and Miscellaneous Alkylating Agents*; International Agency for Research on Cancer: Lyon, 1974; pp. 127–136.
2. International Agency for Research on Cancer. *IARC Monographs on the Evaluation of the Carcinogenic Risk of Chemicals to Humans, Supplement No. 4, Chemicals, Industrial Processes and Industries Associated with Cancer in Humans. IARC Monographs, Volumes 1 to 29*; International Agency for Research on Cancer: Lyon, 1982; pp. 136–138.
3. International Agency for Research on Cancer. *IARC Monographs on the Evaluation of the Carcinogenic Risk of Chemicals to Humans, Supplement No. 7, Overall Evaluations of Carcinogenicity: An Updating of* IARC Monographs *Volumes 1 to 42*; International Agency for Research on Cancer: Lyon, 1987; pp. 223–224.
4. Reference 1, pp. 137–143.
5. Reference 1, pp. 145–152.
6. Toth, B.; Shimizu, H. Methylhydrazine tumorigenesis in Syrian golden hamsters and the morphology of malignant histiocytomas. *Cancer Res.* **1973**, *33*, 2744–2753.
7. Lewis, R.J., Sr. *Sax's Dangerous Properties of Industrial Materials*, 8th ed.; Van Nostrand-Reinhold: New York, 1992; p. 1890.
8. Reference 7, pp. 2759–2760.
9. Kelly, M.G.; O'Gara, R.W.; Yancey, S.T.; Botkin, C. Induction of tumors in rats with procarbazine hydrochloride. *J. Natl. Cancer Inst.* **1968**, *40*, 1027–1051.
10. Kelly, M.G.; O'Gara, R.W.; Yancey, S.T.; Gadekar, K.; Botkin, C.; Oliverio, V.T. Comparative carcinogenicity of *N*-isopropyl-α-(2-methylhydrazino)-*p*-toluamide.HCl (procarbazine hydrochloride), its degradation products, other hydrazines, and isonicotinic acid hydrazide. *J. Natl. Cancer Inst.* **1969**, *42*, 337–344.
11. International Agency for Research on Cancer. *IARC Monographs on the Evaluation of the Carcinogenic Risk of Chemicals to Humans.* Volume 26, *Some Antineoplastic and*

Immunosuppressive Agents; International Agency for Research on Cancer: Lyon, 1981; pp. 311–339.

12. Reference 2, pp. 220–221.
13. Reference 3, pp. 327–328.
14. Reference 2, pp. 146–148.
15. Reference 3, pp. 227–228.
16. Reference 1, pp. 159–172.
17. Toth, B.; Nagel, D.; Patil, K. Carcinogenic effects of 1,1-di-*n*-butylhydrazine in mice. *Carcinogenesis* **1981**, *2*, 651–654.
18. Reference 7, pp. 2892–2893.
19. Reference 7, pp. 2327–2328.
20. Reference 7, pp. 2030–2031.
21. Reference 7, p. 2030.
22. Reference 7, p. 2378.
23. Nielsen, P.A.; Lagersted, A.; Danielsen, S.; Jensen, A.A.; Hart, J.; Larsen, J.C. Mutagenic activity of nine *N,N*-disubstituted hydrazines in the *Salmonella*/mammalian microsome assay. *Mutat. Res.* **1992**, *278*, 215–226.
24. Matsushita, H., Jr.; Endo, O.; Matsushita, H.; Yamamoto, M.; Mochizuki, M. Mutagenicity of alkylhydrazine oxalates in *Salmonella typhimurium* TA100 and TA102 demonstrated by modifying the growth conditions of the bacteria. *Mutat. Res.* **1993**, *301*, 213–222.
25. Fieser, L.F.; Fieser, M. *Reagents for Organic Synthesis*; Wiley: New York, 1967; Vol. 1, p. 434.
26. Reference 7, pp. 1885–1886.
27. Reference 7, pp. 1381–1382.
28. Reference 7, pp. 1397–1398.
29. Castegnaro, M.; Ellen, G.; Lafontaine, M.; van der Plas, H.C.; Sansone, E. B.; Tucker, S.P., Eds., *Laboratory Decontamination and Destruction of Carcinogens in Laboratory Wastes: Some Hydrazines*; International Agency for Research on Cancer: Lyon, 1983 (IARC Scientific Publications No. 54).
30. Other names for this compound are asymmetric dimethylhydrazine, *unsym*-dimethylhydrazine, *N,N*-dimethylhydrazine, Dimazine, and UDMH.
31. Other names for this compound are *N,N'*-dimethylhydrazine, *sym*-dimethylhydrazine, and SDMH. This compound is usually supplied as the dihydrochloride.
32. Another name for this compound is 4-methylphenylhydrazine.
33. Another name for this compound is 2,2'-diphenylcarbonic dihydrazide.
34. Other names for this compound are ibenzmethyzin, IBZ, *N*-isopropyl-α-(2-methylhydrazino)-*p*-toluamide, *N*-isopropyl-*p*-[(2-methylhydrazino)methyl]benzamide, *N*-4-isopropylcarbamoylbenzyl-*N'*-methylhydrazine, 2-(*p*-isopropylcarbamoylbenzyl)-1-methylhydrazine, matulane, MBH, 1-methyl-2-[(isopropylcarbamoyl)benzyl]hydrazine, *N*-(1-methylethyl)-4-[(2-methylhydrazino)methyl]benzamide, *p*-(N^1-methylhydrazinomethyl)-

N-isopropylbenzamide, MIH, Nathulane, natulan, NCI-C01810, NSC-77213, and RO 4-6467. This compound is generally supplied and used as the hydrochloride.

35. Other names for this compound are Amidon, Antimicina, Antituberkulosum, Atcotibine, Azuren, Bacillin, Cedin(Aerosol), Cemidon, Chemiazid, Cotinazin, Defonin, Dibutin, Dinacrin, Ditubin, Eralon, Ertuban, Evalon, Fimaline, Hidranizil, Hidrasonil, Hidrulta, Hycozid, Hydrazid, Hyozid, Hyzyd, Idrazil, INH, Isdonidrin, Isidrina, Ismazide, Isobicina, Isocid, Isocotin, Isolyn, Isonex, Isoniacid, Isonicazide, Isonicid, Isonicotan, isonicotinic acid hydrazide, isonicotinoyl hydrazide, isonicotinoylhydrazine, isonicotinyl hydrazide, isonicotinylhydrazine, Isonide, Isonikazid, Isonilex, Isonin, Isonindon, Isonirit, Isonizide, Isotebezid, Isozide, Isozyd, Laniazid, Mybasan, Neoteben, Neoxin, Neumandin, Nevin, Niadrin, Nicazide, Nicetal, Nicizina, Niconyl, Nicotibina, Nicozide, Nidaton, Nidrazid, Nikozid, Niplen, Nitadon, Niteban, NSC-9659, Nydrazid, Nyscozid, Pelazid, Percin, Pycazide, Pyricidin, 4-pyridinecarboxylic acid hydrazide, Pyrizidin, Raumanon, Retozide, Rimicid, Rimifon, rimitsid, Robisellin, RU-EF-Tb, Sanohydrazina, Sauterazid, Sauterzid, TB-Vis, Tebecid, Tebexin, Teebaconin, Tekazin, Tibazide, Tibinide, Tibivis, Tibizide, Tisin, Tisiodrazida, Tizide, Tubazid, Tubeco, Tubercid, Tubicon, Tubilysin, Tubomel, Tyvid, Tyvid, Unicozyde, USAF CB-2, Vazadrine, Vederon, Zinadon, and Zonazide.

36. Other names for this compound are Euphozid, Fosfazide, IIH, IPN, Iprazid, Iproniazid, Ipronid, Ipronin, isonicotinic acid 2-isopropylhydrazide, 1-isonicotinoyl-2-isopropylhydrazine, 1-isonicotinyl-2-isopropylhydrazine, *N*-isopropyl isonicotinhydrazide, LH, Marsalid, Marsilid, P 887, 4-pyridinecarboxylic acid 2-(1-methylethyl)hydrazide, Rivivol, RO 2-4572, and Yatrozide. This compound is frequently supplied as the phosphate salt.

37. Lunn, G.; Sansone, E.B.; Keefer, L.K. Reductive destruction of hydrazines as an approach to hazard control. *Environ. Sci. Technol.* **1983**, *17*, 240–243.

38. Lunn, G.; Sansone, E.B. Reductive destruction of dacarbazine, procarbazine hydrochloride, isoniazid, and iproniazid. *Am. J. Hosp. Pharm.* **1987**, *44*, 2519–2524.

39. Castegnaro, M.; Brouet, I.; Michelon, J.; Lunn, G.; Sansone, E.B. Oxidative destruction of hydrazines produces *N*-nitrosamines and other mutagenic species. *Am. Ind. Hyg. Assoc. J.* **1986**, *47*, 360–364.

40. Lunn, G.; Sansone, E.B.; Andrews, A.W. Aerial oxidation of hydrazines to nitrosamines. *Environ. Mol. Mutagen.* **1991**, *17*, 59–62.

41. Lunn, G. Unpublished observations.

42. Lunn, G.; Sansone, E.B.; Keefer, L.K. Safe disposal of carcinogenic nitrosamines. *Carcinogenesis* **1983**, *4*, 315–319.

43. Lunn, G.; Sansone, E.B.; De Méo, M.; Laget, M.; Castegnaro, M. Potassium permanganate can be used for degrading hazardous compounds. *Am. Ind. Hyg. Assoc. J.* **1994**, *55*, 167–171.

HYPOCHLORITES

CAUTION! Refer to safety considerations section on page 7 before starting any of these procedures.

Sodium hypochlorite (NaOCl) [7681-52-9],[1-3] calcium hypochlorite [$Ca(OCl)_2$] [7778-54-3],[4-6] and *tert*-butyl hypochlorite [$(CH_3)_3COCl$] [507-40-4][7-8] are oxidizers and are incompatible with a variety of reagents. These compounds are corrosive and irritating, and calcium hypochlorite and *tert*-butyl hypochlorite may explode on storage. *tert*-Butyl hypochlorite decomposes under the influence of light. *tert*-Butyl hypochlorite is readily prepared by the reaction of sodium hypochlorite with *tert*-butyl alcohol.[9] Sodium and calcium hypochlorites have a number of uses, including as bleaches and disinfectants. All of these compounds are used as oxidants in organic chemistry.

Principle of Destruction

Sodium hypochlorite, calcium hypochlorite, and *tert*-butyl hypochlorite can be reduced by addition to sodium metabisulfite solution.[10]

Destruction Procedure

Add 5 mL or 5 g of the hypochlorite to 100 mL of 10% (w/v) sodium metabisulfite solution and stir the mixture. When all the hypochlorite has dissolved and the reaction appears to be over, test for completeness of reaction, and discard the mixture. If the reaction is not complete, add more sodium metabisulfite solution until a negative test is obtained.

Analytical Procedure

Add a few drops of the reaction mixture to an equal volume of 10% (w/v) potassium iodide (KI) solution then acidify with a drop of 1 *M* hydrochloric acid and add a drop of starch solution as an indicator. The deep blue color of the starch–iodine complex indicates that excess oxidant is present.

Assay of Hypochlorites

Hypochlorites in general and sodium hypochlorite solutions in particular tend to deteriorate with time so they should be periodically checked for the amount of active chlorine they contain. A procedure for assaying sodium hypochlorite solutions follows but the procedure for other hypochlorites is similar. Pipette 10 mL of sodium hypochlorite solution into a 100-mL volumetric flask, which is filled to the mark with distilled H_2O. Pipette 10 mL of this solution into a conical flask containing 50 mL of distilled H_2O, 1 g of KI, and 12.5 mL of 2 *M* acetic acid. Titrate this solution against 0.1 *N* sodium thiosulfate solution using starch as an indicator. Each 1 mL of the sodium thiosulfate solution corresponds to 3.545 mg of active chlorine. Commercially available 5.25% sodium hypochlorite solution (Clorox bleach) should contain 45–50 g of active chlorine per liter.

References

1. Other names for this compound are Antiformin, B-K Liquid, Carrel-Dakin solution, Chloros, Chlorox, Clorox, Dakin's solution, Hyclorite, Javex, Milton, Surchlor, and surgical chlorinated soda solution.
2. Lewis, R.J., Sr. *Sax's Dangerous Properties of Industrial Materials*, 8th ed.; Van Nostrand-Reinhold: New York, 1992; pp. 3088–3089.
3. Bretherick, L. *Bretherick's Handbook of Reactive Chemical Hazards*, 4th ed.; Butterworths: London; 1990; pp. 984–986.
4. Other names for this compound are hypochlorous acid calcium salt, B-K Powder, bleach-

ing powder, calcium chlorohydrochlorite, calcium hypochloride, Camporit, caporit, CCH, chloride of lime, chlorinated lime, HTH, Hy-Chlor, lime chloride, Lo-Bax, Losantin, Perchloron, Pittchlor, Pittcide, and Sentry.

5. Reference 2, p. 1967.
6. Reference 3, pp. 918–923.
7. Reference 2, p. 621.
8. Reference 3, p. 473.
9. Mintz, M.J.; Walling, C. *t*-Butyl hypochlorite. In *Organic Syntheses*; Baumgarten, H.E., Ed., Wiley: New York, 1973; Coll. Vol. 5, pp. 184–187.
10. Lunn, G. Unpublished results.

MERCURY

CAUTION! Refer to safety considerations section on page 7 before starting any of these procedures.

Mercury and mercury compounds are poisonous and teratogenic.[1] Although mercury cannot be destroyed it can be removed from dilute aqueous solutions by using ion-exchange resins or by amalgamation with iron. In this way the volume of waste that must be disposed of can be greatly reduced.

Solutions of mercuric acetate [$Hg(CH_3CO_2)_2$] [1600-27-7] and mercury(II) chloride ($HgCl_2$) [7487-94-7][2] were decontaminated using ion exchange resins. Laboratory waste water containing Hg^{2+} was decontaminated using iron powder.

Principles of Decontamination

Positively charged metal ions can be removed from solution by using Amberlite IR-120(plus) or Dowex 50X8-100, strongly acid gel-type resins with a sulfonic acid functionality and solutions of $Hg(CH_3CO_2)_2$ can be decon-

Table 1 Removal of Mercury from Solution Using Ion Exchange Resins (ppm Remaining)

Compound	Resin	1 h	2 h	6 h	24 h
$Hg(CH_3CO_2)_2$	Amberlite IR-120 (plus)	175	183	3.8	<3.8
$Hg(CH_3CO_2)_2$	Dowex 50X8-100	<3.8	<3.8	<3.8	<3.8
$HgCl_2$	Amberlite IRA-400 (Cl)	96	11	<3.8	<3.8
$HgCl_2$	Dowex 1X8-50	62	17	3.8	3.8

taminated in this fashion (Table 1). Solutions of $HgCl_2$ that contain negatively charged species can be decontaminated using Amberlite IRA-400(Cl) and Dowex 1X8-50, strongly basic gel-type resins with a quaternary ammonium functionality.[3] In each case the concentration of mercury was 1000 ppm. Solutions containing other concentrations can be decontaminated by adjusting the solution volume:weight of resin ratio. On a small scale it is most convenient to stir the resin in the solution to be decontaminated but on a larger scale, or for routine use, it might be most convenient to pass the solution through a column packed with the resin. Although the volume of waste that must be disposed of is greatly reduced by using this technique a small amount of waste (i.e., the resin contaminated with the metal) remains and it should be discarded appropriately. The resin can be regenerated by washing with acid but the concentrated metal-containing solution generated by this technique must also be disposed of appropriately.[4] Mercury may also be removed from laboratory waste H_2O using a column of iron powder.[5] The mercury forms mercury amalgam and stays on the column. Some metallic mercury remains in solution but this can be removed by aeration. The final concentration of mercury is less than 5 ppb.

Decontamination Procedures

1. *Solutions containing $Hg(CH_3CO_2)_2$.* For each 200 mL of solution containing no more than 1000 ppm of mercury add 1 g of Amberlite IR-120(plus) or Dowex 50X8-100 resin. Stir the mixture for 24 h, filter, check the filtrate for completeness of decontamination, and discard it. The speed and efficiency of decontamination will depend on factors such as the size

and shape of the flask and the rate of stirring. The beads now contain the metal and they should be discarded appropriately.

2. *Solutions containing $HgCl_2$*. For each 200 mL of solution containing no more than 1000 ppm of mercury add 1 g of Amberlite IRA-400(Cl) or Dowex 1X8-50 resin. Stir the mixture for 24 h, filter, check the filtrate for completeness of decontamination, and discard it. The speed and efficiency of decontamination will depend on factors such as the size and shape of the flask and the rate of stirring. The beads now contain the metal and they should be discarded appropriately.

3. *Waste Water Containing Mercury*. Pass no more than 2 L of the solution containing no more than 2.5 ppm of mercury through a column (6-mm i.d.) packed with 1 g of iron powder (60 mesh) at 250 mL/h. Aerate the resulting effluent to remove metallic mercury and continue aeration for 30 min after the last of the effluent has emerged from the column.

The effluent contains less than 5 ppb of mercury. Use a fresh column for each treatment. Some iron ends up in solution and this can be removed by adjusting the pH to 8 before aeration. Remove the precipitated $Fe(OH)_3$ by filtration. Chloride and glycine do not interfere and bromide, ethylenediaminetetraacetic acid (EDTA), and cysteine do not interfere if the aeration is carried out at pH 8. The presence of iodide or polypeptone may require several treatments to reduce the mercury to an acceptable level. The metallic mercury removed from the solution by aeration can be vented into the fume hood or captured in a mercury trap. Solutions containing a higher concentration of mercury (e.g., 100 ppm) may also be treated but this will result in a higher final concentration of mercury (33 ppb).

Analytical Procedures

Atomic absorption spectroscopy was used to determine the concentrations of mercury in solution in the experiments involving ion-exchange resins. The wavelength was 253.7 nm, the slit width was 0.7 nm, and the limit of detection was 3.8 ppm. A Hiranuma mercury meter model HG-1 was used in the experiments involving iron powder. Methods involving atomic absorption spectroscopy[6,7] and colorimetric procedures[8,9] have been described.

Related Compounds

The choice of resin depends on the nature of the species in solution and is probably best determined by small-scale tests. Ion-exchange resins can be used to decontaminate solutions containing ions of other metals. See also the monographs on Chromium, Heavy Metals, Osmium Tetroxide, Potassium Permanganate, and Uranyl Compounds.

References

1. Lewis, R.J., Sr. *Sax's Dangerous Properties of Industrial Materials*, 8th ed.; Van Nostrand-Reinhold: New York, 1992; pp. 2188–2204.
2. Other names for this compound are mercuric chloride, mercury bichloride, corrosive sublimate, mercury perchloride, and corrosive mercury chloride.
3. Lunn, G. Unpublished observations.
4. Wheaton, R.M.; Lefevre, L.J. Ion exchange. In *Kirk–Othmer Encyclopedia of Chemical Technology*, Vol. 13, 3rd ed.; Wiley: New York, 1981; pp. 678–705.
5. Shirakashi, T.; Nakayama, K.; Kakii, K.; Kuriyama, M. Removal of mercury from laboratory waste water with iron powder. *Nihon Kagaku Gaishi* **1986**, 1352–1356. (*Chem. Abstr.* **1986**, *105*, 213690y.) Translation available upon request.
6. American Public Health Association. *Standard Methods for the Examination of Water and Wastewater*, 16th ed.; American Public Health Association: Washington, 1985; pp. 171–173.
7. Helrich, K., Ed., *Official Methods of Analysis of the Association of Official Analytical Chemists*, 15th ed.; Association of Official Analytical Chemists, Inc.: Arlington, VA, 1990; pp. 262–264.
8. Reference 6, pp. 232–234.
9. Reference 7, pp. 264–266.

METHOTREXATE

CAUTION! Refer to safety considerations section on page 7 before starting any of these procedures.

The degradation of a number of antineoplastic drugs, including methotrexate, was investigated by the International Agency for Research on Cancer (IARC).[1] Methotrexate (**I**) [59-05-2] (L form dihydrate);[2] [15475-56-6] (sodium salt); [51865-79-3] (D form); [60388-53-6] (DL form hydrate)] is a solid (mp of the monohydrate 185–204°C) and is insoluble in H_2O and organic solvents, but it is soluble in dilute acid or base.

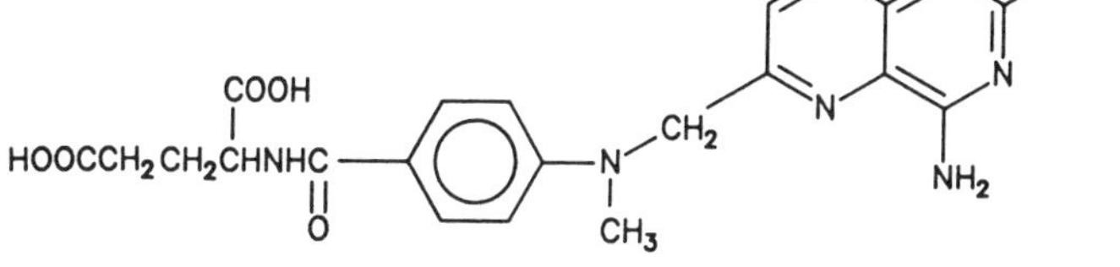

I

Methotrexate is not mutagenic to *Salmonella typhimurium* but it is mutagenic to *Streptococcus faecium*.[3] Methotrexate is a teratogen in humans and laboratory animals and can affect the blood, bone marrow, liver, and other organs.[4] It is used as an antineoplastic drug.

Principles of Destruction

Methotrexate is destroyed by oxidation with potassium permanganate in sulfuric acid ($KMnO_4$ in H_2SO_4), by oxidation with alkaline $KMnO_4$, or by oxidation with sodium hypochlorite (NaOCl). Destruction is greater than 99.5% in all cases. The products of these reactions have not been established.

Destruction Procedures

Destruction of Bulk Quantities of Methotrexate

1. Dissolve in 3 *M* H_2SO_4 so that the concentration does not exceed 5 mg/mL, then add 0.5 g of $KMnO_4$ for each 10 mL of solution and stir for 1 h. Decolorize with sodium metabisulfite, make strongly basic with 10 *M* potassium hydroxide (KOH) solution (**Caution!** Exothermic), dilute with H_2O, filter to remove manganese compounds,[5] neutralize the filtrate, check for completeness of destruction, and discard it.

2. Dissolve in 1 *M* sodium hydroxide (NaOH) solution so that the concentration does not exceed 1 mg/mL. For every 50 mL of this solution add 5.5 mL of a 1% (w/v) aqueous $KMnO_4$ solution and stir. The purple color should persist for at least 30 min; if it does not, add more $KMnO_4$ solution. Decolorize with sodium metabisulfite, dilute with H_2O, filter to remove manganese compounds,[5] neutralize the filtrate, check for completeness of destruction, and discard it.

3. Dissolve in 1 *M* NaOH solution so that the concentration does not exceed 0.5 mg/mL. For every 100 mL of this solution add 10 mL of 5.25% NaOCl solution and stir for 30 min. Treat the reaction mixture with 0.1 *M* sodium bisulfite solution to remove excess oxidant, neutralize with 1 *M* hydrochloric acid (HCl) (**Caution!** Chlorine is evolved), analyze for completeness of destruction, and discard it. Use fresh NaOCl solution (see assay procedure below).

Destruction of Aqueous Solutions of Methotrexate

1. Dilute with H_2O, if necessary, so that the concentration does not exceed 5 mg/mL, then add enough concentrated H_2SO_4 to obtain a 3 *M* solution.

Allow it to cool to room temperature. For each 10 mL of solution add 0.5 g of $KMnO_4$ and stir for 1 h. Decolorize with sodium metabisulfite, make strongly basic with 10 *M* KOH (**Caution!** Exothermic), dilute with H_2O, filter to remove manganese compounds,[5] neutralize the filtrate, check for completeness of destruction, and discard it.

2. Add at least an equal volume of 2 *M* NaOH solution, more if necessary, so that the concentration does not exceed 1 mg/mL. For every 50 mL of this solution add 5.5 mL of a 1% (w/v) aqueous $KMnO_4$ solution and stir. The purple color should persist for at least 30 min; if it does not, add more $KMnO_4$ solution. Decolorize with sodium metabisulfite, dilute with H_2O, filter to remove manganese compounds,[5] neutralizc the filtrate, check for completeness of destruction, and discard it.

3. For every 50 mg of methotrexate add 10 mL of 5.25% NaOCl solution and stir for 30 min. Then treat the reaction mixture with 0.1 *M* sodium bisulfite solution to remove excess oxidant, neutralize with 1 *M* HCl (**Caution!** Chlorine is evolved), analyze for completeness of destruction, and discard it. Use fresh NaOCl solution (see assay procedure below).

Destruction of Injectable Pharmaceutical Preparations of Methotrexate Containing 2–5% Glucose and 0.45% Saline

1. Dilute with H_2O so that the concentration does not exceed 2.5 mg/mL, then add enough concentrated H_2SO_4 to obtain a 3 *M* solution. Allow it to cool to room temperature. For each 10 mL of solution add 1 g of $KMnO_4$, in small portions to avoid frothing, and stir for 1 h. Decolorize with sodium metabisulfite, make strongly basic with 10 *M* KOH (**Caution!** Exothermic), dilute with H_2O, filter to remove manganese compounds,[5] neutralize the filtrate, check for completeness of destruction, and discard it.

2. Add at least an equal volume of 2 *M* NaOH solution, more if necessary, so that the concentration does not exceed 1 mg/mL. For every 50 mL of this solution add 5.5 mL of a 1% (w/v) aqueous $KMnO_4$ solution and stir. The purple color should persist for at least 30 min; if it does not, add more $KMnO_4$ solution. Decolorize with sodium metabisulfite, dilute with H_2O, filter to remove manganese compounds,[5] neutralize the filtrate, check for completeness of destruction, and discard it.

3. For every 50 mg of methotrexate add 10 mL of 5.25% NaOCl solution and stir for 30 min. Treat the reaction mixture with 0.1 *M* sodium bisulfite solution to remove excess oxidant, neutralize with 1 *M* HCl (**Caution!** Chlorine is evolved), analyze for completeness of destruction, and discard it. Use fresh NaOCl solution (see assay procedure below).

Destruction of Solutions of Methotrexate in Volatile Organic Solvents

1. Remove the solvent under reduced pressure using a rotary evaporator and take up the residue in 3 *M* H_2SO_4 so that the concentration does not exceed 5 mg/mL. For each 10 mL of solution add 0.5 g of $KMnO_4$ and stir for 1 h. Decolorize with sodium metabisulfite, make strongly basic with 10 *M* KOH (**Caution!** Exothermic), dilute with H_2O, filter to remove manganese compounds,[5] neutralize the filtrate, check for completeness of destruction, and discard it.

2. Remove the solvent under reduced pressure using a rotary evaporator and take up the residue in 1 *M* NaOH solution so that the concentration does not exceed 1 mg/mL. For every 50 mL of this solution add 5.5 mL of a 1% (w/v) aqueous $KMnO_4$ solution and stir. The purple color should persist for at least 30 min; if it does not, add more $KMnO_4$ solution. Decolorize with sodium metabisulfite, dilute with H_2O, filter to remove manganese compounds,[5] neutralize the filtrate, check for completeness of destruction, and discard it.

3. Remove the solvent under reduced pressure using a rotary evaporator and take up the residue in 1 *M* NaOH solution so that the concentration does not exceed 0.5 mg/mL. For every 100 mL of this solution add 10 mL of 5.25% NaOCl solution and stir for 30 min. Then treat the reaction mixture with 0.1 *M* sodium bisulfite solution to remove excess oxidant, neutralize with 1 *M* HCl (**Caution!** Chlorine is evolved), analyze for completeness of destruction, and discard it. Use fresh NaOCl solution (see assay procedure below).

Destruction of Dimethyl Sulfoxide or N,N-Dimethylformamide Solutions of Methotrexate

Dilute with H_2O so that the concentration of dimethyl sulfoxide (DMSO) or *N,N*-dimethylformamide (DMF) does not exceed 20% and the concentration of the drug does not exceed 2.5 mg/mL, then add enough concentrated H_2SO_4 to obtain a 3 *M* solution. Allow it to cool to room temperature. For each 10 mL of solution add 1 g of $KMnO_4$ and stir for 1 h. Decolorize with sodium metabisulfite, make strongly basic with 10 *M* KOH (**Caution!** Exothermic), dilute with H_2O, filter to remove manganese compounds,[5] neutralize the filtrate, check for completeness of destruction, and discard it.

Decontamination of Glassware Contaminated with Methotrexate

1. Immerse the glassware in a 0.3 *M* solution of $KMnO_4$ in 3 *M* H_2SO_4 for 1 h. Decolorize with sodium metabisulfite and remove and clean the glass-

ware. Make the solution strongly basic with 10 *M* KOH (**Caution!** Exothermic), dilute with H_2O, filter to remove manganese compounds,[5] neutralize the filtrate, check for completeness of destruction, and discard it.

2. Immerse the glassware in a mixture of 50 mL of 1 *M* NaOH solution and 5.5 mL of a 1% (w/v) aqueous $KMnO_4$ solution. After 30 min clean the glassware with 0.1 *M* sodium bisulfite solution. Decolorize the solution with sodium metabisulfite, dilute with H_2O, filter to remove manganese compounds,[5] neutralize the filtrate, check for completeness of destruction, and discard it.

3. Immerse the glassware in 5.25% NaOCl solution for 30 min. Use fresh NaOCl solution (see assay procedure below).

Decontamination of Spills of Methotrexate

1. Allow any organic solvent to evaporate and remove as much of the spill as possible by high efficiency particulate air (HEPA) vacuuming (not sweeping), then rinse the area with 3 *M* H_2SO_4. Take up the rinse with absorbents and allow the rinse and absorbents to react with 0.3 *M* $KMnO_4$ solution in 3 *M* H_2SO_4 for 1 h. If the color fades, add more solution. Check for completeness of decontamination by using a wipe moistened with 0.1 *M* H_2SO_4. Analyze the wipe for the presence of the drug. Decolorize the $KMnO_4$ solution with sodium metabisulfite, make strongly basic with 10 *M* KOH (**Caution!** Exothermic), dilute with H_2O, filter to remove manganese compounds,[5] neutralize the filtrate, check for completeness of destruction, and discard it.

2. Allow any organic solvent to evaporate and remove as much of the spill as possible by HEPA vacuuming (not sweeping), then rinse the area with 1 *M* NaOH solution. Take up the rinse with absorbents and allow the rinse and absorbents to react with a mixture of 50 mL of 1 *M* NaOH solution and 5.5 mL of a 1% (w/v) aqueous $KMnO_4$ solution. Check for completeness of decontamination by using a wipe moistened with 0.1 *M* NaOH solution. Analyze the wipe for the presence of the drug. Decolorize the $KMnO_4$ solution with sodium metabisulfite, dilute with H_2O, filter to remove manganese compounds,[5] neutralize the filtrate, check for completeness of destruction, and discard it.

3. Allow any organic solvent to evaporate and remove as much of the spill as possible by HEPA vacuuming (not sweeping), then rinse the area with 5.25% NaOCl solution, then H_2O. Remove the rinses and discard them. Check for completeness of decontamination by using a wipe moistened with 0.1 *M* NaOH solution. Analyze the wipe for the presence of the drug. Use fresh NaOCl solution (see assay procedure below).

Assay of Sodium Hypochlorite Solution

Sodium hypochlorite solutions tend to deteriorate with time so they should be periodically checked for the amount of active chlorine they contain. Pipette 10 mL of NaOCl solution into a 100-mL volumetric flask and fill to the mark with distilled H_2O. Pipette 10 mL of this solution into a conical flask containing 50 mL of distilled H_2O, 1 g of potassium iodide, and 12.5 mL of 2 *M* acetic acid. Titrate this solution against 0.1 *N* sodium thiosulfate solution using starch as an indicator. Each 1 mL of the sodium thiosulfate solution corresponds to 3.545 mg of active chlorine. The NaOCl solution used in these degradation reactions should contain 45–50 g of active chlorine per liter.

Analytical Procedures

Methotrexate can be analyzed by high-performance liquid chromatography (HPLC) using a 25-cm reverse phase column and ultraviolet (UV) detection at 254 nm. Tetrabutylammonium phosphate (5 m*M*, adjusted to pH 3.5 with phosphoric acid) : methanol (55:45) flowing at 1.5 mL/min; 5 m*M* ammonium formate (adjusted to pH 5 with formic acid) : methanol (60:40) flowing at 1 mL/min; and 0.1 *M* potassium phosphate, monobasic (KH_2PO_4) : methanol (80:20) flowing at 1 mL/min have been recommended as mobile phases.

Mutagenicity Assays

In the IARC study[1] tester strains TA100, TA1530, TA1535, and UTH8414 of *S. typhimurium* were used with and without mutagenic activation (not all strains were used for each procedure). The reaction mixtures were not mutagenic.

Related Compounds

Potassium permanganate in H_2SO_4 is a general oxidative method and should, in principle, be effective in destroying many drugs. However, any new application should be thoroughly validated both for complete destruction of the compound and for the production of nonmutagenic reaction mixtures.

References

1. Castegnaro, M.; Adams, J.; Armour, M-.A.; Barek, J.; Benvenuto, J.; Confalonieri, C.; Goff, U.; Ludeman, S.; Reed, D.; Sansone, E. B.; Telling, G.; Eds., *Laboratory De-*

contamination and Destruction of Carcinogens in Laboratory Wastes: Some Antineoplastic Agents; International Agency for Research on Cancer: Lyon, 1985 (IARC Scientific Publications No. 73).

2. Other names for this compound are *N*-[4-([(2,4-diamino-6-pteridinyl)methyl]-methylamino)benzoyl]-L-glutamic acid, amethopterin, 4-amino-4-deoxy-N^{10}-methylpteroylglutamic acid, 4-amino-10-methylfolic acid, 4-amino-N^{10}-methylpteroylglutamic acid, *N*-bismethylpteroylglutamic acid, methopterin, methotextrate, MTX, Emtexate, A-Methopterin, methylaminopterin, Cl-14377, Rheumatrex, Folex, Mexate, Antifolan, EMT 25,299, HDMTX, NCI-C04671, NSC-740, and R 9985.
3. International Agency for Research on Cancer. *IARC Monographs on the Evaluation of the Carcinogenic Risk of Chemicals to Humans, Supplement No. 4, Chemicals, Industrial Processes and Industries Associated with Cancer in Humans. IARC Monographs, Volumes 1 to 29*; International Agency for Research on Cancer: Lyon, 1982; pp. 157–158.
4. Lewis, R.J., Sr. *Sax's Dangerous Properties of Industrial Materials*, 8th ed.; Van Nostrand-Reinhold: New York, 1992; pp. 2217–2218.
5. Lunn, G.; Sansone, E.B.; De Méo, M.; Laget, M.; Castegnaro, M. Potassium permanganate can be used for degrading hazardous compounds. *Am. Ind. Hyg. Assoc. J.*, **1994**, *55*, 167–171.

2-METHYLAZIRIDINE

CAUTION! Refer to safety considerations section on page 7 before starting any of these procedures.

2-Methylaziridine (**I**) [75-55-8] is a volatile liquid (bp 66–67°C) that has been reported to be a carcinogen in experimental animals.[1] This compound is a severe eye irritant, reacts vigorously with oxidizing materials, and can polymerize explosively.[2] 2-Methylaziridine is used industrially as a chemical intermediate and is miscible with H_2O and soluble in ethanol. Other names for this compound are 2-methylazacyclopropane, 2-methylethyleneimine, and 1,2-propyleneimine.

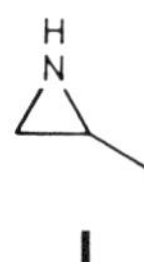

I

Principle of Destruction

2-Methylaziridine is reduced by nickel–aluminum (Ni–Al) alloy in potassium hydroxide (KOH) solution to give a mixture of isopropylamine and *n*-propylamine in an 85:15 ratio.[3] Destruction efficiency is greater than 99.6%.

Destruction Procedure

Take up 2-methylaziridine (50 μL, 40.4 mg) in 10 mL of 1 *M* KOH solution and add 0.5 g of Ni–Al alloy. (If the reaction is done on a bigger scale, add the alloy in portions to avoid frothing.) Stir the reaction mixture for 18 h, then filter through a pad of Celite®. Check the filtrate for completeness of destruction, neutralize, and discard it. Allow the spent nickel to dry on a metal tray for 24 h (away from flammable solvents) and discard it with the solid waste.[4]

Analytical Procedures

Add 100 μL of the solution to be analyzed to 1 mL of a solution of 2 mL of acetic acid in 98 mL of 2-methoxyethanol.[5] Swirl this mixture and add 1 mL of a solution of 5 g of 4-(4-nitrobenzyl)pyridine (4-NBP) in 100 mL of 2-methoxyethanol. Heat the solution at 100°C for 10 min, then cool in ice for 5 min. Add piperidine (0.5 mL) and 2-methoxyethanol (2 mL). Determine the violet color at 560 nm.

Check the efficacy of the analytical procedure by adding a small quantity of 2-methylaziridine in 2-methoxyethanol solution to the solution to be analyzed after the acetic acid-2-methoxyethanol has been added but before the 4-NBP has been added. A positive response indicates that the analytical technique is satisfactory.

Using the analytical procedure described above with 10-mm disposable plastic cuvettes in a Gilford 240 spectrophotometer the limit of detection was 16 mg/L, but this can easily be reduced by using more than 100 μL of the solution to be analyzed.

The products were determined by gas chromatography (GC) using a 1.8-m × 2-mm i.d. packed column filled with 10% Carbowax 20 M + 2% KOH on 80/100 Chromosorb W AW. The oven temperature was 60°C, the injection temperature was 200°C, and the flame ionization detector operated at 300°C. Under these conditions, isopropylamine and *n*-propylamine had retention times of about 1.1 and 1.7 min, respectively. The GC conditions given above are only a guide and the exact conditions would have to be determined experimentally.

Mutagenicity Assays

The mutagenicity assays were carried out as described on page 4 using tester strains TA98, TA100, TA1530, and TA1535. The final reaction mixture (tested at a level corresponding to 0.4 mg of undegraded material) was not mutagenic although it was somewhat toxic to the cells. When it was diluted with an equal volume of pH 7 buffer it was not toxic and it was still not mutagenic. The pure compound was highly mutagenic, whereas the products, isopropylamine and *n*-propylamine, were not mutagenic.

Related Compounds

The procedure should be generally applicable to the destruction of other aziridines but it should be carefully checked to ensure that the compounds are completely degraded.

References

1. International Agency for Research on Cancer. *IARC Monographs on the Evaluation of the Carcinogenic Risk of Chemicals to Man.* Volume 9, *Some Aziridines,* N, S- *and* O-*Mustards and Selenium*; International Agency for Research on Cancer: Lyon, 1975; pp. 61–65.
2. Lewis, R.J., Sr., *Sax's Dangerous Properties of Industrial Materials*, 8th ed.; Van Nostrand-Reinhold: New York, 1992; pp. 2921–2922.
3. Lunn, G.; Sansone, E.B. Validated methods for degrading hazardous chemicals: Some halogenated compounds. *Am. Ind. Hyg. Assoc. J.* **1991**, *52*, 252–257.
4. Lunn, G.; Sansone, E.B.; Andrews, A.W.; Keefer, L.K. Decontamination and disposal of nitrosoureas and related *N*-nitroso compounds. *Cancer Res.* **1988**, *48*, 522–526.
5. This procedure was developed for determining dimethyl sulfate. See Lunn, G.; Sansone, E.B. Validation of techniques for the destruction of dimethyl sulfate. *Am. Ind. Hyg. Assoc. J.* **1985**, *46*, 111–114.

1-METHYL-4-PHENYL-1,2,3,6-TETRAHYDROPYRIDINE (MPTP)

CAUTION! Refer to safety considerations section on page 7 before starting any of these procedures.

1-Methyl-4-phenyl-1,2,3,6-tetrahydropyridine (MPTP) (**I**) [28289-54-5; 23007-85-4 (HCl)] is a colorless crystalline solid (mp 37–40°C), which produces the symptoms of Parkinson's disease in humans.[1] Other toxicological properties have not been established although it does not appear to be mutagenic.[2] It has a tendency to sublime and it should be handled with great care only in a properly functioning chemical fume hood. To avoid volatility problems, MPTP, which is a base, should only be handled in acidic solution. It is usually convenient to handle the compound as a nonvolatile salt. The preparation of MPTP tartrate [124066-53-1] is described

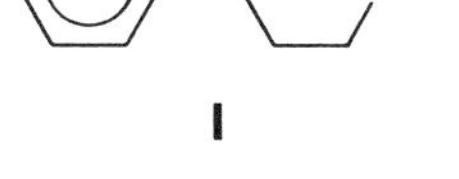

N—

I

below. 1-Methyl-4-phenyl-1,2,3,6-tetrahydropyridine is used in the laboratory in research into Parkinson's disease.

Principles of Destruction

1-Methyl-4-phenyl-1,2,3,6-tetrahydropyridine may be degraded by oxidation with potassium permanganate in 3 *M* sulfuric acid ($KMnO_4$ in H_2SO_4).[2] No detectable pyridinium oxidation product (1-methyl-4-phenylpyridinium ion) was formed.[2] Experiments using labeled MPTP showed that the products of the reaction were polar, H_2O soluble species, probably carboxylic acids.[2] Reduction of MPTP to the physiologically inactive[3] 1-methyl-4-phenylpiperidine with nickel–aluminum alloy gave unsatisfactory results and this procedure could not be recommended for degrading MPTP.[2]

Destruction Procedures

Destruction of Bulk Quantities

For every 25 mg of MPTP add 100 mL of 3 *M* H_2SO_4 and stir the mixture until it is homogeneous. For every 100 mL of solution add 4.7 g of $KMnO_4$ and stir the mixture overnight. The mixture should still be purple. If it is not, add more $KMnO_4$ until it stays purple for at least 1 h. Decolorize with sodium metabisulfite, make strongly basic with 10 *M* KOH (**Caution!** Exothermic), dilute with H_2O, filter to remove manganese compounds,[4] neutralize the filtrate, check for completeness of destruction, and discard it.

Destruction of MPTP in Aqueous Solution

Dilute the solution with H_2O, if necessary, so that the concentration of MPTP does not exceed 0.25 mg/mL (or 0.4 mg/mL of MPTP tartrate), then add an equal volume of 6 *M* H_2SO_4. For every 100 mL of solution add 4.7 g of $KMnO_4$ and stir the mixture overnight. The mixture should still be purple. If it is not, add more $KMnO_4$ until it stays purple for at least 1 h. Decolorize with sodium metabisulfite, make strongly basic with 10 *M* KOH (**Caution!** Exothermic), dilute with H_2O, filter to remove manganese compounds,[4] neutralize the filtrate, check for completeness of destruction, and discard it.

Destruction of MPTP in Ethanol, Methanol, Dimethyl Sulfoxide, and Acetone

Dilute the solution, if necessary, with the same solvent so that the concentration does not exceed 20 mg/mL. For every 1 mL of solution add 200 mL of 3 *M* H_2SO_4. For every 200 mL of this solution add 9.4 g of $KMnO_4$ and stir the mixture overnight. The mixture should still be purple. If it is not, add more $KMnO_4$ until it stays purple for at least 1 h. Decolorize with sodium metabisulfite, make strongly basic with 10 *M* KOH (**Caution!** Exothermic), dilute with H_2O, filter to remove manganese compounds,[4] neutralize the filtrate, check for completeness of destruction, and discard it.

Destruction of MPTP in Acetonitrile:Aqueous Buffer High-Performance Liquid Chromatography Eluant

Note: This procedure applies **only** to acetonitrile : aqueous buffer high-performance liquid chromatography (HPLC) eluant. It has not been tested for, and probably would not work with methanol : aqueous buffer mixtures.

To each volume of solution add two volumes of 6 *M* H_2SO_4 and for every 100 mL of this solution add 4.7 g of $KMnO_4$ and stir the mixture overnight. The mixture should still be purple. If it is not, add more $KMnO_4$ until it stays purple for at least 1 h. Decolorize with sodium metabisulfite, make strongly basic with 10 *M* KOH (**Caution!** Exothermic), dilute with H_2O, filter to remove manganese compounds,[4] neutralize the filtrate, check for completeness of destruction, and discard it.

Preparation of MPTP Tartrate[5]

Add a solution of MPTP (5 g) in absolute ethanol (10 mL) to a solution of *d*-tartaric acid (4.5 g) in absolute ethanol (20 mL). Wash the original MPTP bottle with absolute ethanol (5 mL) and combine the washing with the other reagents. Warm the mixture until a clear solution forms and then allow it to cool to room temperature. Scratch to induce crystallization and filter the crystals, wash with cold ethanol and dry. Typical yields are 97%.

Analytical Procedures

Analysis may be performed by HPLC using a 5 μ C18 reverse phase column (25 cm). The mobile phase was acetonitrile : buffer (60:40) to which was added 0.1% triethylamine. The buffer consisted of 100 m*M* sodium acetate adjusted to pH 5.6 with acetic acid. The flow rate was 1 mL/min. An ultraviolet (UV) detector operating at 230 nm was used. The reaction mixtures were decolorized with sodium ascorbate, then neutralized by the addition of solid sodium bicarbonate before analysis. This mixture may be analyzed directly but problems were encountered when crystallization of the concentrated salt solution occurred. A better procedure was to place 250 μL of the decolorized and neutralized reaction mixture with 250 μL of acetonitrile in a small tube and stir for 5 min. The upper (acetonitrile) layer was then removed and analyzed by HPLC. Any MPTP that was present was extracted into the acetonitrile layer but the salts were left behind.

Mutagenicity Assays

The mutagenicity assays were carried out as described on page 4 using tester strains TA98, TA100, TA1535, TA1537, and TA1538. The final reaction mixtures [tested at a level corresponding to 62 μg of undegraded material (when bulk quantities were treated) per plate] were not mutagenic. The compounds MPTP, MPTP tartrate, 1-methyl-4-phenylpyridinium chloride, 1-methyl-4-phenylpiperidine hydrochloride, and 4-phenylpiperidine were also tested at concentrations of up to 1 mg per plate (2 mg for MPTP tartrate) and were not mutagenic.

Related Compounds

The initial oxidation product is probably the 1-methyl-4-phenylpyridinium ion. This compound can be detected using the same HPLC system used for MPTP. Since no trace of this ion was seen during the course of these oxidations,[2] this method can probably be used to degrade 1-methyl-4-phenylpyridinium chloride and related compounds containing the same cation.

References

1. Markey, S.P.; Schmuff, N.R. The pharmacology of the Parkinsonian syndrome producing neurotoxin MPTP (1-methyl-4-phenyl-1,2,3,6-tetrahydropyridine) and structurally related compounds. *Med. Res. Rev.* **1986**, *6*, 389–429.
2. Yang, S.C.; Markey, S.P.; Bankiewicz, K.S.; London, W.T.; Lunn, G. Recommended safe practices for using the neurotoxin MPTP in animal experiments. *Lab. Animal Sci.* **1988**, *38*, 563–567.
3. Cohen, G.; Mytilineou, C. Studies on the mechanism of action of 1-methyl-4-phenyl-1,2,3,6-tetrahydropyridine (MPTP). *Life Sci.* **1985**, *36*, 237–242.
4. Lunn, G.; Sansone, E.B.; De Méo, M.; Laget, M.; Castegnaro, M. Potassium permanganate can be used for degrading hazardous compounds. *Am. Ind. Hyg. Assoc. J.*, **1994**, *55*, 167–171.
5. Pitts, S.M.; Markey, S.P.; Murphy, D.L.; Weisz, A.; Lunn, G. Recommended practices for the safe handling of MPTP. In *MPTP—A Neurotoxin Producing a Parkinsonian Syndrome*; Markey, S.P., Castagnoli, Jr., N., Trevor, A., Kopin, I., Eds., Academic Press: Orlando, 1986; pp. 703–716.

MISCELLANEOUS PHARMACEUTICALS

CAUTION! Refer to safety considerations section on page 7 before starting any of these procedures.

In this monograph we will report some recent developments, including preliminary results from a study on the degradation of some drugs organized by the International Agency for Research on Cancer (IARC), as well as results obtained from an investigation on the photolytic degradation of some drugs. Additionally, we will discuss the use of Amberlite resins for decontaminating aqueous solutions of some drugs.

The following drugs are considered:

Name	Reference	Structure	Registry Number
Amoxicillin trihydrate	1	**I.3H_2O**	[61336-70-7]
Ampicillin	2	**II**	[69-53-4]
BCNU [*N,N'*-Bis(2-chloroethyl)-*N*-nitrosourea]	3	**III**	[154-93-8]

Name	Reference	Structure	Registry Number
Bleomycin sulfate	4	**IV.HSO_4^-**	[9041-93-4]
CCNU [*N*-(2-Chloroethyl)-*N'*-cyclohexyl-*N*-nitrosourea]	5	**V**	[13010-47-4]
Cephalothin (sodium salt)	6	**VI (Na salt)**	[58-71-9]
Dacarbazine	7	**VII**	[4342-03-4]
Etoposide	8	**VIII**	[33419-42-0]
Metronidazole	9	**IX**	[443-48-1]
Norethindrone	10	**X**	[68-22-4]
Streptozotocin	11	**XI**	[18883-66-4]
Sulfamethoxazole	12,13	**XII**	[723-46-6]
Teniposide	14	**XIII**	[29767-20-2]
Trimethoprim	13,15	**XIV**	[738-70-5]
Verapamil hydrochloride	16	**XV.HCl**	[152-11-4]

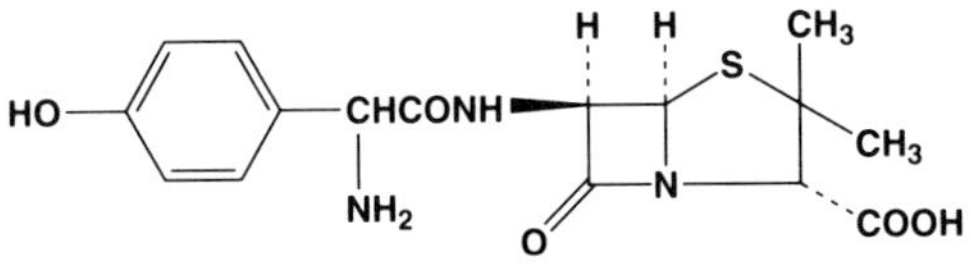

I

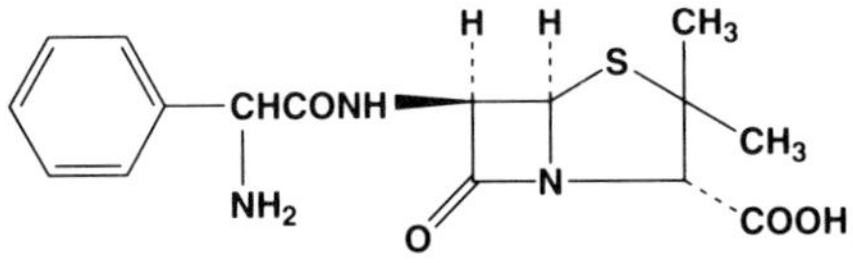

II

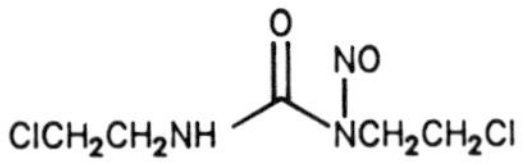

III

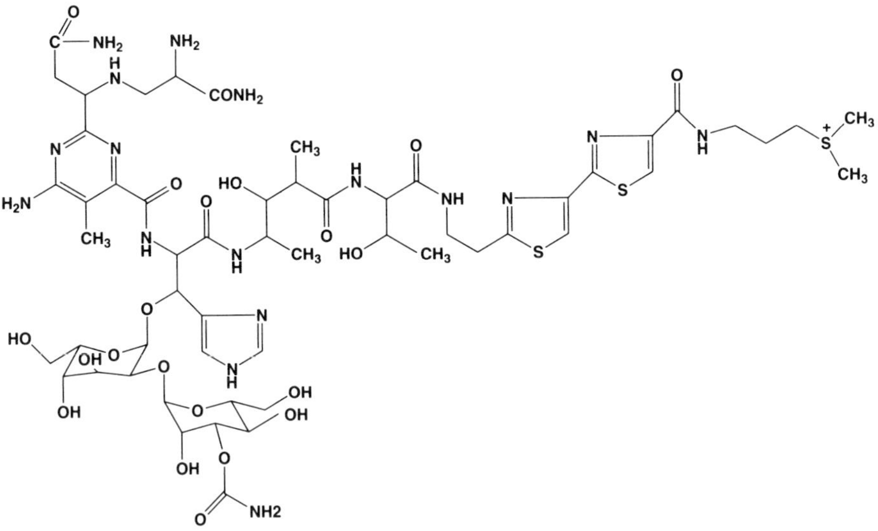

IV

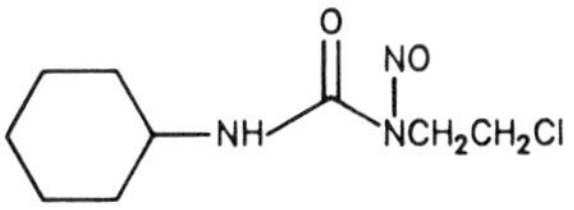

V

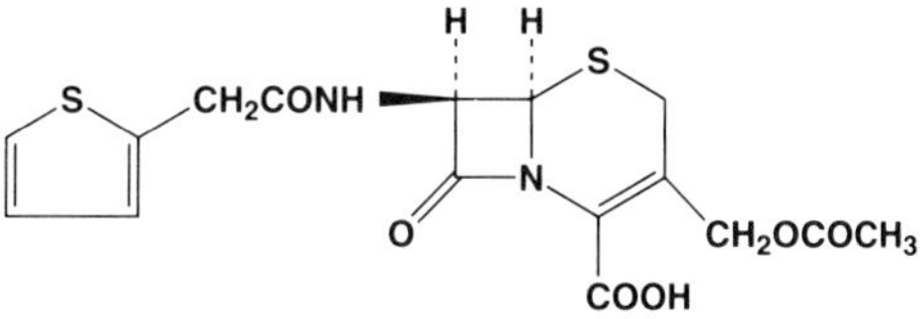

VI

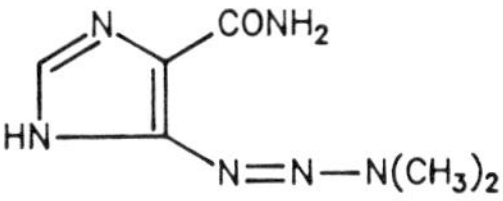

VII

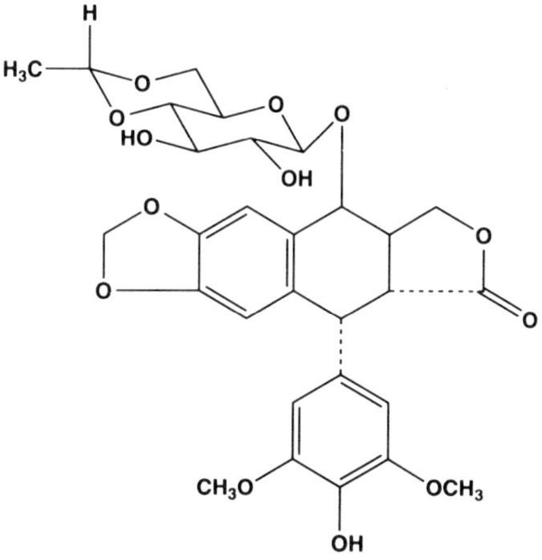

VIII

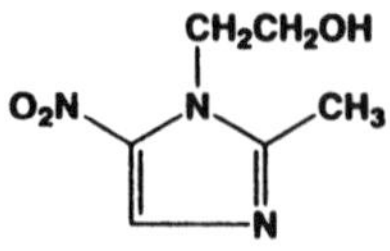

IX 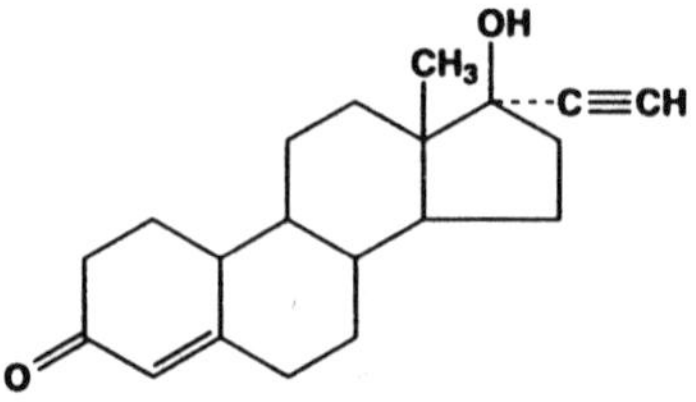 X 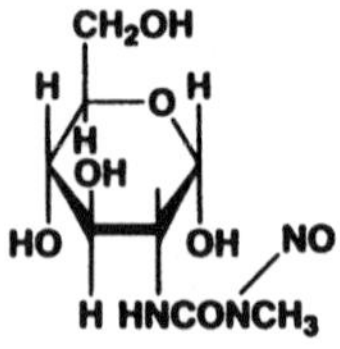 XI

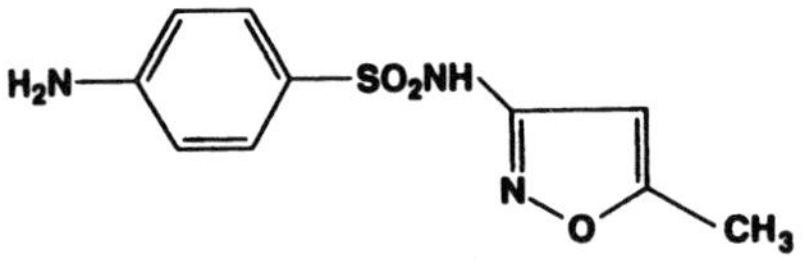

XII 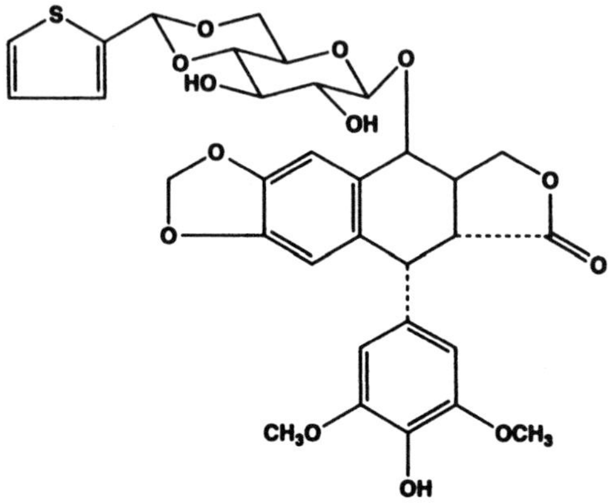 XIII

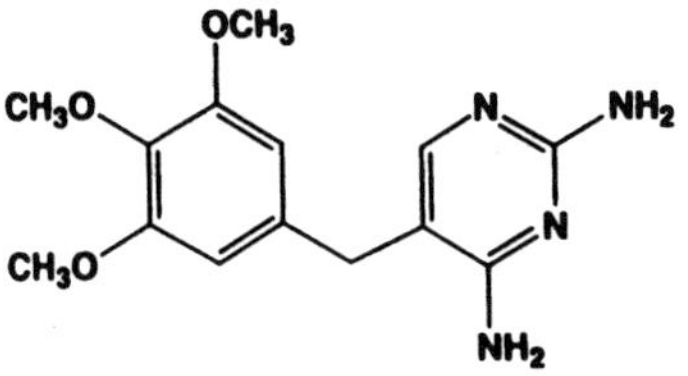

XIV 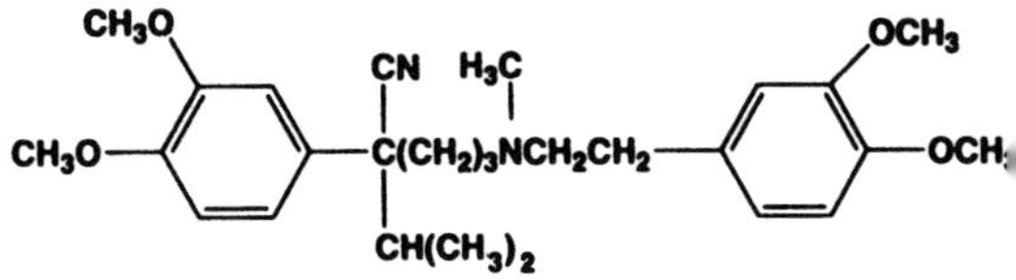 XV

The compounds BCNU, bleomycin, CCNU, dacarbazine, etoposide, streptozotocin, and teniposide are antineoplastics; amoxicillin, ampicillin, sulfamethoxazole, and trimethoprim are antibacterials; metronidazole is an antiprotozoal; norethindrone is used in oral contraceptives; and verapamil is an antianginal.

The compounds BCNU,[17–19] CCNU,[19–21] and streptozotocin[22] are carcinogenic in animals and the IARC has stated that BCNU,[17,19] CCNU,[20] and streptozotocin[22] should be regarded as presenting a carcinogenic risk to humans. The compounds BCNU, CCNU, and streptozotocin are mutagenic.[23] Dacarbazine is carcinogenic in laboratory animals[24–27] and probably carcinogenic to humans.[25] Dacarbazine,[28] etoposide,[29] metronida-

Table 1 Destruction of Pharmaceuticals Using Sodium Hypochlorite

Compound	Solvent	Initial Concentration (mg/mL)	Amount Remaining (%)
BCNU	NaCl (0.9% w/v)	0.5	<0.13
BCNU	Glucose (5% w/v)	0.5	<0.13
Bleomycin	H_2O	1.78 (3 units/mL)	<1.1
CCNU	CH_3OH	5	<0.02
CCNU	CH_3OH	1	<0.1
Dacarbazine	NaCl + Buffer[a]	4	<0.005
Streptozotocin	NaCl (0.9% w/v)	14	<0.07
Streptozotocin	Glucose (5% w/v)	14	<0.07
Sulfamethoxazole	CH_3OH	0.4	<0.1

[a] A 2-mL aliquot of dacarbazine (10 mg/mL in H_2O containing citric acid (10 mg/mL) and mannitol (5 mg/mL)) was added to 3 mL of saline.

zole,[30] and teniposide[29] are mutagenic. Metronidazole,[31,32] norethindrone,[33,34] and sulfamethoxazole[35] are carcinogenic in experimental animals.

Principles of Destruction

These compounds can be degraded using sodium hypochlorite (NaOCl) solution,[29,36] 30% hydrogen peroxide (H_2O_2) solution,[36] potassium permanganate ($KMnO_4$),[29] and photolysis with and without the presence of H_2O_2.[36] Each compound cannot be degraded by each procedure (see Tables 1–4 for details). The products of these reactions are not known but deg-

Table 2 Destruction of Pharmaceuticals Using Hydrogen Peroxide

Compound	Solvent	Concentration (mg/mL)	Amount Remaining (%)
Ampicillin	H_2O	5	<0.2
BCNU	NaCl (0.9% w/v)	0.5	<0.27
BCNU	Glucose (5% w/v)	0.5	<0.13
CCNU	CH_3OH	5	<0.04
CCNU	CH_3OH	1	<0.1
Streptozotocin	NaCl (0.9% w/v)	14	<0.28
Streptozotocin	Glucose (5% w/v)	14	<0.28

Table 3 Destruction of Pharmaceuticals by Photolysis[a]

Compound	Solvent	Concn. (mg/mL)	Reaction Time (h)	H_2O_2 Added (μL/mL)	Amount Remaining (%) Without H_2O_2	With H_2O_2
Amoxicillin	H_2O	1	2	2.75	1.2	<0.05
Ampicillin	H_2O	1	1	3	5.5	<0.2
BCNU	H_2O	0.1	1	0.48	<0.5	<0.5
Bleomycin	H_2O	1	1	7[b]	1.3	<1
CCNU	CH_3OH	10	1	44	0.01	<0.02
Cephalothin	H_2O	1	1	2.5	<0.05	<0.05
Dacarbazine	Buffer[c]	0.1	1	1.25[d]	<0.2	<0.2
Dacarbazine	NaCl (0.9% w/v)	0.1	1	1.25[d]	<0.2	<0.2
Dacarbazine	H_2O	0.1	1	1.25[d]	<0.2	<0.2
Metronidazole	Buffer[e]	5	4	30	16	<0.0005
Norethindrone	CH_3OH	0.1	1	34[f]	0.67	<0.1
Streptozotocin	NaCl (0.9% w/v)	14	2	48	<0.36	<0.36
Streptozotocin	Glucose (5% w/v)	14	2	48	<0.36	<0.36
Sulfamethoxazole	H_2O	0.1	1	0.4	<0.2	<0.2
Verapamil	H_2O	1	1	2.25	<0.1	<0.1

[a] All reactions were carried out in an all quartz apparatus with a 200-W medium-pressure mercury lamp. When H_2O_2 was used 10 moles were employed for each mole of compound present (except where shown).

[b] One hundred moles of H_2O_2 per mole of bleomycin.

[c] The dacarbazine buffer contained 10 mg/mL of citric acid and 5 mg/mL of mannitol.

[d] Twenty two moles of H_2O_2 per mole of dacarbazine.

[e] Metronidazole buffer was NaCl (7.9 mg/mL), NaH_2PO_4 (0.45 mg/mL), and citric acid (0.23 mg/mL).

[f] One thousand moles of H_2O_2 per mole of norethindrone.

radation with NaOCl and 30% H_2O_2 gave a number of unidentified peaks in the high-pressure liquid chromatography (HPLC) chromatogram. The chromatograms obtained after photolysis did not have many unidentified peaks although polar oxidized compounds would be expected to elute with the unretained peak. The final reaction mixtures were not mutagenic. Some of these compounds can also be removed from aqueous solution with Amberlite XAD-16 resin.

Table 4 Removal of Pharmaceuticals from Solution Using Amberlite XAD-16 Resin[a]

Compound	Solvent	Amount Remaining (%)	Volume Decontaminated per Gram of Resin (mL)
Ampicillin	H_2O	<1	20
BCNU	H_2O	<0.17	40
BCNU	NaCl (0.9% w/v)	<0.16	40
BCNU	Glucose (5% w/v)	<0.16	40
Bleomycin	H_2O	<5	20
CCNU	H_2O	<0.5	200
Norethindrone	H_2O	<1.1	80
Streptozotocin	H_2O	<0.5	20
Streptozotocin	NaCl (0.9% w/v)	<0.5	20
Streptozotocin	Glucose (5% w/v)	<0.5	20
Trimethoprim	H_2O	<0.2	40
Verapamil	H_2O	<0.5	80

[a] All solutions were 100 μg/mL except for norethindrone, which was a saturated aqueous solution of 9 μg/mL.

Destruction Procedures

Destruction Using Sodium Hypochlorite

1. Take up bulk quantities of the compound in the indicated solvent at the concentration shown in Table 1.[36] If necessary, dilute solutions so that the indicated concentration is not exceeded. Add an equal volume of 5.25% NaOCl solution (Clorox bleach) and allow to stand for 1 h (4 h for CCNU). Use fresh NaOCl solution (see assay procedure below). At the end of this time neutralize excess oxidizing power by adding sodium metabisulfite, check for completeness of degradation, and discard it.

2. Treat injectable solutions of bleomycin, etoposide, and teniposide by adding 7.0 mL of 5.25% NaOCl solution per milligram of bleomycin, 20.1 mL of 5.25% NaOCl solution per milligram of etoposide, or 28.0 mL of 5.25% NaOCl solution per milligram of teniposide.[29] Use fresh NaOCl solution (see assay procedure below). After the reaction mixture has become homogeneous, neutralize excess oxidizing power by adding sodium metabisulfite, check for completeness of degradation, and discard it.

Table 5 Analytical Conditions

Compound	Mobile Phase	UV (nm)	Retention Time (min)	LOD[a] (ppm)
Amoxicillin	MeCN:5.5 m*M* octanesulfonic acid + 20 m*M* citric acid adjusted to pH 3 18:82	236	6.2	0.5
Ampicillin	MeCN:5.5 m*M* octanesulfonic acid + 20 m*M* citric acid adjusted to pH 3 23:77	230	6.8	1
BCNU	MeOH:0.4 g/L $(NH_4)H_2PO_4$ 55:45	228	8.0	0.16
Bleomycin	MeCN:5.5 m*M* octanesulfonic acid + 20 m*M* citric acid adjusted to pH 3 25:75	230	6.2	10
CCNU	MeOH:0.4 g/L $(NH_4)H_2PO_4$ 75:25	228	6.4	0.5
Cephalothin	MeCN:5.5 m*M* octanesulfonic acid + 20 m*M* citric acid adjusted to pH 3 28:72	245	7.2	0.5
Dacarbazine	MeOH:0.4 g/L $(NH_4)H_2PO_4$ + 0.1% Et_3N 30:70	325	6.4	0.5
Etoposide	MeOH:H_2O 50:50	254	6.9	[b]
Metronidazole	MeCN:45 m*M* KH_2PO_4 (pH 4.5) 15:85	325	7.1	0.05
Norethindrone	MeCN:100 m*M* KH_2PO_4 adjusted to pH 7 50:50	254	8.7	0.1
Streptozotocin	20 m*M* KH_2PO_4	230	9.8	0.5
Sulfamethoxazole	MeCN:5.5 m*M* octanesulfonic acid + 20 m*M* citric acid adjusted to pH 3 35:65	270	6.6	0.2
Teniposide	MeOH:H_2O 60:40	254	4.5	[b]
Trimethoprim	MeCN:5.5 m*M* octanesulfonic acid + 20 m*M* citric acid adjusted to pH 3 30:70	230	6.7	0.2
Verapamil	MeCN:5.5 m*M* octanesulfonic acid + 20 m*M* citric acid adjusted to pH 3 65:35	230	7.1	1

[a] The limits of detection (LOD) shown above represent the best values obtained. Frequently, interferences from other compounds in the reaction mixtures meant that the limits of detection were higher.

[b] Not reported.

Destruction Using 30% Hydrogen Peroxide

Caution! 30% H_2O_2 is a corrosive irritant and a strong oxidizer.

Take up bulk quantities of the compound in the indicated solvent at the concentration shown in Table 2. If necessary, dilute solutions so that the indicated concentration is not exceeded. Add an equal volume of 30% H_2O_2 and allow to stand for 1 h. At the end of this time neutralize excess oxidizing power by adding sodium metabisulfite (**Caution!** Exothermic), check for completeness of degradation, and discard it.

Destruction Using Potassium Permanganate

Treat injectable solutions of bleomycin, etoposide, and teniposide by adding 0.34 mg $KMnO_4$ per milligram of bleomycin, 2.42 mg of $KMnO_4$ per milligram of etoposide, or 2.65 mg of $KMnO_4$ per milligram of teniposide.[29] After the reaction mixture has become homogeneous, add sodium bisulfite solution (1%) until the purple color is discharged. Centrifuge to remove manganese salts, check for completeness of destruction, and discard it.

Destruction Using Photolysis

Caution! Ultraviolet (UV) radiation is harmful to the skin and eyes. Before turning on the lamp cover the apparatus with aluminum foil or other light absorbing material. Turn off the light before removing the covering.

Take up bulk quantities of the compound in the indicated solvent at the concentration shown in Table 3. If necessary, dilute solutions so that the indicated concentration is not exceeded. If necessary, add the required amount of 30% H_2O_2. Pass a slow stream of air through the solution and photolyze in an all quartz apparatus using a 200-W medium-pressure mercury lamp (Ace Glass, Inc., Vineland, NJ, Cat. No. 7825-32 or equivalent) with H_2O cooling for the time shown. At the end of this time neutralize excess oxidizing power by adding sodium metabisulfite, check for completeness of degradation, and discard it.

Photolysis of Concentrated Solutions of Dacarbazine

Caution! Ultraviolet radiation is harmful to the skin and eyes. Before turning on the lamp cover the apparatus with aluminum foil or other light absorbing material. Turn off the light before removing the covering.

Take up bulk quantities of dacarbazine in H_2O containing 10 mg/mL of citric acid and 5 mg/mL of mannitol so that the concentration of dacarbazine does not exceed 10 mg/mL. If necessary, dilute solutions of dacarbazine so that the concentration does not exceed 10 mg/mL. Add 56 μL of 30% H_2O_2 per milliliter of solution. Pass a slow stream of air through the solution and photolyze in an all quartz apparatus using a 200-W medium-pressure mercury lamp (Ace Glass, Inc., Vineland, NJ, Cat. No. 7825-32 or equivalent) with H_2O cooling for 1 h. At the end of this time filter to remove dark red insoluble products that will otherwise stop thc reaction from going to completion. Photolyze the filtrate for 1 h, neutralize excess oxidizing power by adding sodium metabisulfite, check for completeness of degradation, and discard it. (Destruction was >99.9%.)

Decontamination of Aqueous Solutions of Pharmaceuticals

If necessary dilute so that the concentration of the drug does not exceed 100 μg/mL. Add the required quantity of Amberlite XAD-16 resin (see Table 4) and stir for 18 h, filter, check the filtrate for completeness of decontamination, and discard it.

Analytical Procedures

The drugs were determined by reverse phase HPLC with UV detection using the conditions shown in Table 5.[29,36] The flow rate was 1 mL/min except for etoposide (1.5 mL/min) and teniposide (2 mL/min). Before analysis the oxidizing power of the solution should be removed by adding sodium metabisulfite. Afterwards add a few drops of the reaction mixture to an equal volume of 10% (w/v) potassium iodide (KI) solution, then acidify with a drop of 1 *M* hydrochloric acid and add a drop of starch solution as an indicator. The deep blue color of the starch–iodine complex indicates that excess oxidant is present and so more sodium metabisulfite should be added.

Mutagenicity Assays

The reaction mixtures obtained after treatment with NaOCl and H_2O_2 and after photolysis were tested for mutagenicity using tester strains TA97a, TA100, and TA102 of *Salmonella typhimurium*. In all cases in which destruction was complete no mutagenic activity was found.[36] In some cases tester strain TA98 was used but in many cases toxicity problems, which also affected the controls, made these data unusable. Reaction mixtures obtained from the degradation of etoposide and teniposide were assayed using tester strain UTH8413 without S9 activation.[29]

Related Compounds

In principle, these techniques could be applied to other drugs but it was found in the current study that each technique was not applicable to every compound. For example, dacarbazine was not degraded by 30% H_2O_2 and it could not be removed from solution by Amberlite XAD-16 resin. Thus each procedure should be thoroughly validated before it is placed into routine use. Spills of the cephalosporins cefsulodin, cefmenoxime, and cefadroxil can be decontaminated with NaOCl.[37] The mutagenicity of the decontaminated reaction mixtures was not investigated. See also monographs on Antineoplastic Alkylating Agents; Cisplatin; Cycloserine; Dichloromethotrexate, Vincristine, and Vinblastine; Doxorubicin and Daunorubicin; Drugs Containing Hydrazine and Triazene Groups; Methotrexate; Mitomycin C; Nitrosourea Drugs; and 6-Thioguanine and 6-Mercaptopurine.

Assay of Sodium Hypochlorite Solution

Sodium hypochlorite solutions tend to deteriorate with time, so they should be periodically checked for the amount of active chlorine they contain. Pipette 10 mL of the NaOCl solution into a 100-mL volumetric flask and fill it to the mark with distilled H_2O. Pipette 10 mL of this solution into a conical flask containing 50 mL of distilled H_2O, 1 g of KI, and 12.5 mL of 2 *M* acetic acid. Titrate this solution against a 0.1 *N* sodium thiosulfate solution using starch as an indicator. Each 1 mL of the sodium thiosulfate solution corresponds to 3.545 mg of active chlorine. Commercially available NaOCl solution (Clorox bleach) contains 5.25% sodium hypochlorite and should contain about 45–50 g of active chlorine per liter.

References

1. Other names for this compound are [2S-[2α,5α,6β(S*)]]-6-[[amino(4-hydroxyphenyl)acetyl]amino]-3,3-dimethyl-7-oxo-4-thia-1-azabicyclo[3.2.O]heptane-2-carboxylic acid, (−)-6-[2-amino-2-(*p*-hydroxyphenyl)acetamido]-3,3-dimethyl-7-oxo-4-thia-1-azabicyclo[3.2.O]heptane-2-carboxylic acid, 6-[D(−)-α-amino-*p*-hydroxyphenylacetamido]penicillanic acid, α-amino-*p*-hydroxybenzylpenicillin, 6-(*p*-hydroxy-α-aminophenylacetamido)penicillanic acid, *p*-hydroxyampicillin, amoxycillin, AMPC, Alfamox, Almodan, Amocilline, Amolin, Amopenixin, Amoxi, Amoxipen, Anemolin, Aspenil, Betamox, Bristamox, Cabermox, Cuxacillin, Delacillin, Efpenix, Grinsil, Ibiamox, Ospamox, Optium, Piramox, Simoxil, and Sumox. The registry number for the anhydrous form is 26787-78-0. Other names for the trihydrate are BRL 2333, Agram, Amodex, Amoxibiotic, Amoxidal, Amoxidin, Amoxil, Amoxillat, Amoxi-Wolff, Amoxypen, Ardine, AX 250, Clamoxyl, Dura AX, Hiconcil, Infectomycin, Larocin, Larotid, Moxal, Moxaline, Neamoxyl. Pamocil, Pasetocin, Penamox, Polymox, Raylina, Robamox, Sawucillin. Sigamopen, Silamox, Trimox, Uro-Clamoxyl, Utimox, Widecillin, Wymox, and Zamocillin.
2. Other names for this compound are 6-[(aminophenylacetyl)amino]-3,3-dimethyl-7-oxo-4-thia-1-azabicyclo[3.2.O]heptane-2-carboxylic acid, 6-[D(−)-α-aminophenylacetamido]penicillanic acid, D-(−)-α-aminobenzylpenicillin, ampicillin A, Ay 6108, BRL 1341, P 5O, Adobacillin, Alpen, Amblosin, Amfipen, Amipenix S, Ampi-Bol, Ampicin, Ampicina, Ampilar, Ampimed, Ampipenin, Ampi-Tablinen, Amplisom, Amplital, Ampy-

Penyl, Austrapen, Binotal, Bonapicillin, Britacil, Copharcilin, Doktacillin, Grampenil, Guicitrina, Marisilan, Nuvapen, Pen-Bristol, Penbritin, Penbrock, Pénicline, Penstabil, Pentrex, Pentrexyl, Polycillin, Ponecil, QI Damp, Rosampline, Synpenin, Tokiocillin, Totacillin, Totalciclina, Totapen, Ultrabion, Viccillin, and Suractin. Another name for the monohydrate [32388-53-7] is Redicilin. Another name for the potassium salt [23277-71-6] is Suractin. Other names for the trihydrate [7177-48-2] are Acillin, Amcap, Amcill, Amperil, Ampichel, Ampikel. Ampinova, Amplin, Cetampin, Cymbi, Divercillin, Lifeampil, Morepen, Pen A, Pensyn, Princillin, Principen, Ro-Ampen, Trafarbiot, Ukopen, and Vidopen. Other names for the anhydrous [69-53-4] form are Ampicillin B, Omnipen, and Orbicilina. Other names for the sodium salt [69-52-3] are sodium ampicillin, Alpen-N, Amcill-S, Ampilag, Cilleral, Domicillin, Omnipen-N, Pen A/N, Penbritin-S, Penialmen, Polycillin-N, and Principen/N.

3. Other names for this compound are Carmustine, BiCNU, Nitrumon, NSC-409962, SK 27702, NCI-C04733, SRI 1720, FDA 0345, Becenun, and Carmubris.
4. Other names for this compound are Blenoxane and Blexane. Other names for the free base [11056-06-7] are Bleo, Bleocin, BLM, and NSC-125066. Bleomycin A_1 [58995-26-9] is N^1-[3-(methylsulfinyl)propyl]bleomycinamide. Bleomycin A_2 [11116-31-7] is N^1-[3-(dimethylsulfonio)propyl]bleomycinamide. Bleomycin A_5 [11116-32-8] is N^1-[3-[(4-aminobutyl)amino]propyl]bleomycinamide. Bleomycin A_6 [37293-17-7] is N^1-[3-[[4-[(3-aminopropyl)amino]butyl]amino]propyl]bleomycinamide. Bleomycin B_1 [41138-54-9] is bleomycinamide. Bleomycin B_2 [9060-10-0] is N^1-[4-[(aminoiminomethyl)amino]butyl]bleomycinamide. Bleomycin B_4 [9060-11-1] is N^1-[4-[(aminoiminomethyl)[4-[(aminoiminomethyl)amino]butyl]amino]butyl]bleomycinamide. Bleomycin B_6 [73666-80-5] is N^1-(20-amino-6,13,20-triimino-5,7,12,14,19-pentaazaeicosyl)bleomycinamide. Other variants are known.
5. Other names for this compound are Lomustine, 1-(2-chloroethyl)-3-cyclohexyl-1-nitrosourea, Belustine, Cecenu, CeeNU, CiNU, ICIG 1109, NCI-C04740, NSC-79037, and RB 1509.
6. Other names for this compound are 3-[(acetyloxy)methyl]-8-oxo-7-[(2-thienylacetyl)amino]-5-thia-1-azabicyclo[4.2.0]oct-2-ene-2-carboxylic acid, 3-(hydroxymethyl)-8-oxo-7-[2-(2-thienyl)acetamido]-5-thia-1-azabicyclo[4.2.0]oct-2-ene-2-carboxylic acid acetate, 7-(2-thienylacetamido)cephalosporanic acid, and 7-(thiophene-2-acetamido)cephalosporanic acid. The registry number of the free acid is [153-61-7]. Other names for the sodium salt are Averon-1, Cefalotin, Cephation, Ceporacin, Cepovenin, Chephalotin, Coaxin, Keflin, Lospoven, Microtin, Synclotin, and Toricelocin.
7. Other names for this compound are 5-(3,3-dimethyl-1-triazenyl)-1*H*-imidazole-4-carboxamide, 4-(3,3-dimethyl-1-triazeno)imidazole-5-carboxamide, 4-(5)-(3,3-dimethyl-1-triazeno)imidazole-5(4)-carboxamide, 5-(3,3-dimethyl-1-triazeno)imidazole-4-carboxamide, DIC, DTIC, DTIC-Dome, Deticene, NSC-45388, Dacatic, and NCI-C04717.
8. Other names for this compound are 9[(4,6-*O*-ethylidene-β-D-glucopyranosyl)oxy]-5,8,8a,9-tetrahydro-5-(4-hydroxy-3,5-dimethoxyphenyl)furo[3′,4′:6,7]naphtho[2,3-d]-1,3-dioxol-6(5a*H*)-one, 4′-demethylepipodophyllotoxin 9-[4,6-*O*-ethylidene-β-D-glucopyranoside], demethylepipodophyllotoxin ethylidene glucoside, EPEG, NK 171, NSC-141540, VP-16-213, Lastet, and Vepesid.
9. Other names for this compound are 2-methyl-5-nitroimidazole-l-ethanol, 1-(2-hydroxyethyl)-2-methyl-5-nitroimidazole, 1-(β-ethylol)-2-methyl-5-nitro-3-azapyrrole, Bayer 5630, RP 8823, Arilin, Clont, Cont, Danizol, Deflamon, Elyzol, Flagyl, Fossyol, Gineflavir, Klion, MetroGel, Metrolag, Metrolyl, Nidazol, Orvagil, Rathimed N, Sanatri-

chom, Trichazol, Trichocide, Tricho Cordes, Tricho-Gynaedron, Tricocet, Trivazol, Vagilen, Vagimid, and Zadstat. Other names for the hydrochloride [69198-10-3] are SC-32642, and Flagyl I.V.

10. Other names for this compound are (17α)-17-hydroxy-19-norpregn-4-en-20-yn-3-one, 19-nor-17α-ethynyltestosterone, 17α-ethynyl-19-nortestosterone, 19-nor-17α-ethynyl-17β-hydroxy-4-androsten-3-one, 19-nor-17α-ethynylandrosten-17β-ol-3-one, anhydrohydroxynorprogesterone, 19-norethisterone, norpregneninolone, "mini-pill", Conceplan, Conludag, Micronor, Micronett, Mini-Pe, Norcolut, Noriday, Micronovum, Milligynon, Nor-QD, Norluten, Ostro-Primolut, Primolut N, Norluton, Norlutin, Noralutin, and Utovlan.
11. Other names for this compound are 2-deoxy-2-([(methylnitrosoamino)carbonyl]amino)-D-glucopyranose, streptozocin, 2-deoxy-2-(3-methyl-3-nitrosoureido)-D-glucopyranose, *N*-D-glucosyl-(2)-*N*′-nitrosomethylurea, Zanosar, NCI-C03167, NSC-85998, STR, STZ, STRZ, and U-9889.
12. Other names for this compound are 4-amino-*N*-(5-methyl-3-isoxazolyl)benzensulfonamide, N^1-(5-methyl-3-isoxazolyl)sulfanilamide, 5-methyl-3-sulfanilamidoisoxazole, 3-sulfanilamido-5-methylisoxazole, 3-(*p*-aminophenylsulfonamido)-5-methylisoxazole, sulfisomezole, sulfamethylisoxazole, sulfamethoxizole, Gantanol, and Sinomin.
13. Other names for sulfamethoxazole/trimethoprim (usually 5:1) mixtures are co-trimoxazoe, Abacin, Apo-Sulfatrim, Bactramin, Bactrim, Bactromin, Baktar, Chemotrim, Comox, Cotrim-Puren, Drylin, Duratrimet, Eltrianyl, Eusaprim, Fectrim, Gantaprim, Gantrim, Helveprim, Imexim, Kepinol, Laratrim, Linaris, Microtrim, Momentol, Nopil, Omsat, Septra, Septrim, Sigaprim, Sulfotrim, Sulfotrimin, Sulprim, Sumetrolim, Supracombin, Suprim, Tacumil, Teleprim, Thiocuran, TMS 480, Trigonyl, Trimesulf, Trimforte, Uroplus, and Uro-Septra.
14. Other names for this compound are [5R-[5α,5aβ,8aα,9β(R*)]]-5,8,8a,9-tetrahydro-5-(4-hydroxy-3,5-dimethoxyphenyl)-9-[[4,6-*O*-(2-thienylmethylene)-β-D-glucopyranosyl]-oxy]furo[3′,4′:6,7]naphtho[2,3-d]-1,3-dioxol-6(5a*H*)-one, 4′-demethylepipodophyllotoxin 9-(4,6-*O*-2-thenylidene-β-D-glucopyranoside), 4′-demethylepipodophyllotoxin-β-D-thenylidene glucoside, ETP, EPT, PTG, Veham-Sandoz, NSC-122819, VM-26, Vehem, and Vumon.
15. Other names for this compound are 5-[(3,4,5-trimethoxyphenyl)methyl]-2,4-pyrimidinediamine, 2,4-diamino-5-(3,4,5-trimethoxybenzyl)pyrimidine, Monotrim, Proloprim, Syraprim, Tiempe, Trimanyl, Trimogal, Trimopan, Trimpex, Uretrim, and Wellcoprim.
16. Other names for this compound are α-[3-[[2-(3,4-dimethoxyphenyl)ethyl]methylamino]-propyl]-3,4-dimethoxy-α-(1-methylethyl)benzeneacetonitrile, 5-[(3,4-dimethoxyphenethyl)methylamino]-2-(3,4-dimethoxyphenyl)-2-isopropylvaleronitrile, α-isopropyl-α-[(*N*-methyl-*N*-homoveratryl)-γ-aminopropyl]-3,4-dimethoxyphenylacetonitrile, and iproveratril. The registry number of the free base is [52-53-9]. Other names for the hydrochloride are Berkatens, Calan, Cardiagutt, Cardibeltin, Cordilox, Dignover, Drosteakard, Geangin, Isoptin, Securon, Univer, Vasolan, Veramex, Veraptin, Verelan, and Verexamil.
17. International Agency for Research on Cancer. *IARC Monographs on the Evaluation of the Carcinogenic Risk of Chemicals to Humans.* Volume 26, *Some Antineoplastic and Immunosuppressive Agents*; International Agency for Research on Cancer: Lyon, 1981; pp. 79–95.
18. International Agency for Research on Cancer. *IARC Monographs on the Evaluation of*

the Carcinogenic Risk of Chemicals to Humans, Supplement No. 4, Chemicals, Industrial Processes and Industries Associated with Cancer in Humans. IARC Monographs, Volumes 1 to 29; International Agency for Research on Cancer: Lyon, 1982; pp. 63–64.

19. International Agency for Research on Cancer. *IARC Monographs on the Evaluation of the Carcinogenic Risk of Chemicals to Humans, Supplement No. 7, Overall Evaluations of Carcinogenicity: An Updating of* IARC Monographs *Volumes 1 to 42*; International Agency for Research on Cancer: Lyon, 1987; pp. 150–152.
20. Reference 15, pp. 137–149.
21. Reference 16, pp. 83–84.
22. International Agency for Research on Cancer. *IARC Monographs on the Evaluation of the Carcinogenic Risk of Chemicals to Humans.* Volume 17, *Some* N-*Nitroso Compounds*; International Agency for Research on Cancer: Lyon, 1978; pp. 337–349.
23. Lunn, G.; Sansone, E.B.; Andrews, A.W.; Hellwig, L.C. Degradation and disposal of some antineoplastic drugs. *J. Pharm. Sci.* **1989**, *78*, 652–659.
24. Skibba, J.L.; Ertürk, E.; Bryan, G.T. Induction of thymic lymphosarcoma and mammary adenocarcinomas in rats by oral administration of the antitumor agent, 4(5)-(3,3-dimethyl-1-triazeno)imidazole-5(4)-carboxamide. *Cancer* **1970**, *26*, 1000–1005.
25. Reference 15, pp. 203–215.
26. Reference 16, pp. 103–104.
27. Reference 17, pp. 184–185.
28. Lunn, G.; Sansone, E.B. Reductive destruction of dacarbazine, procarbazine hydrochloride, isoniazid, and iproniazid. *Am. J. Hosp. Pharm.* **1987**, *44*, 2519–2524.
29. Benvenuto, J.A.; Connor, T.H.; Monteith, D.K.; Laidlaw, J.L.; Adams, S.C.; Matney, T.S.; Theiss, J.C. Degradation and inactivation of antitumor drugs. *J. Pharm. Sci.* **1993**, *82*, 988–991.
30. Melo, M.E.; Ferreira, L.C. Screening the mutagenic activities of commonly used antiparasite drugs by the Simultest, a simplified Salmonella/microsome plate incorporation assay. *Rev. Inst. Med. Trop. Sao Paolo* **1990**, *32*, 269–274.
31. International Agency for Research on Cancer. *IARC Monographs on the Evaluation of the Carcinogenic Risk of Chemicals to Man.* Volume 13, *Some Miscellaneous Pharmaceutical Substances*; International Agency for Research on Cancer: Lyon, 1977; pp. 113–122.
32. Reference 17, pp. 250–252.
33. International Agency for Research on Cancer. *IARC Monographs on the Evaluation of the Carcinogenic Risk of Chemicals to Man.* Volume 6, *Sex Hormones*; International Agency for Research on Cancer: Lyon, 1974; pp. 179–189.
34. International Agency for Research on Cancer. *IARC Monographs on the Evaluation of the Carcinogenic Risk of Chemicals to Humans.* Volume 21, *Sex Hormones (II)*; International Agency for Research on Cancer: Lyon, 1979; pp. 441–460.
35. International Agency for Research on Cancer. *IARC Monographs on the Evaluation of the Carcinogenic Risk of Chemicals to Humans.* Volume 24, *Some Pharmaceutical Drugs*; International Agency for Research on Cancer: Lyon, 1980; pp. 285–295.
36. Lunn, G. Unpublished observations.
37. Gorski, R.J.; Plasz, A.C.; Elrod, L., Jr.; Yoder, J.; White, L.B. Determination of cefsulodin, cefmenoxime, and cefadroxil as residues on surfaces. *Pharm. Res.*, **1991**, *8*, 1525–1527.

MITOMYCIN C

CAUTION! Refer to safety considerations section on page 7 before starting any of these procedures.

Mitomycin C (**I**) (mp >360°C) [50-07-7][1] is a solid soluble in H_2O, methanol, acetone, butyl acetate, and cyclohexanone. Mitomycin C is carcinogenic in animals,[2] mutagenic,[3] and a teratogen[4] and is used as an antineoplastic drug.

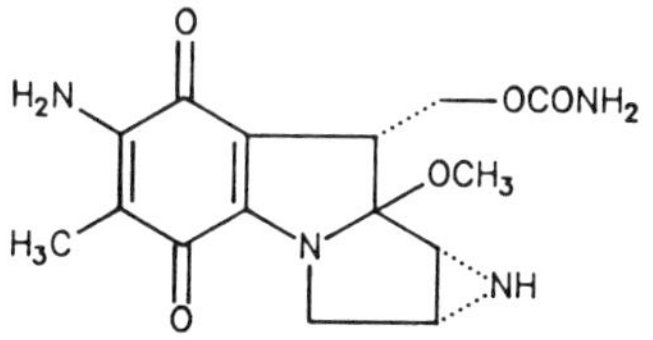

O
H_2N OCONH$_2$
OCH$_3$
H_3C N NH
O

I

Principle of Destruction

Mitomycin C in urine can be destroyed by using sodium hypochlorite (NaOCl)[3,5] or potassium permanganate ($KMnO_4$).[5]

Destruction Procedure

Destruction of Mitomycin C in Aqueous Solution[5]

1. Dilute, if necessary, so that the concentration of Mitomycin C does not exceed 0.5 mg/mL. For each 10 mL of solution add 15 mL of 5.25% NaOCl solution. Destruction is rapid. Destroy the excess NaOCl by adding excess 1% sodium bisulfite solution. Neutralize the reaction mixture, check for completeness of destruction, and discard it. Use fresh NaOCl solution (see assay procedure below).

2. Dilute, if necessary, so that the concentration of Mitomycin C does not exceed 0.5 mg/mL. For each 10 mL of solution add 4.55 mg $KMnO_4$. When the mixture has become homogeneous add excess 1% sodium bisulfite solution until the color of the $KMnO_4$ has disappeared. Centrifuge to remove manganese salts, check for completeness of destruction, and discard it.

Destruction of Mitomycin C in Urine[3]

For each 10 mL of urine add 1 mL of 5.25% NaOCl solution. Destruction is rapid. Destroy the excess NaOCl by adding 140 mg of sodium bisulfite. Neutralize the reaction mixture, check for completeness of destruction, and discard it. Use fresh NaOCl solution (see assay procedure below).

Assay of Sodium Hypochlorite Solution

Sodium hypochlorite solutions tend to deteriorate with time so they should be periodically checked for the amount of active chlorine they contain. Pipette 10 mL of NaOCl solution into a 100-mL volumetric flask and fill to the mark with distilled H_2O. Pipette 10 mL of this solution into a conical flask containing 50 mL of distilled H_2O, 1 g of potassium iodide, and 12.5 mL of 2 *M* acetic acid. Titrate this solution against 0.1 *N* sodium thiosulfate solution using starch as an indicator. Each 1 mL of the sodium thiosulfate solution corresponds to 3.545 mg of active chlorine. The NaOCl solution used in these degradation reactions should contain 45–50 g of active chlorine per liter.

Analytical Procedures

Mitomycin C can be analyzed by high-performance liquid chromatography (HPLC) using a μBondapak C18 column and UV detection at 254 nm. The mobile phase was methanol : H_2O 50:50 flowing at 1 mL/min.[5]

Mutagenicity Assays

Tester strains TA100, TA102, UTH8413, and UTH8414 of *Salmonella typhimurium* were used without S9 activation to test the mutagenicity of the reaction mixtures obtained when Mitomycin C in urine was degraded.[3] No mutagenic activity was observed. When Mitomycin C in aqueous solution was degraded tester strain TA102 was used without activation.[5] No mutagenic activity was observed. Mitomycin C in urine was mutagenic to TA102. The final reaction mixtures were not mutagenic to any of the strains.

Related Compounds

Although oxidation with NaOCl works with Mitomycin C, the procedure should be thoroughly validated, including testing the final reaction mixtures for mutagenicity, before it is used for any other compound. For example, treatment of daunorubicin with NaOCl gave unacceptable results.[6]

References

1. Other names for this compound are [1aR]-6-amino-8-([(aminocarbonyl)oxy]methyl)-1,1a,2,8,8a,8b-hexahydro-8a-methoxy-5-methylazirino[2′,3′:3,4]pyrrolo[1,2-*a*]indole-4,7-dione, Ametycin, Ametycine, MMC, Mitocin-C, Mutamycin, 7-amino-9-α-methoxymitosane, MIT-C, Mito-C, Mitomycinum, NCI-C04706, and NSC-26980.
2. International Agency for Research on Cancer. *IARC Monographs on the Evaluation of Carcinogenic Risk of Chemicals to Man*. Volume 10, *Some Naturally Occurring Substances*; International Agency for Research on Cancer: Lyon, 1975; pp. 171–179.
3. Monteith, D.K.; Connor, T.H.; Benvenuto, J.A.; Fairchild, E.J.; Theiss, J.C. Stability and inactivation of mutagenic drugs and their metabolites in the urine of patients administered antineoplastic therapy. *Environ. Mol. Mutagen.* **1987**, *10*, 341–356.
4. Lewis, R.J., Sr. *Sax's Dangerous Properties of Industrial Materials*, 8th ed.; Van Nostrand-Reinhold: New York, 1992; pp. 135–136.
5. Benvenuto, J.A.; Connor, T.H.; Monteith, D.K.; Laidlaw, J.L.; Adams, S.C.; Matney, T.S.; Theiss, J.C. Degradation and inactivation of antitumor drugs. *J. Pharm. Sci.* **1993**, *82*, 988–991.
6. Castegnaro, M.; Adams, J.; Armour, M-.A.; Barek, J.; Benvenuto, J.; Confalonieri, C.; Goff, U.; Ludeman, S.; Reed, D.; Sansone, E. B.; Telling, G., Eds., *Laboratory Decontamination and Destruction of Carcinogens in Laboratory Wastes: Some Antineoplastic Agents*; International Agency for Research on Cancer: Lyon, 1985 (IARC Scientific Publications No. 73).

4-NITROBIPHENYL

CAUTION! Refer to safety considerations section on page 7 before starting any of these procedures.

4-Nitrobiphenyl (4-NBP) [92-93-3] is a crystalline solid (mp 113–114 °C), which is insoluble in H_2O, slightly soluble in alcohol, and soluble in organic solvents. Other names for 4-NBP are 4-nitro-1,1′-biphenyl, PNB, 4-nitrodiphenyl, and 4-phenylnitrobenzene. 4-Nitrobiphenyl causes cancer in laboratory animals.[1] This compound is used as an intermediate in the chemical industry and in organic synthesis.

Principle of Destruction

4-Nitrobiphenyl is reduced by zinc in acid solution to 4-aminobiphenyl (4-ABP), which is then oxidized by potassium permanganate in sulfuric acid ($KMnO_4$ in H_2SO_4).[2] Less than 0.2% of 4-NBP or 4-ABP was left in the final reaction mixture.[3]

Destruction Procedure[2]

Dissolve bulk quantities of 4-NBP in glacial acetic acid so that the concentration does not exceed 1 mg/mL; dilute solutions in glacial acetic acid, if necessary, so that the concentration does not exceed 1 mg/mL, and evaporate organic solvents and take up the residue in glacial acetic acid so that the concentration does not exceed 1 mg/mL. To each of these solutions add an equal volume of 2 *M* H_2SO_4. Treat aqueous solutions by adding an equal volume of glacial acetic acid, then cautiously add, with stirring, 53 mL of concentrated H_2SO_4 per liter of solution. To each 20 mL of these acidified solutions add 165 mg of zinc powder with stirring. Stir the mixture overnight; then, for every 20 mL of solution, add 10 mL of 0.2 *M* $KMnO_4$ solution. Stir this mixture for 10 h, decolorize with sodium metabisulfite, make strongly basic with 10 *M* potassium hydroxide solution (**Caution!** Exothermic), dilute with H_2O, filter to remove manganese compounds,[4] neutralize the filtrate, check for completeness of destruction, and discard it.

Analytical Procedures

The following high-performance liquid chromatography (HPLC) analysis has been recommended.[2] A 250 × 4.6-mm i.d. reverse phase column was used and the mobile phase was MeCN : MeOH : buffer (10:30:20) flowing at 1.5 mL/min. The buffer was 1.5 m*M* in K_2HPO_4 and 1.5 m*M* in KH_2PO_4. If a variable wavelength UV detector is available, use 280 nm for 4-NBP and 275 nm for 4-ABP. Otherwise, a fixed wavelength detector operating at 254 nm should be satisfactory.

Mutagenicity Assays

The reaction mixtures were tested for mutagenicity using *Salmonella typhimurium* strains TA97, TA98, and TA100.[2] No mutagenic activity was seen.

References

1. International Agency for Research on Cancer. *IARC Monographs on the Evaluation of Carcinogenic Risk of Chemicals to Man.* Volume 4, *Some Aromatic Amines, Hydrazine and Related Substances,* N-*Nitroso Compounds and Miscellaneous Alkylating Agents*; International Agency for Research on Cancer: Lyon, 1974; pp. 113–117.
2. Castegnaro, M.; Barek, J.; Dennis, J.; Ellen, G.; Klibanov, M.; Lafontaine, M.; Mitchum, R.; van Roosmalen, P.; Sansone, E.B.; Sternson, L.A.; Vahl, M., Eds. *Laboratory Decontamination and Destruction of Carcinogens in Laboratory Wastes: Some Aromatic Amines and 4-Nitrobiphenyl*; International Agency for Research on Cancer: Lyon, 1985 (IARC Scientific Publications No. 64).

3. Barek, J.; Berka, A.; Müller, M.; Procházka, M.; Zima, J. Chemical destruction of 4-nitrobiphenyl in laboratory waste and its monitoring by differential pulse polarography and voltammetry and by high-performance liquid chromatography. *Coll. Czech. Chem. Commun.* **1986**, *51*, 1604–1608.
4. Lunn, G.; Sansone, E.B.; De Méo, M.; Laget, M.; Castegnaro, M. Potassium permanganate can be used for degrading hazardous compounds. *Am. Ind. Hyg. Assoc. J.* **1994**, 55, 167–171.

N-NITROSO COMPOUNDS: NITROSAMIDES

CAUTION! Refer to safety considerations section on page 7 before starting any of these procedures.

Nitrosamides are compounds of the general form (**I**), where R is usually alkyl (e.g., methyl or ethyl) and the X—N bond is labile. Compounds in which the X—N bond is not labile are termed nitrosamines. Since the chemistry of their decomposition is quite different, these compounds are dealt with in a separate monograph.

$$\begin{matrix} R \searrow \\ \quad N—NO \\ X \nearrow \end{matrix}$$

I

A particular problem with nitrosamides is base-induced decomposition to give diazoalkanes. For example, treatment of an ethanolic solution of *N*-methyl-*N*-nitroso-*p*-toluenesulfonamide with 1 *M* potassium hydroxide

(KOH) solution generated diazomethane in 29% yield. Diazoalkanes are toxic, explosive, and carcinogenic and their generation should be avoided. All of the methods described below, with the exception of potassium permanganate in sulfuric acid ($KMnO_4$ in H_2SO_4), have been carefully checked for the generation of diazoalkanes; none were found. Because of the acid nature of the reaction medium, diazoalkane generation should not be a problem with $KMnO_4$ in H_2SO_4. The following nitrosamides are compounds whose decomposition has been extensively studied:

Abbreviation	Compound Name	Reference	mp or bp (°C)	Registry Number
MNTS	*N*-Methyl-*N*-nitroso-*p*-toluenesulfonamide	1	mp 61–62	[80-11-5]
MNU	*N*-Methyl-*N*-nitrosourea		mp 126	[684-93-5]
ENU	*N*-Ethyl-*N*-nitrosourea		mp 104	[759-73-9]
MNUT	*N*-Methyl-*N*-nitrosourethane	2	bp 62–64/12 mm Hg	[615-53-2]
ENUT	*N*-Ethyl-*N*-nitrosourethane	3	bp 75/16 mm Hg	[614-95-9]
MNNG	*N*-Methyl-*N*′-nitro-*N*-nitrosoguanidine		mp 123	[70-25-7]
ENNG	*N*-Ethyl-*N*′-nitro-*N*-nitrosoguanidine		mp 118–120	[4245-77-6]

Many nitrosamides cause cancer in laboratory animals and these compounds should be regarded as potential human carcinogens. The International Agency for Research on Cancer (IARC) has determined that MNU[4,5] and ENU[6,7] cause cancer in laboratory animals and should be regarded for practical purposes as carcinogenic to humans. The compounds

MNNG,[8,9] MNUT,[10] ENNG,[11] and ENUT[12] cause cancer in laboratory animals while MNU,[13] MNNG,[14] MNUT,[15] ENU,[16] and ENUT[12] are teratogens. All the nitrosamides discussed in this monograph are mutagenic.[17] The compound MNNG will explode when heated or under impact,[14] MNUT explodes when heated,[15] and MNU can detonate[13] and has exploded on storage.[18] Nitrosamides should all be stored below -10°C. Some nitrosamides are solids and some are volatile liquids. These compounds are generally soluble in polar organic solvents and only very sparingly soluble in H_2O.

As commercially supplied these nitrosamides may come in varying degrees of purity and this should be checked before commencing any experiments. For example, MNU and ENU are frequently supplied dampened with a little dilute acid to preserve their stability; purities in the 50–80% range are typical. **Unless absolutely necessary** these compounds should be used as supplied and purification should not be attempted because of the explosion hazard. Purity is best assessed by running an ultraviolet (UV) spectrum and comparing the observed absorbance with reported values.[19,20] If the purity is unacceptably low, obtain a fresh supply.

As far as possible these compounds should be handled in dilute solution using disposable equipment. Any reactions that may generate diazoalkanes should be done behind a safety shield using smooth glass apparatus with rubber stoppers and plastic tubing.

Nitrosamides are generally used in laboratories for the induction of tumors in experimental animals, although they also find some use in organic chemistry for the generation of diazoalkanes.

Principles of Destruction

Many methods have been investigated for the destruction of nitrosamides. The method of choice depends on the nitrosamide and the matrix in which it is found. The methods we will discuss are:

Oxidation by potassium permanganate in sulfuric acid ($KMnO_4$ in H_2SO_4).[19] The products of the reaction have not been determined. Degradation efficiency was greater than 99.5%.

Reaction with sulfamic acid in hydrochloric acid solution (HCl).[19] The strong hydrochloric acid causes displacement of the nitroso group. The nitrosyl chloride formed reacts with the sulfamic acid to form nitrogen and

H_2SO_4. This reaction prevents any reformation of the nitrosamide. The products of the reaction are the corresponding amides produced by simple removal of the nitroso group. Degradation efficiency was greater than 99.5%.

Reaction with iron filings in HCl solution.[19] The strong HCl causes displacement of the nitroso group. The nitrosyl chloride formed is reduced by the iron filings in the acid to ammonia. This reaction prevents any reformation of the nitrosamide. The products of the reaction are the corresponding amides produced by simple removal of the nitroso group except for MNNG and ENNG, where reductive removal of the nitro group causes the major products to be methylguanidine and ethylguanidine, respectively. Degradation efficiency was greater than 99%.

Reaction with sodium bicarbonate solution ($NaHCO_3$).[21] This weak base causes a slow, base-mediated decomposition. The rate of reaction is sufficiently slow so that any diazoalkanes that are formed react with the solvent before escaping from the solution. The products of the reaction have not been definitely identified but they probably include methanol from MNU and MNUT, ethanol from ENU, MNUT,and ENUT, and cyanate from MNU and ENU. Degradation efficiency was greater than 99.99% for MNU, ENU, MNUT, and ENUT. The method is not suitable for MNNG, ENNG, or MNTS.

Reaction with $NaHCO_3$ solution, then nickel–aluminum (Ni–Al) alloy and sodium carbonate (Na_2CO_3) solution, then potassium hydroxide (KOH) solution.[22] The slow increase in pH of the solution produced by sequential addition of the bases causes a slow degradation of the nitrosamide. The degradation rate is sufficiently slow so that any diazoalkanes that are formed have time to react with the solvent before escaping from the solution. The products from this reaction have been discussed.[22] Degradation efficiency was greater than 99.9%.

In all cases, using the procedures described below, destruction was complete, no diazoalkanes were detected, and the final reaction mixtures were not mutagenic.[17,19,21,22] Not all nitrosamides can be degraded by each procedure and these limitations are indicated at the start of each section. Provided that these restrictions are observed, the final reaction mixtures will not be mutagenic.

Destruction Procedures

Destruction of Bulk Quantities of Nitrosamides

1. Take up 1 g of the nitrosamide in 30 mL of methanol and add 30 mL of saturated aqueous $NaHCO_3$ solution. Stir the mixture at room temperature for 24 h, then add 30 mL of 1 *M* Na_2CO_3 solution and 10 g of Ni–Al alloy and stir for 24 h. At the end of this time add 30 mL of 1 *M* KOH solution (more if the solution has become too thick for easy stirring) and stir this mixture for 24 h, then filter it through a pad of Celite®. Neutralize the filtrate, check for completeness of destruction, and discard it. Allow the solid to dry on a metal tray away from flammable solvents for 24 h, then discard it with the solid waste. For larger amounts of nitrosamide increase the quantities given here proportionately, but do this procedure **at least** on this scale. Dissolve small quantities of nitrosamides in **at least** 30 mL of methanol.

2. Dissolve the nitrosamide in 3 *M* H_2SO_4 so that the concentration does not exceed 5 g/L and add 47.4 g of $KMnO_4$ for each liter of solution. Stir the mixture at room temperature overnight, then decolorize with sodium metabisulfite, make strongly basic with 10 *M* KOH (**Caution!** Exothermic), dilute with H_2O, filter to remove manganese compounds,[23] neutralize the filtrate, check for completeness of destruction, and discard it.

3. For ENUT, MNUT, MNU, and ENU only. This method should **not** be used to degrade MNTS, MNNG, and ENNG. Take up 15 g of the nitrosamide in 1 L of ethanol and add 1 L of saturated $NaHCO_3$ solution. Stir the mixture for 24 h, neutralize, check for completeness of destruction, and discard it.

4. For MNTS, MNUT, MNU, ENU, MNNG, and ENNG only. This method should **not** be used to degrade ENUT. Take up 30 g of the nitrosamide in 1 L of methanol and slowly add 1 L of 6 *M* HCl with stirring. Add 70 g of sulfamic acid and stir the mixture for 24 h, neutralize, check for completeness of destruction, and discard it.

5. For MNTS, MNUT, MNU, and ENU only. This method should **not** be used to degrade ENUT, MNNG, and ENNG. Take up 30 g of the nitrosamide in 1 L of methanol and slowly add 1 L of 6 *M* HCl with stirring. Add 70 g of iron filings and stir the mixture for 24 h, neutralize, check for completeness of destruction, and discard it.

Destruction of Nitrosamides in Methanol

1. Dilute the solution,if necessary, with methanol so that the concentration of the nitrosamide does not exceed 1 g in 30 mL. For each 30 mL of this

solution add 30 mL of saturated aqueous $NaHCO_3$ solution. Stir the mixture at room temperature for 24 h, then add 30 mL of 1 *M* Na_2CO_3 solution and 10 g of Ni–Al alloy and stir for 24 h. At the end of this time add 30 mL of 1 *M* KOH solution (more if the solution has become too thick for easy stirring) and stir this mixture for 24 h, then filter it through a pad of Celite®. Neutralize the filtrate, check for completeness of destruction, and discard it. Allow the solid to dry on a metal tray away from flammable solvents for 24 h, then discard it with the solid waste. For larger amounts of nitrosamide increase the quantities given here proportionately, but do this procedure **at least** on this scale. Dissolve small quantities of nitrosamides in **at least** 30 mL of methanol.

2. Dilute the solution with H_2O so that the concentration of methanol does not exceed 20% and the concentration of the nitrosamide does not exceed 0.5%, then cautiously add 160 mL of concentrated H_2SO_4, with stirring, to each liter of solution. After cooling add 47.4 g of $KMnO_4$ for each liter of solution. Stir the mixture overnight at room temperature, decolorize with sodium metabisulfite, make strongly basic with 10 *M* KOH (**Caution!** Exothermic), dilute with H_2O, filter to remove manganese compounds,[23] neutralize the filtrate, check for completeness of destruction, and discard it.

3. For MNTS, MNUT, MNU, ENU, MNNG, and ENNG only. This method should **not** be used to degrade ENUT. Dilute the solution, if necessary, with methanol so that the concentration of the nitrosamide does not exceed 30 g/L. For each liter slowly add 1 L of 6 *M* HCl with stirring. Add 70 g of sulfamic acid, stir the mixture for 24 h, neutralize, check for completeness of destruction, and discard it.

4. For MNTS, MNUT, MNU, and ENU only. This method should **not** be used to degrade ENUT, MNNG, and ENNG. Dilute the solution, if necessary, with methanol so that the concentration of the nitrosamide does not exceed 30 g/L. For each liter slowly add 1 L of 6 *M* HCl with stirring. Add 70 g of iron filings, stir the mixture for 24 h, neutralize, check for completeness of destruction, and discard it.

Destruction of Nitrosamides in Ethanol

1. Dilute the solution,if necessary, with ethanol so that the concentration of the nitrosamide does not exceed 1 g in 30 mL. For each 30 mL of this solution add 30 mL of saturated aqueous $NaHCO_3$ solution. Stir the mixture at room temperature for 24 h, then add 30 mL of 1 *M* Na_2CO_3 solution

and 10 g of Ni–Al alloy and stir for 24 h. At the end of this time add 30 mL of 1 *M* KOH solution (more if the solution has become too thick for easy stirring) and stir this mixture for 24 h, then filter it through a pad of Celite®. Neutralize the filtrate, check for completeness of destruction, and discard it. Allow the solid to dry on a metal tray away from flammable solvents for 24 h, then discard it with the solid waste. For larger amounts of nitrosamide increase the quantities given here proportionately, but do this procedure **at least** on this scale. Dissolve small quantities of nitrosamides in **at least** 30 mL of ethanol.

2. Dilute the solution with H_2O so that the concentration of methanol does not exceed 20% and the concentration of the nitrosamide does not exceed 0.5%, then cautiously add 160 mL of concentrated H_2SO_4, with stirring, to each liter of solution. After cooling add 47.4 g of $KMnO_4$ for each liter of solution. Stir the mixture overnight at room temperature, decolorize with sodium metabisulfite, make strongly basic with 10 *M* KOH (**Caution!** Exothermic), dilute with H_2O, filter to remove manganese compounds,[23] neutralize the filtrate, check for completeness of destruction, and discard it.

3. For MNTS, MNU, ENU, MNNG, and ENNG only. This method should **not** be used to degrade MNUT and ENUT. Dilute the solution, if necessary, with ethanol so that the concentration of the nitrosamide does not exceed 30 g/L. For each liter slowly add 1 L of 6 *M* HCl with stirring. Add 70 g of sulfamic acid, stir the mixture for 24 h, neutralize, check for completeness of destruction, and discard it.

4. Dilute the solution, if necessary, with ethanol so that the concentration of the nitrosamide does not exceed 30 g/L. For each liter of solution slowly add 1 L of 6 *M* HCl with stirring. Add 70 g of iron filings, stir the mixture for 24 h, neutralize, check for completeness of destruction, and discard it.

Destruction of Nitrosamides in Dimethyl Sulfoxide

1. Dilute the solution,if necessary, with dimethyl sulfoxide (DMSO) so that the concentration of the nitrosamide does not exceed 1 g in 30 mL. For each 30 mL of this solution add 30 mL of saturated aqueous $NaHCO_3$ solution. Stir the mixture at room temperature for 24 h, then add 30 mL of 1 *M* Na_2CO_3 solution and 10 g of Ni–Al alloy and stir for 24 h. At the end of this time add 30 mL of 1 *M* KOH solution (more if the solution has become too thick for easy stirring) and stir this mixture for 24 h, then

filter it through a pad of Celite®. Neutralize the filtrate, check for completeness of destruction, and discard it. Allow the solid to dry on a metal tray away from flammable solvents for 24 h, then discard it with thc solid waste. For larger amounts of nitrosamide increase the quantities given here proportionately, but do this procedure **at least** on this scale. Dissolve small quantities of nitrosamides in **at least** 30 mL of DMSO.

2. Dilute the solution with H_2O so that the concentration of DMSO does not exceed 20% and the concentration of the nitrosamide does not exceed 0.5%, then cautiously add 160 mL of concentrated H_2SO_4, with stirring, to each liter of solution. After cooling add 47.4 g of $KMnO_4$ for each liter of solution. Stir the mixture overnight at room temperature, decolorize with sodium metabisulfite, make strongly basic with 10 *M* KOH (**Caution!** Exothermic), dilute with H_2O, filter to remove manganese compounds,[23] neutralize the filtrate, check for completeness of destruction, and discard it.

3. For MNTS, MNUT, MNU, ENU, MNNG, and ENNG only. This method should **not** be used to degrade ENUT. Dilute the solution, if necessary, with DMSO so that the concentration of the nitrosamide does not exceed 30 g/L. For each liter slowly add 1 L of 6 *M* HCl with stirring. Add 70 g of sulfamic acid, stir the mixture for 24 h, neutralize, check for completeness of destruction, and discard it.

4. Dilute the solution, if necessary, with DMSO so that the concentration of the nitrosamide does not exceed 30 g/L. For each liter slowly add 1 L of 6 *M* HCl with stirring. Add 70 g of iron filings, stir the mixture for 24 h, neutralize, check for completeness of destruction, and discard it.

Destruction of Nitrosamides in Acetone

Dilute the solution, if necessary, with acetone so that the concentration of the nitrosamide does not exceed 1 g in 30 mL. For each 30 mL of this solution add 30 mL of saturated aqueous $NaHCO_3$ solution. Stir the mixture at room temperature for 24 h, then add 30 mL of 1 *M* Na_2CO_3 solution and 10 g of Ni–Al alloy and stir for 24 h. At the end of this time add 30 mL of 1 *M* KOH solution (more if the solution has become too thick for easy stirring) and stir this mixture for 24 h, then filter it through a pad of Celite®. Neutralize the filtrate, check for completeness of destruction, and discard it. Allow the solid to dry on a metal tray away from flammable solvents for 24 h, then discard it with the solid waste. For larger amounts of nitrosamide increase the quantities given here proportionately, but do this procedure **at least** on this scale. Dissolve small quantities of nitrosamides in **at least** 30 mL of acetone.

Destruction of Nitrosamides in Water

1. For each 30 mL of solution add 30 mL of saturated aqueous $NaHCO_3$ solution. Stir the mixture at room temperature for 24 h, then add 30 mL of 1 *M* Na_2CO_3 solution and 10 g of Ni–Al alloy and stir for 24 h. At the end of this time add 30 mL of 1 *M* KOH solution (more if the solution has become too thick for easy stirring) and stir this mixture for 24 h, then filter it through a pad of Celite®. Neutralize the filtrate, check for completeness of destruction, and discard it. Allow the solid to dry on a metal tray away from flammable solvents for 24 h, then discard it with the solid waste. For larger amounts of nitrosamide increase the quantities given here proportionately, but do this procedure **at least** on this scale. Dissolve small quantities of nitrosamides in **at least** 30 mL of H_2O.

2. Dilute the solution, if necessary, with H_2O so that the concentration of the nitrosamide does not exceed 0.5%, then cautiously add 160 mL of concentrated H_2SO_4, with stirring, to each liter of solution. After cooling add 47.4 g of $KMnO_4$ for each liter of solution. Stir the mixture overnight at room temperature, decolorize with sodium metabisulfite, make strongly basic with 10 *M* KOH (**Caution!** Exothermic), dilute with H_2O, filter to remove manganese compounds,[23] neutralize the filtrate, check for completeness of destruction, and discard it.

Destruction of Nitrosamides in Ethyl Acetate

Dilute the solution, if necessary, with ethyl acetate so that the concentration of the nitrosamide does not exceed 0.5%. Then, for each 30 mL of this solution, add 20 mL of *N,N*-dimethylformamide and 50 mL of a 0.3 *M* solution of $KMnO_4$ in H_2O. Cautiously add 16 mL of concentrated H_2SO_4, with stirring, and shake the mixture for about 1 min, then add 2.5 g of $KMnO_4$. Stir the mixture overnight at room temperature, decolorize with sodium metabisulfite, make strongly basic with 10 *M* KOH (**Caution!** Exothermic), dilute with H_2O, filter to remove manganese compounds,[23] neutralize the filtrate, check for completeness of destruction, and discard it.

Decontamination of Glassware

As far as possible drain the glassware of all the nitrosamide solution and treat this solution using one of the methods detailed above. Treat the drained glassware with one of the following methods.

1. Immerse the glassware overnight in a solution containing 3 *M* H_2SO_4 and 0.3 *M* $KMnO_4$. Clean the glassware by immersion in a solution of sodium metabisulfite. Decolorize the $KMnO_4$ solution with sodium metabisulfite, make strongly basic with 10 *M* KOH (**Caution!** Exothermic), dilute with H_2O, filter to remove manganese compounds,[23] neutralize the filtrate, check for completeness of destruction, and discard it.

2. For glassware contaminated with ENUT, MNUT, MNU, and ENU only. This method should **not** be used to degrade MNTS, MNNG, and ENNG. Soak the glassware in **either** equal volumes of saturated $NaHCO_3$ solution and ethanol **or** the clear filtrate obtained when equal volumes of ethanol and saturated $NaHCO_3$ solution are mixed and filtered. After 2 (MNU), 4 (ENU), or 24 h (MNUT and ENUT), clean the glassware.

3. For glassware contaminated with MNTS, MNNG, and ENNG only. This method should **not** be used to degrade ENUT, MNUT, MNU, and ENU. Soak the glassware in a mixture of equal volumes of methanol and sulfamic acid in 2 *M* HCl (70 g/L). After 2 (MNTS), 6 (MNNG), or 24 h (ENNG), clean the glassware.

Treatment of Spills[17]

Remove as much material as possible by using absorbents or by high efficiency particulate air (HEPA) vacuuming. Treat the material that is removed by one of the methods detailed above. Treat the contaminated area with one of the following methods.

1. For spills of ENUT, MNUT, MNU, and ENU only. This method should **not** be used for MNTS, MNNG, and ENNG. Add ethanol until all the nitrosamide appears to be dissolved, then add an equal volume of saturated $NaHCO_3$ solution. Alternatively, treat the area with the clear filtrate obtained when equal volumes of ethanol and saturated $NaHCO_3$ solution are mixed and filtered. After 2 (MNU), 4 (ENU), or 24 h (MNUT and ENUT), clean the area. At the end of the procedure check for completeness of destruction by using a wipe soaked in methanol and analyzing the wipe for the presence of the compound.

2. For spills of MNTS, MNNG, and ENNG only. This method should **not** be used for ENUT, MNUT, MNU, and ENU. Soak the area in methanol until the nitrosamide appears to be dissolved, then add an approximately equal volume of sulfamic acid in 2 *M* HCl (70 g/L). After 2 (MNTS), 6 (MNNG), or 24 h (ENNG), clean the area. At the end of the procedure check for completeness of destruction by using a wipe soaked in methanol and analyzing the wipe for the presence of the compound.

3. Cover the area overnight with a solution containing 3 *M* H_2SO_4 and 0.3 *M* $KMnO_4$. Then treat the area with sodium metabisulfite and remove the liquid with absorbents. Squeeze out the absorbents, make the solution strongly basic with 10 *M* KOH (**Caution!** Exothermic), dilute with H_2O, filter to remove manganese compounds,[23] neutralize the filtrate, check for completeness of destruction, and discard it. At the end of the procedure check for completeness of destruction by using a wipe soaked in methanol to wipe the cleaned area and analyze the wipe for the presence of the compound.

Analytical Procedures[22]

The nitrosamides may be determined by HPLC using a UV detector operating at 254 nm and a 250 × 4.6-mm i.d. reverse phase column. The mobile phase is a mixture of methanol and 3.5 m*M* $(NH_4)H_2PO_4$ buffer flowing at 1 mL/min (except 2 mL/min for MNTS). The methanol : buffer ratios are 50:50 for MNTS, 40:60 for ENNG, MNUT, and ENUT, 12:88 for ENU and MNNG, and 7:93 for MNU.

To determine diazoalkanes, use smooth glass apparatus with plastic tubing and rubber stoppers behind a safety shield. Use a fresh reaction flask for each determination since scoured glass surfaces can cause decomposition of the diazoalkanes. Sweep the reaction mixture with nitrogen flowing at 50–55 mL/min for a total of 2 h during and after the addition of the reagent to the nitrosamide. Allow the nitrogen to bubble through 75 mL of ether containing 2.5 mL of valeric acid. At the end of the reaction wash the ether solution twice with saturated aqueous $NaHCO_3$ solution and dry it over anhydrous magnesium sulfate. Add 100 μL of 1-butanol as an internal standard and analyze the mixture by gas chromatography (GC) using a 1.8-m × 2-mm i.d. glass column packed with 28% Pennwalt 223 + 4% KOH on 80/100 Gas Chrom R at 170°C. Methyl or ethyl valerate, produced by diazoalkane alkylation of the valeric acid, indicates the presence of diazomethane or diazoethane, respectively.

Toluene produced by the decomposition of MNTS interferes with this determination so butyric acid is substituted for valeric acid and 1-propanol is substituted for 1-butanol in the above procedure when analyzing reaction mixtures containing MNTS. Methyl butyrate is detected with the GC oven temperature set at 130°C. From time to time a positive control consisting of 1 g of MNTS in 30 mL of ethanol to which is added 30 mL of 1 *M* KOH solution should be used.

Reagents

Prepare saturated $NaHCO_3$ solution by occasionally shaking a mixture of $NaHCO_3$ and H_2O. Add solid $NaHCO_3$ until a precipitate persists. The supernatant is saturated $NaHCO_3$ solution.

Prepare 6 *M* HCl by cautiously adding concentrated HCl to an equal volume of H_2O. This reaction is exothermic.

Mutagenicity Assays

The mutagenicity assays were carried out as described on page 4 using tester strains TA98, TA100, TA1530, and TA1535. The final reaction mixtures were not mutagenic.[17,21,22] The pure compounds were mutagenic,[17] whereas the products that could be identified were not.[21]

Related Compounds

In theory, the methods described above should be applicable to other nitrosamides, for example, *N*-methyl-*N*-nitrosoacetamide, but the chemistry of these compounds varies widely and the destruction procedure chosen **must** be fully validated before being employed on a routine basis. Validation includes determining that the nitrosamide is completely degraded, no diazoalkanes are generated, the final reaction mixtures are nonmutagenic, and, if possible, the products are nontoxic. The destruction procedure should be validated for the nitrosamide dissolved in each solvent in which it is likely to be encountered. As an example of the problems that may arise, the mixed base/Ni–Al alloy method produced no mutagenic residues when applied to the seven nitrosamides whose physical characteristics are listed at the beginning of this monograph but did produce mutagens when applied to nitrosoureas containing β-chloroethyl groups, for example, 1,3-bis(2-chloroethyl)-1-nitrosourea (BCNU).[22]

References

1. Other names for this compound are *N*,4-dimethyl-*N*-nitrosobenzenesulfonamide, *p*-tolylsulfonyl methyl nitrosamine, *p*-tolylsulfonylmethylnitrosamide, and Diazald.
2. Another name for this compound is *N*-methyl-*N*-nitrosoethylcarbamate.
3. Another name for this compound is *N*-ethyl-*N*-nitrosocarbamic acid ethyl ester.
4. International Agency for Research on Cancer. *IARC Monographs on the Evaluation of Carcinogenic Risk of Chemicals to Man.* Volume 1; International Agency for Research on Cancer: Lyon, 1971; pp. 125–134.

5. International Agency for Research on Cancer. *IARC Monographs on the Evaluation of the Carcinogenic Risk of Chemicals to Humans*. Volume 17, *Some* N-*Nitroso Compounds*; International Agency for Research on Cancer: Lyon, 1978; pp. 227–255.
6. Reference 4, pp. 135–140.
7. Reference 5, pp. 191–215.
8. International Agency for Research on Cancer. *IARC Monographs on the Evaluation of the Carcinogenic Risk of Chemicals to Humans, Supplement No. 7, Overall Evaluations of Carcinogenicity: An Updating of* IARC Monographs *Volumes 1 to 42*; International Agency for Research on Cancer: Lyon, 1987; pp. 248–250.
9. International Agency for Research on Cancer. *IARC Monographs on the Evaluation of Carcinogenic Risk of Chemicals to Man*. Volume 4, *Some Aromatic Amines, Hydrazine and Related Substances,* N-*Nitroso Compounds and Miscellaneous Alkylating Agents*; International Agency for Research on Cancer: Lyon, 1974; pp. 183–195.
10. Reference 9, pp. 211–220.
11. Nakamura, T.; Matsuyama, M.; Kishimoto, H. Tumors of the esophagus and duodenum induced in mice by oral administration of *N*-ethyl-*N'*-nitro-*N*-nitrosoguanidine. *J. Natl. Cancer Inst.* **1974**, *52*, 519–522.
12. Lewis, R.J., Sr. *Sax's Dangerous Properties of Industrial Materials*, 8th ed.; Van Nostrand-Reinhold: New York, 1992; p. 2567.
13. Reference 12, p. 2365.
14. Reference 12, pp. 2356–2357.
15. Reference 12, pp. 2362–2363.
16. Reference 12, pp. 1645–1646.
17. Lunn, G.; Sansone, E.B. Dealing with spills of hazardous chemicals: Some nitrosamides. *Food Chem. Toxic.* **1988**, *26*, 481–484.
18. Sparrow, A.H. Hazards of chemical carcinogens and mutagens. *Science* **1973**, *181*, 700–701.
19. Castegnaro, M.; Benard, M.; van Broekhoven, L. W.; Fine, D.; Massey, R.; Sansone, E.B.; Smith, P.L.R.; Spiegelhalder, B.; Stacchini, A.; Telling, G.; Vallon, J.J., Eds., *Laboratory Decontamination and Destruction of Carcinogens in Laboratory Wastes: Some* N-*Nitrosamides*; International Agency for Research on Cancer: Lyon, 1983 (IARC Scientific Publications No. 55).
20. Extinction coefficients in dichloromethane are MNU 8500 (231 nm); ENU 7100 (237 nm); MNUT 6800 (236 nm); ENUT 7100 (236 nm); and, in methanol, MNNG 18197 (275 nm) (reported in Reference 19). Also, in methanol, ENNG 13798 (277 nm) (Reference 21) and MNTS 14412 (222 nm) (Sadtler Standard Ultraviolet Spectrum 3516).
21. Lunn, G. Unpublished results.
22. Lunn, G.; Sansone, E.B.; Andrews, A.W.; Keefer, L.K. Decontamination and disposal of nitrosoureas and related *N*-nitroso compounds. *Cancer Res.* **1988**, *48*, 522–526.
23. Lunn, G.; Sansone, E.B.; De Méo, M.; Laget, M.; Castegnaro, M. Potassium permanganate can be used for degrading hazardous compounds. *Am. Ind. Hyg. Assoc. J.* **1994**, *55*, 167–171.

N-NITROSO COMPOUNDS: NITROSAMINES

CAUTION! Refer to safety considerations section on page 7 before starting any of these procedures.

Nitrosamines are compounds of the general form (**I**), where R and R′ are usually aryl or alkyl, although other variants where R or R′ are heteroatoms (e.g., oxygen) are known. Compounds in which R or R′ is a good leaving group are termed nitrosamides. Since the chemistry of their decomposition is quite different they are dealt with in a separate monograph.

$$R(R^1)N{-}NO$$

I

Many of the lower molecular weight nitrosamines are volatile liquids and they should all be regarded as carcinogenic in humans. For example, NDMA (*N*-methyl-*N*-nitrosomethanamine, *N*-nitrosodimethylamine, or dimethylnitrosamine) [62-75-9] is a yellow-green water-soluble liquid (bp 149–150°C), which has been shown to cause cancer in mice at the part per billion level in drinking H_2O. Nitrosamines are soluble in organic solvents and to varying degrees in H_2O.

Other commonly used nitrosamines whose degradation has been investigated include:

NDEA	*N*-Ethyl-*N*-nitrosoethanamine, *N*-nitrosodiethylamine, or diethylnitrosamine bp 177°C; soluble 106 mg/mL in H_2O	[55-18-5]
NDPA	*N*-Nitrosodipropylamine or dipropylnitrosamine bp 81°C/5 mm Hg; soluble 9.8 mg/mL in H_2O	[621-64-7]
NDiPA	*N*-Nitrosodiisopropylamine or diisopropylnitrosamine mp 48°C; slightly soluble in H_2O	[601-77-4]
NDBA	*N*-Nitrosodibutylamine[1] bp 116°C/14 mm Hg; soluble 1.2 mg/mL in H_2O	[924-16-3]
NPYR	*N*-Nitrosopyrrolidine or NO-PYR bp 98°C/12 mm Hg; miscible with H_2O	[930-55-2]
NPIP	*N*-Nitrosopiperidine bp 100°C/14 mm Hg; soluble 77 mg/mL in H_2O	[100-75-4]
NMOR	*N*-Nitrosomorpholine or 4-nitrosomorpholine bp 96°C/6 mm Hg; miscible with H_2O	[59-89-2]
di-NPZ	*N,N'*-Dinitrosopiperazine mp 160°C; soluble 5.7 mg/mL in H_2O	[140-79-4]
NMA	*N*-Nitroso-*N*-methylaniline[2] mp 15°C, bp 121°C/13 mm Hg; insoluble in H_2O	[614-00-6]
NDPhA	*N*-Nitrosodiphenylamine mp 66.5°C	[86-30-6]

The International Agency for Research on Cancer (IARC) has determined that NDMA,[3,4] NDEA,[5,6] NDPA,[7] NDBA,[8,9] NPYR,[10] NPIP,[11] and

NMOR[12] cause cancer in laboratory animals and should be regarded for practical purposes as carcinogenic to humans. The compounds NDiPA,[13] di-NPZ,[14] NMA,[15] and NDPhA[16] cause cancer in laboratory animals. *N*-Nitrosodimethylamine causes fatal liver disease and a number of systemic effects upon ingestion.[17] *N*-Nitrosodiphenylamine is an eye irritant and can react vigorously with oxidizing materials.[18] The compounds NDMA,[17] NDEA,[19] NDPA,[20] NDBA,[21] NPIP,[22] di-NPZ,[23] and NMA[24] are teratogens.

Nitrosamines are generally used in laboratories for the induction of tumors in experimental animals although they find some use as intermediates in organic chemistry, for example, for the preparation of hydrazines or in the α-alkylation of amines. These compounds are also found as unwanted byproducts of industrial processes, for example, in the rubber industry.

Principles of Destruction

Nitrosamines may be reduced to the corresponding amine by using nickel–aluminum (Ni–Al) alloy in dilute base.[25] The nitrosamines were completely degraded (99.9%) and only the amines (RR′NH) were found in the final reaction mixtures. No traces (generally <0.1%) of the corresponding, possibly carcinogenic, hydrazines ($RR'NNH_2$) were found in the final reaction mixtures.

Nitrosamines may be oxidized by potassium permanganate in 3 *M* sulfuric acid ($KMnO_4$ in H_2SO_4).[26] The nitrosamines were completely destroyed (>99.5%). The products of this reaction have not been determined.

Nitrosamines may be destroyed by using hydrogen bromide (HBr) in glacial acetic acid.[26] The nitrosamines were completely destroyed (>99%) and the products were presumably the corresponding amines.

All of these procedures were validated by an international collaborative study.[26]

Destruction Procedures[25,26]

Destruction of Bulk Quantities of Nitrosamines

1. Dissolve the nitrosamine in H_2O so that the concentration does not exceed 10 mg/mL. If the nitrosamine is not sufficiently soluble in H_2O, use methanol instead. Add an equal volume of potassium hydroxide (KOH) solution (1 *M*) and stir the mixture magnetically. For every 100 mL of this

solution add 5 g of Ni–Al alloy at such a rate that excessive frothing does not occur. The reaction can be quite exothermic. Do it in a reaction vessel whose volume is at least three times that of the final reaction mixture. Cover the reaction mixture and stir for 24 h, then filter it through a pad of Celite®. Neutralize the filtrate, check for completeness of destruction, and discard it. Allow the spent nickel to dry on a metal tray for 24 h (away from flammable solvents) and discard it.

2. Dissolve the nitrosamine in H_2SO_4 (3 *M*) so that the nitrosamine concentration does not exceed 6 mg/L and then add 47.4 g of $KMnO_4$ to each liter. Stir the reaction mixture overnight. If the reaction mixture is no longer purple, add more $KMnO_4$ until the purple color is maintained for at least 1 h. Decolorize the reaction mixture with sodium metabisulfite, make strongly basic with 10 *M* KOH solution (**Caution!** Exothermic), dilute with H_2O, filter to remove manganese compounds,[27] neutralize the filtrate, check for completeness of destruction, and discard it.

Destruction of Nitrosamines in Aqueous Solution

1. Dilute the mixture, if necessary, with H_2O so that the concentration does not exceed 10 mg/mL. Add an equal volume of KOH solution (1 *M*) and stir the mixture magnetically. For every 100 mL of this solution add 5 g of Ni–Al alloy at such a rate that excessive frothing does not occur. The reaction can be quite exothermic. Do it in a reaction vessel whose volume is at least three times that of the final reaction mixture. Cover the reaction mixture and stir for 24 h, then filter it through a pad of Celite®. Neutralize the filtrate, check for completeness of destruction, and discard it. Allow the spent nickel to dry on a metal tray for 24 h (away from flammable solvents) and discard it with the solid waste.

2. Stir the aqueous solution and slowly add concentrated H_2SO_4 **to the aqueous solution** so that the H_2SO_4 concentration is 3 *M*. If necessary, use a cold H_2O bath. Add 3 *M* H_2SO_4, if necessary, so that the nitrosamine concentration does not exceed 6 mg/L. Add 47.4 g of $KMnO_4$ to each liter of solution and stir the mixture overnight. If the reaction mixture is no longer purple, add more $KMnO_4$ until the purple color is maintained for at least 1 h. Decolorize the reaction mixture with sodium metabisulfite, make strongly basic with 10 *M* KOH solution (**Caution!** Exothermic), dilute with H_2O, filter to remove manganese compounds,[27] neutralize the filtrate, check for completeness of destruction, and discard it.

Destruction of Nitrosamines in Aprotic Organic Solvents (For Example, Dichloromethane)

1. Dilute the solution, if necessary, so that the nitrosamine concentration does not exceed 10 mg/mL. Stir the reaction mixture and add one volume of KOH solution (2 *M*) and three volumes of methanol (i.e., dichloromethane: KOH: methanol 1:1:3). For every 100 mL of this solution add 5 g of Ni–Al alloy at such a rate that excessive frothing does not occur. The reaction can be quite exothermic. Do it in a reaction vessel whose volume is at least three times that of the final reaction mixture. Cover the reaction mixture, stir for 24 h, then filter it through a pad of Celite®. Neutralize the filtrate, check for completeness of destruction, and discard it. Allow the spent nickel to dry on a metal tray for 24 h (away from flammable solvents) and discard it with the solid waste.

2. Dry the solution, if necessary, with sodium sulfate. If necessary, dilute the mixture with more solvent so that the nitrosamine concentration does not exceed 1 mg/mL. For every volume of nitrosamine solution add 10 volumes of a 3% solution of HBr in glacial acetic acid (obtained by diluting the commercially available 30% solution). After 2 h dilute the reaction mixture with H_2O (using ice cooling if necessary), neutralize, check for completeness of destruction, and discard it.

Destruction of Nitrosamines in Alcohols

Dilute the solution, if necessary, so that the nitrosamine concentration does not exceed 10 mg/mL. Add an equal volume of KOH solution (1 *M*) and stir the mixture magnetically. For every 100 mL of this solution add 5 g of Ni–Al alloy at such a rate that excessive frothing does not occur. The reaction can be quite exothermic. Do it in a reaction vessel whose volume is at least three times that of the final reaction mixture. Cover the reaction mixture, stir for 24 h, then filter it through a pad of Celite®. Neutralize the filtrate, check for completeness of destruction, and discard it. Allow the spent nickel to dry on a metal tray for 24 h (away from flammable solvents) and discard it.

Destruction of Nitrosamines in Dimethyl Sulfoxide

Dilute the mixture with methanol, if necessary, so that the nitrosamine concentration does not exceed 10 mg/mL. Add an equal volume of KOH solution (1 *M*) and stir the mixture magnetically. For every liter of this solution add 100 g of Ni–Al alloy at such a rate that excessive frothing does not occur. The reaction can be quite exothermic. Do it in a reaction vessel whose volume is at least four times that of the final reaction mixture. Cover the reaction mixture, stir for 24 h, then filter it through a pad of Celite®. Neutralize the filtrate, check for completeness of destruction, and discard it. Allow the spent nickel to dry on a metal tray for 24 h (away from flammable solvents) and discard it.

Destruction of Nitrosamines in Olive Oil and Mineral Oil

Dilute olive oil solutions with olive oil and mineral oil solutions with hexane, if necessary, so that the nitrosamine concentration does not exceed 10 mg/mL. Add an equal volume of KOH solution (1 *M*) and stir the resulting two-phase mixture magnetically. For every 100 mL of this solution add 5 g of Ni–Al alloy at such a rate that excessive frothing does not occur. The reaction can be quite exothermic. Do it in a reaction vessel whose volume is at least four times that of the final reaction mixture. Cover the reaction mixture, stir for 24 h, then filter it through a pad of Celite®. Neutralize the filtrate, check for completeness of destruction, and discard it. Allow the spent nickel to dry on a metal tray for 24 h (away from flammable solvents) and discard it.

Destruction of Nitrosamines in Agar Gel

Add the contents of the agar plate (~ 17 g) to 75 mL of KOH solution (1 *M*) and stir and warm the mixture until the agar dissolves. Note that many nitrosamines are liable to volatilize under these conditions. When cool add 2 g of Ni–Al alloy slowly enough that excessive frothing does not occur. (If more than 100 mg of nitrosamine is present, increase the weight of alloy and volume of KOH solution proportionately.) Stir the mixture for 24 h, then filter it through a pad of Celite®. Neutralize the filtrate, check for completeness of destruction, and discard it. Allow the spent nickel to dry on a metal tray away from flammable solvents for 24 h, then discard it.

Dealing with Spills of Nitrosamines

Remove as much of the spill as possible with an absorbent and cover the remainder with a 47.4 g/L solution of $KMnO_4$ in 3 *M* H_2SO_4. Leave this mixture overnight, then decolorize with sodium metabisulfite, and remove with paper towels. Squeeze out the liquid, make strongly basic with 10 *M* KOH solution (**Caution!** Exothermic), dilute with H_2O, filter to remove manganese compounds,[27] neutralize the filtrate, check for completeness of destruction, and discard it. If possible, check for completeness of decontamination by taking wipe samples and analyze these samples. Note that this procedure may damage painted surfaces or Formica. Place the absorbent material in a beaker and decontaminate it by one of the methods described above.

Decontamination of Equipment Contaminated with Nitrosamines

1. Rinse five times with an appropriate solvent, then decontaminate this solution using one of the methods described above. To minimize the scale of the final reaction keep the volume of these rinses as small as practicable.

2. Fill the equipment with a 47.4 g/L solution of $KMnO_4$ in 3 *M* H_2SO_4 (or cover the contaminated surface with this solution). After leaving overnight the solution should still be purple. If it is not, replace with fresh solution, which should stay purple for at least 1 h. At the end of the reaction decolorize with sodium metabisulfite, make strongly basic with 10 *M* KOH solution (**Caution!** Exothermic), dilute with H_2O, filter to remove manganese compounds,[27] neutralize the filtrate, check for completeness of destruction, and discard it. Finally, clean the equipment in a conventional fashion.

Analytical Procedures

Many procedures have been published for the analysis of nitrosamines. The thermal energy analysis (TEA) detector is specific for nitrosamines and great sensitivity can be achieved using this detector. For many applications the flame ionization detector (FID) provides sufficient sensitivity.

For analysis by gas chromatography (GC)[25] a 1.8-m × 2-mm i.d. packed column can be used. For *N*-nitrosodimethylamine (80°C), *N*-nitrosodiethylamine (120°C), *N*-nitrosodiisopropylamine (120°C), *N*-nitrosodibutylamine (150°C), *N*-nitrosopyrrolidine (140°C), *N*-nitrosopiperidine (150°C), and *N*-nitrosomorpholine (150°C) the packing is 10% Carbowax 20 M + 2% KOH in 80/100 Chromosorb W AW, for *N,N'*-dinitrosopiperazine (180°C) the packing is 2% Carbowax 20 M + 1% KOH on 80/100 Supelcoport and for *N*-methyl-*N*-nitrosoaniline (110°C) the packing is 3% SP 2401-DB on 100/120 Supelcoport. The oven temperatures shown above in parentheses are only a guide and the exact conditions would have to be determined experimentally. High-performance liquid chromatography (HPLC) conditions for nitrosamines have been described using a 250 × 4.6-mm i.d. column of 10 μ Lichrosorb Si 60 with acetone : isooctane (7:93)[28] or a 250 × 4.6-mm i.d. column of 5 μ Lichrosorb Si 100 with an acetone : *n*-hexane gradient increasing from 1% to 40% acetone in 15 min.[29] In each case the flow rate was 2 mL/min and a TEA detector was used.

Mutagenicity Assays

The mutagenicity assays were carried out as described on page 4 using tester strains TA98, TA100, TA1530, and TA1535. The reaction mixtures obtained from the Ni–Al alloy degradation of NDMA, NDBA, NPYR, and NPIP were neutralized, diluted with pH 7 buffer in an attempt to avoid toxicity problems (probably from aluminum compounds), and tested. None of the reaction mixtures was mutagenic. Each of the four nitrosamines was mutagenic when tested as the pure compound, whereas the corresponding amines were not mutagenic.[30]

Related Compounds

The procedures listed above should be generally applicable to nitrosamines. Limitations were usually imposed by the matrices in which the nitrosamines were found rather than the nature of the nitrosamines themselves. The only instance where the nature of the nitrosamine made a difference was the use of HBr in glacial acetic acid. In this case, *N*-nitrosopyrrolidine took significantly longer to degrade than the other nitrosamines tested. The conditions given above should, however, degrade *N*-nitrosopyrrolidine.

Nickel–aluminum alloy has also been shown to reduce the related nitramines *N*-nitromorpholine [4164-32-3] and diisopropylnitramine [4164-30-1] to the corresponding amines.[31] Destruction of the nitramines was greater than 99.9% and in the final reaction mixtures less than 0.1% of the theoretical amounts of the related nitrosamines or hydrazines were found.[30] The reaction should be generally applicable to nitramines.

References

1. Other names for this compound are dibutylnitrosamine and *N*-butyl-*N*-nitroso-1-butanamine.

2. Other names for this compound are methylphenylnitrosamine, *N*-methyl-*N*-nitroso-aniline, and *N*-methyl-*N*-nitrosobenzenamine.
3. International Agency for Research on Cancer. *IARC Monographs on the Evaluation of Carcinogenic Risk of Chemicals to Man.* Volume 1; International Agency for Research on Cancer: Lyon, 1971; pp. 95–106.
4. International Agency for Research on Cancer. *IARC Monographs on the Evaluation of the Carcinogenic Risk of Chemicals to Humans.* Volume 17, *Some* N-*Nitroso Compounds*; International Agency for Research on Cancer: Lyon, 1978; pp. 125–175.
5. Reference 3, pp. 107–124.
6. Reference 4, pp. 83–124.
7. Reference 4, pp. 177–189.
8. International Agency for Research on Cancer. *IARC Monographs on the Evaluation of Carcinogenic Risk of Chemicals to Man.* Volume 4, *Some Aromatic Amines, Hydrazine and Related Substances,* N-*Nitroso Compounds and Miscellaneous Alkylating Agents*; International Agency for Research on Cancer: Lyon, 1974; pp. 197–210.
9. Reference 4, pp. 51–75.
10. Reference 4, pp. 313–326.
11. Reference 4, pp. 287–301.
12. Reference 4, pp. 263–280.
13. Lewis, R.J., Sr. *Sax's Dangerous Properties of Industrial Materials*, 8th ed.; Van Nostrand-Reinhold: New York, 1992; pp. 2563–2564.
14. Hadidian, Z.; Fredrickson, T.N.; Weisburger, E.K.; Weisburger, J.H.; Glass, R.M.; Mantel, N. Tests for chemical carcinogens. Report on the activity of derivatives of aromatic amines, nitrosamines, quinolines, nitroalkanes, amides, epoxides, aziridines, and purine antimetabolites. *J. Natl. Cancer. Inst.* **1968**, *41*, 985–1036.
15. Michejda, C.J.; Kroeger-Koepke, M.B.; Kovatch, R.M. Carcinogenic effects of sequential administration of two nitrosamines in Fischer 344 rats. *Cancer Res.* **1986**, *46*, 2252–2256.
16. International Agency for Research on Cancer. *IARC Monographs on the Evaluation of the Carcinogenic Risk of Chemicals to Humans.* Volume 27, *Some Aromatic Amines, Anthraquinones and Nitroso Compounds, and Inorganic Fluorides Used in Drinking-Water and Dental Preparations*; International Agency for Research on Cancer: Lyon, 1982; pp. 213–225.
17. Reference 13, p. 2564.
18. Reference 13, pp. 1463–1464.
19. Reference 13, pp. 2562–2563.
20. Reference 13, pp. 2565–2566.
21. Reference 13, p. 628.
22. Reference 13, p. 2581.
23. Reference 13, pp. 1438–1439.
24. Reference 13, pp. 2360–2361.

25. Lunn, G.; Sansone, E.B.; Keefer, L.K. Safe disposal of carcinogenic nitrosamines. *Carcinogenesis* **1983**, *4*, 315–319.

26. Castegnaro, M.; Eisenbrand, G.; Ellen, G.; Keefer, L.; Klein, D.; Sansone, E. B.; Spincer, D.; Telling, G.; Webb K., Eds., *Laboratory Decontamination and Destruction of Carcinogens in Laboratory Wastes: Some* N-*Nitrosamines*; International Agency for Research on Cancer: Lyon, 1982 (IARC Scientific Publications No. 43).

27. Lunn, G.; Sansone, E.B.; De Méo, M.; Laget, M.; Castegnaro, M. Potassium permanganate can be used for degrading hazardous compounds. *Am. Ind. Hyg. Assoc. J.* **1994**, *55*, 167–171.

28. Goff, U. High-performance liquid chromatography of volatile nitrosamines. In *Environmental Carcinogens Selected Methods of Analysis.* Volume 6, N-*Nitroso Compounds*; Egan, H., Preussmann, R., Eisenbrand, G., Spiegelhalder, B., O'Neill, I.K., Bartsch, H., Eds.; International Agency for Research on Cancer: Lyon, 1983 (IARC Scientific Publications No. 45); pp. 389–394.

29. Sen, N.P.; Seaman, S.W.; Kushwaha, S.C. Determination of non-volatile N-nitrosamines in baby bottle rubber nipples and pacifiers by high-performance liquid chromatography-thermal energy analysis. *J. Chromatogr.* **1989**, *463*, 419–428.

30. Lunn, G. Unpublished results.

31. Lunn, G.; Sansone, E.B.; Keefer, L.K. General cleavage of N—N and N—O bonds using nickel/aluminum alloy. *Synthesis* **1985**, 1104–1108.

NITROSOUREA DRUGS

> **CAUTION!** Refer to safety considerations section on page 7 before starting any of these procedures.

The nitrosourea drugs considered in this monograph are all antineoplastic agents and these compounds all have an N—NO functionality in common. These nitrosoureas are all crystalline solids and are moderately soluble in alcohols. The H_2O solubility of these compounds varies.

The following compounds are considered:

BCNU (**I**)	*N,N′*-Bis(2-chloroethyl)-*N*-nitrosourea;[1] mp 30–32°C; solubility in H_2O 4 mg/mL	[154-93-8]
CCNU (**II**)	*N*-(2-Chloroethyl)-*N′*-cyclohexyl-*N*-nitroso-urea;[2] mp 90°C; solubility in H_2O <0.05 mg/mL	[13010-47-4]

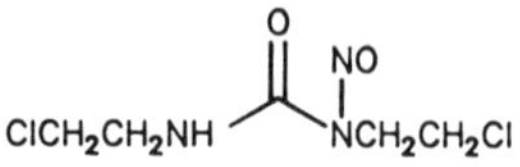

I

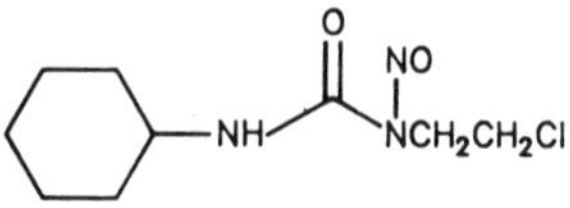

II

STZ (**III**, R = CH_3) — Streptozotocin;[3] mp 115°C; soluble in H_2O and ketones, not soluble in other organic solvents [18883-66-4]

CTZ (**III**, R = CH_2CH_2Cl) — Chlorozotocin;[4] mp 140–141°C; soluble in H_2O [54749-90-5]

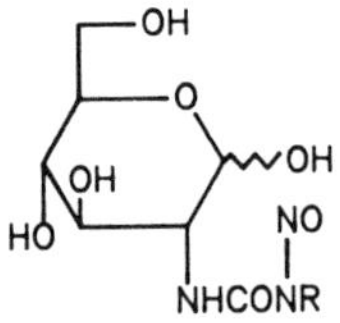

III

PCNU (**IV**) — *N*-(2-Chloroethyl)-*N*′-(2,6-dioxo-3-piperidinyl)-*N*-nitrosourea; NSC-95466; mp 130–131°C; solubility in H_2O <1 mg/mL, soluble in acetone [13909-02-9]

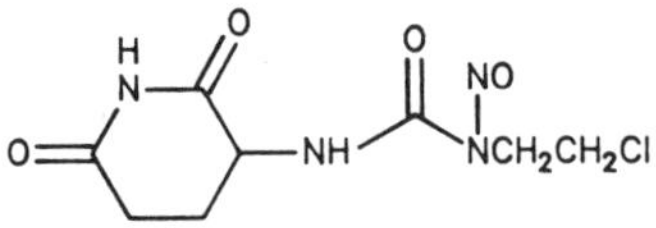

IV

Methyl CCNU (**V**) *N*-(2-Chloroethyl)-*N*′-(4-methylcyclohexyl)-*N*-nitrosourea;[5]
mp 68–69°C; solubility in H_2O 0.09 mg/mL, soluble in dimethyl sulfoxide (DMSO) [13909-09-6]

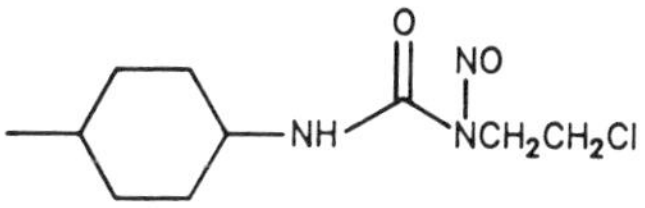

V

The compounds BCNU,[6–8] CCNU,[8–10] methyl CCNU,[8,11] CTZ,[12] and STZ[13] are carcinogenic in experimental animals and the International Agency for Research on Cancer (IARC) has stated that BCNU,[6,8] CCNU,[9] methyl CCNU,[8] and STZ[13] should be regarded as presenting a carcinogenic risk to humans. In addition, BCNU[14] and CCNU[15] are teratogens and have toxic effects on the blood, STZ is a teratogen and affects the liver and kidneys,[16] and methyl CCNU can affect the kidneys.[17] All of these compounds are mutagenic.[18]

Principles of Destruction

Streptozotocin is degraded with saturated sodium bicarbonate ($NaHCO_3$) solution (<0.012% remains) to give methanol.[18] The other drugs are degraded by reduction with nickel–aluminum alloy (Ni–Al) in potassium hydroxide (KOH) solution (<0.2% remains).[18] To obtain nonmutagenic products from the reduction of BCNU it is necessary to increase the ratio of reductant to substrate.[18] The products obtained were ethanol and cyclohexylamine (from CCNU) and 4-methylcyclohexylamine (from methyl CCNU). No trace of diazoethane or 2-chlorodiazoethane were found from CCNU and no trace of diazomethane was found from STZ. The use of hydrogen bromide in glacial acetic acid has been proposed for the destruction of these compounds,[19] but in a number of cases the product was found to be the denitrosated compound,[18] which was also a mutagen, so this method cannot be recommended as a destruction procedure.

Destruction Procedures

Destruction of Bulk Quantities of CCNU, CTZ, PCNU, and Methyl CCNU

Dissolve the nitrosourea in methanol so that the concentration does not exceed 10 mg/mL, then add an equal volume of 2 *M* KOH solution. For every 20 mL of this basified solution add 1 g of Ni–Al alloy. Add quantities of Ni–Al alloy in excess of 5 g in portions to prevent the reaction from frothing too violently. Perform the reaction in a vessel at least three times larger than the final volume as some foaming may occur. Stir the mixture overnight, then filter through a pad of Celite®. Neutralize the filtrate, check for completeness of destruction, and discard it. Allow the spent nickel, which is filtered off, to dry on a metal tray away from flammable solvents for 24 h, then discard it with the solid waste.

Destruction of Bulk Quantities of BCNU

Dissolve 100 mg of BCNU in 3 mL of ethanol and add 27 mL of H_2O. To this solution add 30 mL of 2 *M* KOH solution and 3 g of Ni–Al alloy. Add quantities of Ni–Al alloy in excess of 5 g in portions to prevent the reaction from frothing too violently. Perform the reaction in a vessel at least three times larger than the final volume as some foaming may occur. Stir the mixture overnight, then filter through a pad of Celite®. Neutralize the filtrate, check for completeness of destruction, and discard it. Allow the spent nickel, which is filtered off, to dry on a metal tray away from flammable solvents for 24 h, then discard it with the solid waste.

Destruction of Pharmaceutical Preparations of BCNU, CCNU, and CTZ

BCNU The pharmaceutical preparation consists of 100 mg of drug in 3 mL of ethanol to which is added 27 mL of H_2O.

CCNU Open the capsules and allow the shells to remain in the reaction vessel. For every capsule (100 mg) add 10 mL of methanol.

CTZ The pharmaceutical preparation consists of 50 mg of drug in 5 mL of saline solution.

To each of these solutions add an equal volume of 2 *M* KOH solution. For every 20 mL of this basified solution add 1 g of Ni–Al alloy. Add quantities of more than 5 g in portions to prevent the reaction from frothing too violently. Perform the reaction in a vessel at least three times larger than the final volume as some foaming may occur. Stir the mixture overnight, then filter through a pad of Celite®. Neutralize the filtrate, check for completeness of destruction, and discard it. Allow the spent nickel, which is filtered off, to dry on a metal tray away from flammable solvents for 24 h, then discard it with the solid waste.

Destruction of STZ

Take up bulk quantities in H_2O so that the concentration does not exceed 100 mg/mL. If necessary, dilute pharmaceutical preparations with H_2O so that they do not exceed 100 mg/mL. For each volume of STZ solution add five times the volume of saturated $NaHCO_3$ solution, allow the mixture to stand overnight, check for completeness of destruction, and discard it. Prepare saturated $NaHCO_3$ solution by mixing $NaHCO_3$ and H_2O in a container. Shake the container occasionally. If solid persists, the solution is saturated; if not, add more $NaHCO_3$.

Analytical Procedures

Analysis was by high-performance liquid chromatography (HPLC) using a 250 × 4.6-mm i.d. column of Microsorb C8. The injection volume was 20 μL and the mobile phase flowed at 1 mL/min. Ultraviolet (UV) detection at 254 nm was used. For BCNU, CCNU, PCNU, and methyl CCNU methanol and a 3.5 m*M* $(NH_4)H_2PO_4$ buffer were used; for BCNU and PCNU the methanol:buffer ratio was 50:50 and for CCNU and methyl CCNU it was 75:25. For CTZ methanol:20 m*M* KH_2PO_4 buffer (4:96) was used and for STZ pure 20 m*M* KH_2PO_4 buffer was used. On our equipment these mobile phase combinations were found to give reasonable retention times (4–14 min). It was frequently advantageous to add some KH_2PO_4 buffer to an aliquot of the neutralized reaction mixture and to centrifuge before analysis. This technique removed salts that could clog the chromatograph.

Gas chromatography (GC) using a 1.8-m × 2-mm i.d. glass column packed with 10% Carbowax 20 M + 2% KOH on 80/100 Chromosorb W AW was used to determine the products of these reactions. The injection temperature was 200°C, except where shown, and the flame ionization detector operated at 300°C. The oven temperature was 60°C and approximate retention times were methanol (1.7 min), ethanol (2.2 min), cyclohexylamine (13 min), and 4-methylcyclohexylamine (17 min).

To check for diazoalkanes the reaction mixture was swept with nitrogen into ether containing either valeric acid or acetic acid. The ether solution was washed with $NaHCO_3$ solution, then analyzed by GC with the equipment described above for the presence of esters generated by the reaction of diazoalkanes with the carboxylic acids. Thus, CCNU might give 2-chlorodiazoethane, which would produce 2-chloroethyl acetate (retention time 5 min, oven temperature 110°C, injection temperature 110°C) or diazoethane (produced by Ni–Al dechlorination of the starting material or 2-chlorodiazoethane), which would produce ethyl acetate (retention time 2.2 min, oven temperature 60°C), and STZ might give diazomethane, which would produce methyl valerate (retention time 4 min, oven temperature 70°C). No diazoalkanes were detected.

Mutagenicity Assays[18]

The mutagenicity assays were carried out as described on page 4 using tester strains TA98, TA100, TA1530, and TA1535. To avoid cell toxicity problems it was generally necessary to mix an aliquot of the neutralized reaction mixture with an equal volume of pH 7 buffer before testing. The final reaction mixtures [tested at a level corresponding to 0.25 mg (0.085 mg for BCNU) undegraded material per plate] were not mutagenic. The reaction mixtures from the STZ degradations were tested without using buffer at a level corresponding to 1.7 mg of undegraded product and were not mutagenic. All of the drugs were tested in DMSO solution and were found to be mutagenic. None of the products detected were found to be mutagenic.

Related Compounds

The Ni–Al alloy technique described above should be applicable to compounds of the general form R—NH—CO—N(NO)—CH_2CH_2Cl, but the procedure should be thoroughly validated. The problems encountered with BCNU indicate that sometimes a higher reductant : substrate ratio may be required.

References

1. Other names for this compound are Carmustine, BiCNU, Nitrumon, NSC-409962, SK 27702, NCI-C04733, SRI 1720, FDA 0345, Becenun, and Carmubris.
2. Other names for this compound are Lomustine, 1-(2-chloroethyl)-3-cyclohexyl-1-nitrosourea, Belustine, Cecenu, CeeNU, CiNU, ICIG 1109, NCI-C04740, NSC-79037, and RB 1509.
3. Other names for this compound are 2-deoxy-2-([(methylnitrosoamino)carbonyl]amino)-D-glucopyranose, streptozocin, 2-deoxy-2-(3-methyl-3-nitrosoureido)-D-glucopyranose, *N*-D-glucosyl-(2)-*N'*-nitrosomethylurea, Zanosar, NCI-C03167, NSC-85998, STR, STRZ, and U-9889.
4. Other names for this compound are 2-[([(2-chloroethyl)nitrosoamino]carbonyl)amino]-2-deoxy-D-glucose, 2-[3-(2-chloroethyl)-3-nitrosoureido]-2-deoxy[D-glucosopyranose, 1-(2-chloroethyl)-1-nitroso-3-(D-glucos-2-yl)urea, DCNU, CHLZ, NSC-178248, and NSC D 254157.
5. Other names for this compound are Semustine, Me-CCNU, Methyl-Lomustine, NCI-C04955, and NSC-95441.
6. International Agency for Research on Cancer. *IARC Monographs on the Evaluation of the Carcinogenic Risk of Chemicals to Humans.* Volume 26, *Some Antineoplastic and Immunosuppressive Agents*; International Agency for Research on Cancer: Lyon, 1981; pp. 79–95.
7. International Agency for Research on Cancer. *IARC Monographs on the Evaluation of the Carcinogenic Risk of Chemicals to Humans, Supplement No. 4, Chemicals, Industrial Processes and Industries Associated with Cancer in Humans. IARC Monographs, Volumes 1 to 29*; International Agency for Research on Cancer: Lyon, 1982; pp. 63–64.
8. International Agency for Research on Cancer. *IARC Monographs on the Evaluation of the Carcinogenic Risk of Chemicals to Humans, Supplement No. 7, Overall Evaluations of Carcinogenicity: An Updating of* IARC Monographs *Volumes 1 to 42*; International Agency for Research on Cancer: Lyon, 1987; pp. 150–152.
9. Reference 6, pp. 137–149.
10. Reference 7, pp. 83–84.
11. Weisburger, E.K. Bioassay program for carcinogenic hazards of cancer chemotherapeutic agents. *Cancer* **1977**, *40*, 1935–1949.
12. Habs, M.; Eisenbrand, G.; Schmähl, D. Carcinogenic activity in Sprague-Dawley rats of 2-[3-(2-chloroethyl)-3-nitrosoureido]-D-glucopyranose (chlorozotocin). *Cancer Lett.* **1979**, *8*, 133–137.
13. International Agency for Research on Cancer. *IARC Monographs on the Evaluation of the Carcinogenic Risk of Chemicals to Humans.* Volume 17, *Some* N-*Nitroso Compounds*; International Agency for Research on Cancer: Lyon, 1978; pp. 337–349.
14. Lewis, R.J., Sr. *Sax's Dangerous Properties of Industrial Materials*, 8th ed.; Van Nostrand-Reinhold: New York, 1992; pp. 461–462.
15. Reference 14, p. 804.
16. Reference 14, p. 3139.

17. Reference 14, p. 810.

18. Lunn, G.; Sansone, E.B.; Andrews, A.W.; Hellwig, L.C. Degradation and disposal of some antineoplastic drugs. *J. Pharm. Sci.* **1989**, *78*, 652–659.

19. Castegnaro, M.; Adams, J.; Armour, M-.A.; Barek, J.; Benvenuto, J.; Confalonieri, C.; Goff, U.; Ludeman, S.; Reed, D.; Sansone, E.B.; Telling, G., Eds., *Laboratory Decontamination and Destruction of Carcinogens in Laboratory Wastes: Some Antineoplastic Agents*; International Agency for Research on Cancer: Lyon, 1985 (IARC Scientific Publications No. 73).

OCHRATOXIN A

CAUTION! Refer to safety considerations section on page 7 before starting any of these procedures.

Ochratoxin A (I) [303-47-9][1] is a fungal metabolite from *Aspergillus ochraceus*. Ochratoxin A is teratogenic and carcinogenic to experimental animals[2–6] but it does not appear to be mutagenic.[7–9] Ochratoxin A is a white crystalline solid (mp 169°C). The solid compound may become electrostatically charged and cling to glassware or protective clothing. In a recent collaborative study organized by the International Agency for Research on Cancer (IARC) the safe disposal of ochratoxin A was investigated.[2]

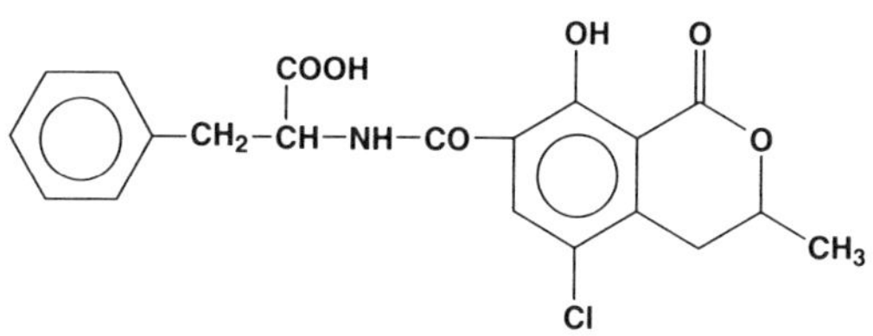

OH O
COOH
—CH$_2$—CH—NH—CO—
O
CH$_3$
Cl

I

Principle of Destruction

Ochratoxin A may be degraded using dilute sodium hypochlorite (NaOCl) solution. The efficiency of degradation was greater than 99%. It may also be degraded using potassium permanganate in sodium hydroxide solution ($KMnO_4$ in NaOH). The efficiency of degradation was greater than 99.9%.

Destruction Procedures[2]

Destruction of Bulk Quantities of Ochratoxin A

1. Prepare a dilute solution of NaOCl by adding 100 mL of commercial 5.25% NaOCl solution (Clorox bleach) to 200 mL of H_2O. Dissolve each 1 mg of ochratoxin A in 1 mL of ethanol. For each 1 mL of solution add 50 mL of the dilute NaOCl solution. Sonicate to improve solubilization, allow to react for at least 30 min, check for completeness of destruction, and discard the reaction mixture. Use fresh NaOCl solution (see assay procedure below).

2. Prepare a 0.3 *M* solution of $KMnO_4$ in 2 *M* NaOH solution by stirring the mixture for at least 30 min but no more than 2 h. Dissolve 2 mg of ochratoxin A in 5 mL of acetonitrile and add 10 mL of $KMnO_4$ in NaOH. Stir for at least 3 h. The color should be either green or purple. If it is not, add more $KMnO_4$ in NaOH until the green or purple color persists for at least 1 h. For each 10 mL of $KMnO_4$ in NaOH add 0.8 g of sodium metabisulfite (more if necessary for complete decolorization), dilute with an equal volume of water, filter to remove the manganese salts,[10] check for completeness of destruction, and discard the solid and filtrate appropriately.

Destruction of Ochratoxin A in Aqueous Solution

1. Prepare a dilute solution of NaOCl by adding 100 mL of commercial 5.25% NaOCl solution (Clorox bleach) to 200 mL of H_2O. If necessary adjust the pH of the ochratoxin A solution to neutral or alkaline. For each 1 mg of ochratoxin A present add 1 mL of ethanol. For each 1 mg of ochratoxin A add 50 mL of the dilute NaOCl solution. Sonicate to improve solubilization, allow to react for at least 30 min, check for completeness of destruction, and discard the reaction mixture. Use fresh NaOCl solution (see assay procedure below).

2. Dilute with H_2O, if necessary, so that the concentration of ochratoxin A does not exceed 200 μg/mL. Add sufficient NaOH, with stirring, to

make the concentration 2 M then add sufficient solid $KMnO_4$ to make the concentration 0.3 M. Stir for at least 3 h. The color should be either green or purple. If it is not, add more $KMnO_4$ in NaOH until the green or purple color persists for at least 1 h. For each 10 mL of $KMnO_4$ in NaOH add 0.8 g of sodium metabisulfite (more if necessary for complete decolorization), dilute with an equal volume of water, filter to remove the manganese salts,[10] check for completeness of destruction, and discard the solid and filtrate appropriately.

Destruction of Ochratoxin A in Volatile Organic Solvents

1. Prepare a dilute solution of NaOCl by adding 100 mL of commercial 5.25% NaOCl solution (Clorox bleach) to 200 mL of H_2O. Remove the solvent under reduced pressure using a rotary evaporator. Add enough ethanol to wet the glass adding at least 1 mL of ethanol for each 1 mg of ochratoxin A present. For each 1 mL of solution add 50 mL of the dilute NaOCl solution. Sonicate to improve solubilization, allow to react for at least 30 min, check for completeness of destruction, and discard the reaction mixture. Use fresh NaOCl solution (see assay procedure below).

2. Prepare a 0.3 M solution of $KMnO_4$ in 2 M NaOH solution by stirring the mixture for at least 30 min but no more than 2 h. Remove the organic solvent under reduced pressure using a rotary evaporator. Dissolve 2 mg of ochratoxin A in 5 mL of acetonitrile and add 10 mL of $KMnO_4$ in NaOH. Stir for at least 3 h. The color should be either green or purple. If it is not, add more $KMnO_4$ in NaOH until the green or purple color persists for at least 1 h. For each 10 mL of $KMnO_4$ in NaOH add 0.8 g of sodium metabisulfite (more if necessary for complete decolorization), dilute with an equal volume of H_2O, filter to remove the manganese salts,[10] check for completeness of destruction, and discard the solid and filtrate appropriately.

Destruction of Ochratoxin A in Dimethyl Sulfoxide or N,N-Dimethylformamide

Prepare a dilute solution of NaOCl by adding 100 mL of commercial 5.25% NaOCl solution (Clorox bleach) to 200 mL of H_2O. Use fresh NaOCl solution (see assay procedure below). Dilute the dimethyl sulfoxide or *N,N*-dimethylformamide solution with 2 volumes of H_2O and extract three times with equal volumes of dichloromethane, pool the extracts, and dry them over anhydrous sodium sulfate. Remove the sodium sulfate by filtration and wash it with one volume of dichloromethane. Evaporate to dryness and make sure that all the dichloromethane is removed under reduced pressure using a rotary evaporator. Add enough ethanol to wet the glass adding at least 1 mL of ethanol for each 1 mg of ochratoxin A present. For each 1 mL of solution add 50 mL of the dilute NaOCl solution. Sonicate to improve solubilization, allow to react for at least 30 min, check for completeness of destruction, and discard the reaction mixture.

Decontamination of Glassware

1. Prepare a dilute solution of NaOCl by adding 100 mL of commercial 5.25% NaOCl solution (Clorox bleach) to 200 mL of H_2O. Use fresh NaOCl solution (see assay procedure below). Add enough ethanol to wet the glassware and immerse it in the dilute NaOCl solution for at least 30 min, check for completeness of destruction, and discard the decontaminating solution.

2. Prepare a 0.3 *M* solution of $KMnO_4$ in 2 *M* NaOH solution by stirring the mixture for at least 30 min but no more than 2 h. Rinse the glassware 5 times with small portions of dichloromethane. Combine the rinses and evaporate the dichloromethane under reduced pressure using a rotary evaporator. For each 2 mg of ochratoxin A that is present add 5 mL of acetonitrile and swirl until it is dissolved. Add 10 mL of the solution of $KMnO_4$ in NaOH. Stir for at least 3 h. The color should be either green or purple. If it is not, add more $KMnO_4$ in NaOH until the green or purple color persists for at least 1 h. For each 10 mL of $KMnO_4$ in NaOH add 0.8 g of sodium metabisulfite (more if necessary for complete decolorization), dilute with an equal volume of H_2O, filter to remove the manganese salts,[10] check for completeness of destruction, and discard the solid and filtrate appropriately.

Decontamination of Protective Clothing

Prepare a dilute solution of NaOCl by adding 100 mL of commercial 5.25% NaOCl solution (Clorox bleach) to 200 mL of H_2O. Use fresh NaOCl solution (see assay procedure below). Add enough ethanol to wet the protective clothing and immerse it in the dilute NaOCl solution for at least 30 min, check for completeness of destruction, and discard the decontaminating solution.

Decontamination of Spills

1. Prepare a dilute solution of NaOCl by adding 100 mL of commercial 5.25% NaOCl solution (Clorox bleach) to 200 mL of H_2O. Use fresh NaOCl solution (see assay procedure below). Collect spills of liquid with a dry cloth and spills of solid with a tissue wetted with sodium bicarbonate solution (5% w/v). Wipe the area with a cloth wetted with sodium bicarbonate solution (5% w/v). Immerse all cloths in the dilute NaOCl solution, allow to react for at least 30 min, check for completeness of decontamination, and discard the decontaminating solution.

Cover the spill area with the dilute NaOCl solution. After at least 30 min absorb the liquid with cloths and discard it. Check the surface for completeness of decontamination by using a wipe moistened with methanol and analyzing the wipe for the presence of ochratoxin A.

2. Prepare a 0.3 *M* solution of $KMnO_4$ in 2 *M* NaOH solution by stirring the mixture for at least 30 min but no more than 2 h. Collect spills of liquid with a dry tissue and spills of solid with a tissue wetted with dichloromethane. Immerse all tissues in the $KMnO_4$ in NaOH solution. Allow to react for at least 3 h. The color should be either green or purple. If it is not, add more $KMnO_4$ in NaOH until the green or purple color persists for at least 1 h. For each 10 mL of $KMnO_4$ in NaOH add 0.8 g of sodium metabisulfite (more if necessary for complete decolorization), dilute with an equal volume of water, filter to remove manganese salts,[10] check for completeness of destruction, and discard the solid and filtrate appropriately.

Cover the spill area with an excess of the $KMnO_4$ in NaOH solution and allow to react for 3 h. Collect the solution on a tissue and immerse the tissue in 2 *M* sodium metabisulfite solution. If the pH of this solution is acidic, make it alkaline with sodium hydroxide. Rinse the spill area with a 2 *M* solution of sodium metabisulfite. Check the surface for completeness of decontamination by using a wipe moistened with methanol and analyzing the wipe for the presence of ochratoxin A.

Decontamination of Thin-Layer Chromatography Plates

Prepare a dilute solution of NaOCl by adding 100 mL of commercial 5.25% NaOCl solution (Clorox bleach) to 200 mL of H_2O. Use fresh NaOCl solution (see assay procedure below). Spray the plate with the dilute NaOCl solution and allow to react for at least 30 min. Check for completeness of destruction by scraping the plate and eluting any remaining ochratoxin A with a suitable solvent.

Analytical Procedures

1. For reaction mixtures from the NaOCl decontaminations acidify an aliquot of the final reaction mixture to pH 3–4 with concentrated hydrochloric acid (HCl) and pass nitrogen through the mixture for at least 1 min to remove chlorine. Analyze by reverse phase high-performance liquid chromatography (HPLC) using acetonitrile:0.25 *M* aqueous phosphoric acid 75:25 flowing at 1.5 mL/min and a UV detector set at 254 nm or a spectrofluorometric detector using 340 nm for excitation and 465 nm for emission.[2] An alternative mobile phase is methanol:2% acetic acid 70:30.[11]

2. For reaction mixtures obtained using the $KMnO_4$ procedures, acidify an aliquot to pH 2–3 using concentrated HCl. Extract this mixture three times with an equal volume of dichloromethane, pool the extracts, and dry them over anhydrous sodium sulfate. Remove the sodium sulfate by filtration, evaporate to dryness and take up the residue in 0.5 mL of acetonitrile:water 75:25. Analyze by HPLC as above.

Mutagenicity Assays

The residues from these degradation procedures were tested using tester strains TA97, TA98, TA100, and TA102 of *Salmonella typhimurium* with and without metabolic activation. No mutagenic activity was found.[2] Ochratoxin A has not been found to be mutagenic in testing.[7–9]

Related Compounds

The above techniques were investigated for ochratoxin A but they may also be applicable to some other mycotoxins. However, these techniques should be thoroughly investigated before being applied to other compounds. See also the monographs on Aflatoxins, Citrinin, Patulin, and Sterigmatocystin.

Assay of Sodium Hypochlorite Solution

Sodium hypochlorite solutions tend to deteriorate with time, so they should be periodically checked for the amount of active chlorine they contain. Pipette 10 mL of the NaOCl solution into a 100-mL volumetric flask and fill it to the mark with distilled H_2O. Pipette 10 mL of this solution into a conical flask containing 50 mL of distilled H_2O, 1 g of potassium iodide, and 12.5 mL of 2 *M* acetic acid. Titrate this solution against a 0.1 *N* sodium thiosulfate solution using starch as an indicator. Each 1 mL of the sodium thiosulfate solution corresponds to 3.545 mg of active chlorine. Commercially available NaOCl solution (Clorox bleach) contains 5.25% NaOCl and should contain 45–50 g of active chlorine per liter.

References

1. Other names for this compound are (*R*)-*N*-[(5-chloro-3,4-dihydro-8-hydroxy-3-methyl-1-oxo-1*H*-2-benzopyran-7-yl)carbonyl]-L-phenylalanine, *N*-[[(3*R*)-5-chloro-8-hydroxy-3-methyl-1-oxo-7-isochromanyl]carbonyl]-3-phenyl-L-alanine, and (−)-*N*-[(5-chloro-8-hydroxy-3-methyl-1-oxo-7-isochromanyl)carbonyl]-3-phenylalanine.
2. Castegnaro, M.; Barek, J.; Frémy, J.-M.; Lafontaine, M.; Miraglia, M.; Sansone, E.B.; Telling, G.M., Eds., *Laboratory Decontamination and Destruction of Carcinogens in Laboratory Wastes: Some Mycotoxins*; International Agency for Research on Cancer: Lyon, 1991 (IARC Scientific Publications No. 113).
3. Kanisawa, M. Synergistic effect of citrinine on hepatorenal carcinogenesis of ochratoxin A in mice. In *Toxic Fungi, Their Toxins and Health Hazard*; Kurata, H., Ueno, Y., Eds., Elsevier/North Holland: Amsterdam, 1984; pp. 245–254.
4. Bendele, A.M.; Carlton, W.W.; Krogh, P.; Lillehoj, E.B. Ochratoxin A carcinogenesis in the (C57BL/6JXC3H)F_1 mouse. *J. Natl. Cancer Inst.* **1985**, *75*, 733–739.
5. Boorman, G.A. *Toxicology and carcinogenesis studies of ochratoxin A (CAS No. 303-47-9) in F344/N rats*; National Toxicology Program Technical Report 358: NIH Publication No. 89-2813, 1989.
6. International Agency for Research on Cancer. *IARC Monographs on the Evaluation of Carcinogenic Risk of Chemicals to Man.* Volume 31, *Some Food Additives, Feed Additives and Naturally Occurring Substances*; International Agency for Research on Cancer: Lyon, 1983; pp. 191–206.
7. Hayes, A.W. *Mycotoxin Teratogenicity and Mutagenicity*; CRC Press: Boca Raton, FL, 1981.
8. Bendele, A.M.; Neal, S.B.; Oberley, T.J.; Thompson, C.Z.; Bewsey, B.J.; Hill, L.E.; Rexroat, M.A.; Carlton, W.W.; Probst, G.S. Evaluation of ochratoxin A for mutagenicity in a battery of bacterial and mammalian cell assays. *Fd. Chem. Toxicol.* **1985**, *23*, 911–918.
9. Würgler, F.E.; Friedrich, U.; Schlatter, J. Lack of mutagenicity of ochratoxin A and B, citrinin, patulin and cnestine in *Salmonella typhimurium* TA102. *Mutat. Res.* **1991**, *261*, 209–216.
10. Lunn, G.; Sansone, E.B.; De Méo, M.; Laget, M.; Castegnaro, M. Potassium permanganate can be used for degrading hazardous compounds. *Am. Ind. Hyg. Assoc. J.* **1994**, *55*, 167–171.
11. Lunn, G. Unpublished observations.

ORGANIC NITRILES

> **CAUTION!** Refer to safety considerations section on page 7 before starting any of these procedures.

Little work has been done on the chemical degradation of these compounds but some results have been obtained for simple compounds. The following compounds were degraded with nickel–aluminum (Ni–Al) alloy in dilute base:[1] acetonitrile [75-05-8],[2] 3-ethoxypropionitrile [2141-62-0], benzonitrile [100-47-0],[3] and benzyl cyanide [140-29-4].[4] Acetonitrile (bp 82°C), 3-ethoxypropionitrile (bp 172°C), benzonitrile (bp 188°C), and benzyl cyanide (bp 233–234°C) are volatile liquids. Acetonitrile reacts exothermically with sulfuric acid, is moderately toxic, produces convulsions, nausea, and vomiting, and is a teratogen.[5] Benzyl cyanide has been known to explode with sodium hypochlorite.[6]

Principle of Destruction

These compounds were reduced with Ni–Al alloy in dilute base to give the corresponding amine in 67–86% yield. Destruction was greater than 99% in all cases.

Destruction Procedure

Take up 0.5 g of the nitrile in 50 mL of H_2O, then add 50 mL of 1 *M* potassium hydroxide (KOH) solution. Stir this mixture and add 5 g of Ni–Al alloy in portions to avoid frothing. Stir the reaction mixture overnight, then filter through a pad of Celite®. Neutralize the filtrate, check for completeness of destruction, and discard it. Place the spent nickel on a metal tray, allow it to dry away from flammable solvents for 24 h, and discard it.

Analytical Procedures

For analysis by gas chromatography (GC)[1] a 1.8-m × 2-mm i.d. packed column can be used together with flame ionization detection. The injection temperature was 200°C, the detector temperature was 300°C, and the carrier gas was nitrogen flowing at 30 mL/min. For acetonitrile (100°C) and 3-ethoxypropionitrile (100°C) the packing was 10% Carbowax 20 M + 2% KOH on 80/100 Chromosorb W AW and for benzonitrile (120°C) and benzyl cyanide (120°C) the packing was 2% Carbowax 20 M + 1% KOH on 80/100 Supelcoport. The oven temperatures shown above in parentheses are only a guide and the exact conditions would have to be determined experimentally.

Related Compounds

Reduction with Ni–Al alloy does not degrade inorganic cyanides.[1] These compounds should be degraded as described in the Cyanides and Cyanogen Bromide monograph. A number of aryl nitriles have been reduced to the corresponding amines with Ni–Al alloy,[7,8] but complete destruction of the starting material has not been established. Full validation should be carried out before this process is used on a routine basis.

References

1. Lunn, G. Unpublished observations.
2. Other names for this compound are ethyl nitrile, methanecarbonitrile, cyanomethane, methyl cyanide, ethanenitrile, NCI-C60822, and USAF EK-488.
3. Other names for this compound are benzoic acid nitrile, cyanobenzene, phenyl cyanide, and benzenenitrile.

4. Other names for this compound are benzeneacetonitrile, benzyl nitrile, α-cyanotoluene, ω-cyanotoluene, phenylacetonitrile, α-tolunitrile, (cyanomethyl)benzene, and USAF K-21.

5. Lewis, R.J., Sr. *Sax's Dangerous Properties of Industrial Materials*, 8th ed.; Van Nostrand-Reinhold: New York, 1992; p. 24.

6. Reference 5, p. 2737.

7. Staskun, B.; van Es, T. Reductions with Raney alloy in alkaline solution. *J. Chem. Soc. (C)* **1966**, 531–532.

8. Kametani, T.; Nomura, Y. Studies on a catalyst. II. Reduction of nitrogen compounds by Raney nickel alloy and alkali solution. 2. Synthesis of amines by reduction of nitriles. *J. Pharm. Soc. Jpn.* **1954**, *74*, 889–891; *Chem. Abstr.* **1956**, *50*, 2467f.

OSMIUM TETROXIDE

CAUTION! Refer to safety considerations section on page 7 before starting any of these procedures.

Osmium tetroxide [OsO_4, osmic acid, osmium(VIII) oxide] [20816-12-0] is a low-melting solid, mp 39.5–41°C, bp 130°C. This compound is quite volatile and its vapors irritate and burn the eyes severely[1] and affect the lungs.[2] Osmium tetroxide is widely used in synthetic organic chemistry and in electron microscopy laboratories.

Principle of Decontamination

Osmium tetroxide reacts with double bonds to form a very stable diester. In this form the OsO_4 is no longer volatile. Corn oil contains a large proportion of double bonds and it is an effective agent for the neutralization of OsO_4.[3] Commercially available Mazola corn oil was used in the tests. Although this procedure eliminates hazards due to the volatility of OsO_4, the material still contains osmium and it should be disposed of as waste containing heavy metals. It has been reported that OsO_4 can be reduced to the dioxide by reacting it with an alkene, bubbling hydrogen sulfide through the solution and removing the osmium dioxide by filtration.[4]

Disposal Procedures

Bulk Quantities and Residues in Containers

Place corn oil in the container. Test for completeness of reaction.

Aqueous Solutions (2%)

Allow to react with twice the volume of corn oil. Test for completeness of reaction.

Spills

A 200-g sample of absorbent granules (e.g., cat litter) absorbs 100 mL of corn oil. Use this mixture to neutralize a spill of 50 mL of 2% OsO_4. Test for completeness of reaction then remove the granules. The corn oil absorbent granules mixture can be kept in a tightly sealed plastic bag for at least 1 month with no loss of effectiveness. Laboratories in which OsO_4 is in routine use should keep at least one bag on hand.

Analytical Procedures

Either a glass cover slip coated in corn oil or a piece of filter paper soaked in corn oil was suspended over the solution. Blackening indicated that OsO_4 was still present.

Related Compounds

This method is specific for OsO_4.

References

1. Bretherick, L., Ed., *Hazards in the Chemical laboratory*, 4th ed.; Royal Society of Chemistry: London, 1986; p. 434.
2. Lewis, R.J., Sr. *Sax's Dangerous Properties of Industrial Materials*, 8th ed.; Van Nostrand-Reinhold: New York, 1992; p. 2637.
3. Cooper, K. Neutralization of osmium tetroxide in case of accidental spillage and for disposal. *Bull. Microscop. Soc. Canada* **1980**, *8*, 24–28.
4. Armour, M-.A.; Browne, L.M.; Weir, G.L., Eds., *Hazardous Chemicals. Information and Disposal Guide*, 3rd ed.; University of Alberta: Edmonton, Alberta, 1987; p. 263.

PATULIN

CAUTION! Refer to safety considerations section on page 7 before starting any of these procedures.

Patulin (I) [149-29-1][1] is an antibiotic isolated from a number of fungi (e.g., *Aspergillus clavatus*).[2] Patulin is a white solid (mp 111°C), which is soluble in H_2O and common organic solvents except petroleum ether.[2] Patulin may be carcinogenic in experimental animals[3] and a teratogen[4] but it does not appear to be mutagenic.[5,6] The solid compound may become electrostatically charged and cling to glassware or protective clothing. In a recent collaborative study organized by the International Agency for Research on Cancer (IARC) the safe disposal of this compound was investigated.[7]

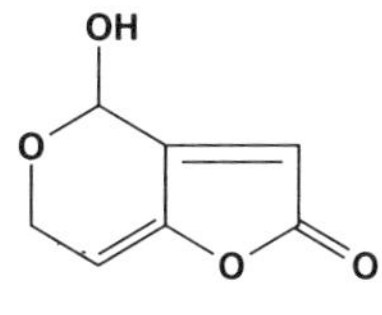

OH

O

O O

I

Principles of Destruction

Patulin may be degraded using potassium permanganate in sodium hydroxide solution ($KMnO_4$ in NaOH) or by heating with ammonia (NH_3) in an autoclave. The degradation efficiency was greater than 99.9% using $KMnO_4$ and greater than 99.5% using NH_3.

Destruction Procedures[7]

Destruction of Bulk Quantities of Patulin

1. Prepare a 0.3 *M* solution of $KMnO_4$ in 2 *M* NaOH solution by stirring the mixture for at least 30 min but no more than 2 h. Dissolve 400 μg of patulin in 5 mL of acetonitrile and add 10 mL of $KMnO_4$ in NaOH. Stir for at least 3 h. The color should be either green or purple. If it is not, add more $KMnO_4$ in NaOH until the green or purple color persists for at least 1 h. For each 10 mL of $KMnO_4$ in NaOH add 0.8 g of sodium metabisulfite (more if necessary for complete decolorization), dilute with an equal volume of H_2O, filter to remove the manganese salts,[8] check for completeness of destruction, and discard the solid and filtrate appropriately.

2. For each 100 μg of patulin add at least 10 mL of 5% (w/w) NH_3 solution. Sonicate for 1 min to improve solubilization. Cover the container with aluminum foil but do not close the container tightly. Place the container in an autoclave and heat under pressure to 120 °C for 15 min. Allow to cool, test for completeness of degradation, and discard the reaction mixture.

Destruction of Patulin in Aqueous Solution

1. Dilute with H_2O, if necessary, so that the concentration of patulin does not exceed 200 μg/mL. Add sufficient NaOH, with stirring, to make the concentration 2 *M* then add sufficient solid $KMnO_4$ to make the concentration 0.3 *M*. Stir for at least 3 h. The color should be either green or purple. If it is not, add more $KMnO_4$ in NaOH until the green or purple color persists for at least 1 h. For each 10 mL of $KMnO_4$ in NaOH add 0.8 g of sodium metabisulfite (more if necessary for complete decolorization), dilute with an equal volume of H_2O, filter to remove the manganese salts,[8] check for completeness of destruction, and discard the solid and filtrate appropriately.

2. Dilute with H_2O, if necessary, so that the concentration of patulin does not exceed 20 μg/mL then add an equal volume of 10% (w/w) NH_3 solution. Sonicate for 1 min to improve solubilization. Cover the container with aluminum foil but do not close the container tightly. Place the container in an autoclave and heat under pressure to 120°C for 15 min. Allow to cool, test for completeness of degradation, and discard the reaction mixture.

Destruction of Patulin in Volatile Organic Solvents

1. Prepare a 0.3 *M* solution of $KMnO_4$ in 2 *M* NaOH solution by stirring the mixture for at least 30 min but no more than 2 h. Remove the organic solvent under reduced pressure using a rotary evaporator. Dissolve 400 μg of patulin in 5 mL of acetonitrile and add 10 mL of $KMnO_4$ in NaOH. Stir for at least 3 h. The color should be either green or purple. If it is not, add more $KMnO_4$ in NaOH until the green or purple color persists for at least 1 h. For each 10 mL of $KMnO_4$ in NaOH add 0.8 g of sodium metabisulfite (more if necessary for complete decolorization), dilute with an equal volume of H_2O, filter to remove the manganese salts,[8] check for completeness of destruction, and discard the solid and filtrate appropriately.

2. Evaporate the solvent under reduced pressure using a rotary evaporator. For each 100 μg of patulin add at least 10 mL of 5% (w/w) NH_3 solution. Sonicate for 1 min to improve solubilization. Cover the container with aluminum foil but do not close the container tightly. Place the container in an autoclave and heat under pressure to 120°C for 15 min. Allow to cool, test for completeness of degradation, and discard the reaction mixture.

Decontamination of Animal Litter

Spread the contaminated litter on a suitable tray to a maximum depth of 5 cm and, for each 10 g of litter, add 16 mL of 5% ammonia solution (w/w). Autoclave at 128–130°C for 20 min but do not preevacuate the autoclave, which would remove the ammonia. Cool to room temperature, analyze the litter for completeness of decontamination, and discard it.

Decontamination of Glassware

1. Prepare a 0.3 *M* solution of $KMnO_4$ in 2 *M* NaOH solution by stirring the mixture for at least 30 min but no more than 2 h. Rinse the glassware

five times with small portions of dichloromethane. Combine the rinses and evaporate the dichloromethane under reduced pressure using a rotary evaporator. Dissolve 400 μg of patulin in 5 mL of acetonitrile and add 10 mL of $KMnO_4$ in NaOH. Stir for at least 3 h. The color should be either green or purple. If it is not, add more $KMnO_4$ in NaOH until the green or purple color persists for at least 1 h. For each 10 mL of $KMnO_4$ in NaOH add 0.8 g of sodium metabisulfite (more if necessary for complete decolorization), dilute with an equal volume of H_2O, filter to remove the manganese salts,[8] check for completeness of destruction, and discard the solid and filtrate appropriately.

2. Rinse the glassware five times with ethyl acetate and evaporate the ethyl acetate under reduced pressure using a rotary evaporator. For each 100 μg of patulin add at least 10 mL of 5% (w/w) NH_3 solution. Sonicate for 1 min to improve solubilization. Cover the container with aluminum foil but do not close the container tightly. Place the container in an autoclave and heat under pressure to 120°C for 15 min. Allow to cool, test for completeness of degradation, and discard the reaction mixture.

Decontamination of Spills

Prepare a 0.3 *M* solution of $KMnO_4$ in 2 *M* NaOH solution by stirring the mixture for at least 30 min but no more than 2 h. Collect spills of liquid with a dry tissue and spills of solid with a tissue wetted with dichloromethane. Immerse all tissues in the $KMnO_4$ in NaOH solution. Allow to react for at least 3 h. The color should be either green or purple. If it is not, add more $KMnO_4$ in NaOH until the green or purple color persists for at least 1 h. For each 10 mL of $KMnO_4$ in NaOH add 0.8 g of sodium metabisulfite (more if necessary for complete decolorization), dilute with an equal volume of H_2O, filter to remove the manganese salts,[8] check for completeness of destruction, and discard the solid and filtrate appropriately.

Cover the spill area with an excess of the $KMnO_4$ in NaOH solution and allow to react for 3 h. Collect the solution on a tissue and immerse the tissue in 2 *M* sodium metabisulfite solution. If the pH of this solution is acidic, make it alkaline with NaOH. Rinse the spill area with a 2 *M* solution of sodium metabisulfite. Check the surface for completeness of decontamination by using a wipe moistened with methanol and analyzing the wipe for the presence of patulin.

Analytical Procedures

1. For reaction mixtures obtained using the $KMnO_4$ procedures, acidify an aliquot to pH 2–3 using concentrated hydrochloric acid (HCl). Extract this mixture three times with an equal volume of dichloromethane, pool the extracts, and dry them over anhydrous sodium sulfate. Remove the sodium sulfate by filtration, evaporate to dryness, and take up the residue in 0.5 mL of acetonitrile: water 90:10. Analyze by reverse phase high-performance liquid chromatography (HPLC) using acetonitrile:water 90:10 flowing at 1 mL/min and an ultraviolet (UV) detector set at 275 nm.

2. For decontaminated animal litter stir 10 g of cooled litter with 50 mL of dichloromethane for 30 min and filter. Separate the layers and dry the lower (dichloromethane) layer over anhydrous sodium sulfate. Remove the solvent under reduced pressure using a rotary evaporator and take up the residue in about 2 mL of methanol. Analyze by HPLC as above.

3. For decontamination using aqueous NH_3 acidify an aliquot to pH 5–6 using concentrated HCl and extract this solution three times with equal volumes of ethyl acetate. Pool the extracts, dry over anhydrous sodium sulfate, and evaporate to near dryness under reduced pressure using a rotary evaporator. Evaporate to complete dryness under a gentle stream of nitrogen then take up the residue in 0.5 mL of acetonitrile:water 90:10 and analyze by HPLC as above.

Mutagenicity Assays[7]

The residues from the decontamination of animal litter were not mutagenic to tester strains TA98, TA100, TA1530, and TA1535 of *Salmonella typhimurium* with and without metabolic activation. The residues from the other ammonia degradations were not mutagenic to TA97, TA98, TA100, and TA102. Patulin itself has not been found to be mutagenic.[5,6]

Related Compounds

The above techniques were investigated for patulin but they may also be applicable to some other mycotoxins. However, they should be thoroughly investigated before being applied to other compounds. See also the monographs for Aflatoxins, Citrinin, Ochratoxin A, and Sterigmatocystin.

References

1. Other names for this compound are 4-hydroxy-4*H*-furo[3,2-*c*]pyran-2(6*H*)-one, anhydro-3-hydroxymethylene-tetrahydro-γ-pyrene-2-carboxylic acid, clairformin, clavacin, clavatin, claviformin, [2,4-dihydroxy-2*H*-pyran 3 (6*H*)ylidene] acetic acid 3,4-lactone, 2,4-dihydroxy-2*H*-pyran-Δ-3(6*H*)-α-acetic acid-3,4-lactone, expansin, expansine, mycoin, mycoin C, mycoin C_3, penicidin, terinin, tercinin, gigantin, leucopin, mycoine C3, mycosin, and penatin.
2. Budavari, S., Ed., *The Merck Index*, 11th ed; Merck & Co., Inc.: Rahway, NJ, 1989, p. 1116.
3. International Agency for Research on Cancer. *IARC Monographs on the Evaluation of Carcinogenic Risk of Chemicals to Man*. Volume 10, *Some Naturally Occurring Substances*; International Agency for Research on Cancer: Lyon, 1975; pp. 205–210.
4. Lewis, R.J., Sr. *Sax's Dangerous Properties of Industrial Materials*, 8th ed.; Van Nostrand-Reinhold: New York, 1992; pp. 920–921.
5. Hayes, A.W. *Mycotoxin Teratogenicity and Mutagenicity*; CRC Press: Boca Raton, FL, 1981.
6. Würgler, F.E.; Friedrich, U.; Schlatter, J. Lack of mutagenicity of ochratoxin A and B, citrinin, patulin and cnestine in *Salmonella typhimurium* TA102. *Mutat. Res.* **1991**, *261*, 209–216.
7. Castegnaro, M.; Barek, J.; Frémy, J-.M.; Lafontaine, M.; Miraglia, M.; Sansone, E.B.; Telling, G.M., Eds., *Laboratory Decontamination and Destruction of Carcinogens in Laboratory Wastes: Some Mycotoxins*; International Agency for Research on Cancer: Lyon, 1991 (IARC Scientific Publications No. 113).
8. Lunn, G.; Sansone, E.B.; De Méo, M.; Laget, M.; Castegnaro, M. Potassium permanganate can be used for degrading hazardous compounds. *Am. Ind. Hyg. Assoc. J.* **1994**, 55, 167–171.

PERACIDS

CAUTION! Refer to safety considerations section on page 7 before starting any of these procedures.

Peracids, for example, peracetic acid ($CH_3C(O)OOH$) [79-21-0][1] and *m*-chloroperbenzoic acid ($3\text{-}ClC_6H_4C(O)OOH$) [937-14-4],[2] have an extra oxygen. These compounds are powerful oxidizing agents and are incompatible with a variety of organic and inorganic compounds.[3,4] Peracetic acid[3] and *m*-chloroperbenzoic acid[4] may be carcinogenic. These compounds are used in organic chemistry.

Destruction Procedure[5]

Add 5 mL or 5 g of the compound to 100 mL of 10% (w/v) sodium metabisulfite solution and stir the mixture at room temperature. Test for completeness of destruction by adding a few drops of the reaction mixture to an equal volume of 10% (w/v) potassium iodide solution, acidifying with 1 *M* hydrochloric acid solution, and adding a drop of starch as an indicator. A deep blue color indicates the presence of excess oxidant. If destruction is complete, discard the mixture. If destruction is not complete, add more sodium metabisulfite solution until a negative test is obtained.

Related Compounds

This technique should be generally applicable to other peracids.

References

1. Other names for this compound are peroxyacetic acid, acetyl hydroperoxide, and ethaneperoxoic acid.
2. Other names for this compound are MCPBA, 3-chlorobenzenecarboperoxoic acid, *m*-chlorobenzoyl hydroperoxide, and 3-chloroperoxybenzoic acid.
3. Lewis, R.J., Sr. *Sax's Dangerous Properties of Industrial Materials*, 8th ed.; Van Nostrand-Reinhold: New York, 1992; p. 2709.
4. Reference 3, pp. 846–847.
5. Lunn, G. Unpublished observations.

PEROXIDES AND HYDROPEROXIDES

> **CAUTION!** Refer to safety considerations section on page 7 before starting any of these procedures.

CAUTION! Peroxides are frequently formed in certain organic solvents, such as ethers. These compounds are dangerously unstable and actions such as removing the container cap may cause them to explode. The help of people specially trained to deal with explosives should be sought in these cases.

Some relatively stable peroxides and hydroperoxides are used in organic chemistry, particularly for the initiation of polymerization reactions. If care is exercised, the use of these compounds should not be particularly hazardous although they may explode when subjected to shock or exposed to heat. These compounds are powerful oxidizers and may react violently with reducing agents.[1] Hydrogen peroxide (H_2O_2) [7722-84-1][2–4] has a wide variety of uses in organic and inorganic chemistry, as a bleach, and as a disinfectant. This compound may be carcinogenic.[5] Sodium peroxide (Na_2O_2) [1313-60-6][6–8] and *tert*-butyl hydroperoxide [$(CH_3)_3COOH$] [75-91-2][9–11] are also commonly used in the laboratory. *tert*-Butyl hydroperoxide may explode on distillation and on contact with molecular sieve and it may cause fatal respiratory arrest at high concentrations.[10–11] All of these compounds are severe skin, eye, and mucous membrane irritants and are incompatible with a wide variety of organic and inorganic compounds.

Principles of Destruction

Hydrogen peroxide, *tert*-butyl hydroperoxide, and sodium peroxide can be reduced with sodium metabisulfite and diacyl peroxides can be reduced with sodium or potassium iodide (NaI or KI).

Destruction Procedures

Hydrogen Peroxide[12]

Add 5 mL of 30% H_2O_2 to 100 mL of 10% (w/v) sodium metabisulfite solution and stir the mixture at room temperature. Test for completeness of destruction by adding a few drops of the reaction mixture to an equal volume of 10% (w/v) KI, acidifying with 1 *M* hydrochloric acid (HCl) solution, and adding a drop of starch as an indicator. A deep blue color indicates the presence of excess oxidant. If destruction is complete, discard the mixture. If destruction is not complete, add more sodium metabisulfite solution until a negative test is obtained.

tert-*Butyl Hydroperoxide*[12]

Add 5 mL of *tert*-butyl hydroperoxide to 100 mL of 10% (w/v) sodium metabisulfite solution and stir the mixture at room temperature. Test for completeness of destruction by adding a few drops of the reaction mixture to an equal volume of 10% (w/v) KI solution, acidifying with 1 *M* HCl solution, and adding a drop of starch as an indicator. A deep blue color indicates the presence of excess oxidant. If destruction is complete, discard the mixture. If destruction is not complete, add more sodium metabisulfite solution until a negative test is obtained.

Sodium Peroxide[12]

Add 1 g of sodium peroxide to 100 mL of 10% (w/v) sodium metabisulfite solution and stir the mixture at room temperature. Test for completeness of destruction by adding a few drops of the reaction mixture to an equal volume of 10% (w/v) KI solution, acidifying with 1 *M* HCl solution, and adding a drop of starch as an indicator. A deep blue color indicates the presence of excess oxidant. If destruction is complete, discard the mixture. If destruction is not complete, add more sodium metabisulfite solution until a negative test is obtained.

Diacyl Peroxides[13]

Dissolve 3.3 g of NaI or 3.65 g of KI in 70 mL of glacial acetic acid. Stir this mixture at room temperature and slowly add 0.01 mol of the peroxide. The mixture darkens because iodine is formed. After 30 min discard the mixture.

References

1. Lewis, R.J., Sr. *Sax's Dangerous Properties of Industrial Materials*, 8th ed.; Van Nostrand-Reinhold: New York, 1992; pp. 2708–2709.
2. Other names for this compound are Albone, dihydrogen dioxide, Hioxyl, hydrogen dioxide, hydroperoxide, Inhibine, Oxydol, Perhydrol, Perone, Peroxan, Peroxide, Superoxol, and T-Stuff.
3. Reference 1, pp. 1901–1902.
4. Bretherick, L. *Bretherick's Handbook of Reactive Chemical Hazards*, 4th ed.; Butterworths: London, 1990; pp. 1198–1215.
5. International Agency for Research on Cancer. *IARC Monographs on the Evaluation of the Carcinogenic Risk of Chemicals to Humans*. Volume 36, *Allyl Compounds, Aldehydes, Epoxides and Peroxides*; International Agency for Research on Cancer. Lyon, 1985; pp. 285–314.
6. Other names for this compound are disodium dioxide, disodium peroxide, Flocool 180, sodium dioxide, sodium oxide, sodium superoxide, and Solozone.
7. Reference 1, pp. 3104–3105.
8. Reference 4, pp. 1380–1384.
9. Other names for this compound are Cadox TBH, 1,1-dimethylethyl hydroperoxide, 2-hydroperoxy-2-methylpropane, Perbutyl H, and TBHP-70.
10. Reference 1, p. 619.
11. Reference 4, pp. 488–489.
12. Lunn, G. Unpublished observations.
13. National Research Council, Committee on Hazardous Substances in the Laboratory. *Prudent Practices for Disposal of Chemicals from Laboratories;* National Academy Press: Washington, DC, 1983; p. 76.

PHOSGENE

CAUTION! Refer to safety considerations section on page 7 before starting any of these procedures.

Phosgene ($COCl_2$) [75-44-5][1] is a toxic gas (bp 8°C) that is most conveniently handled in toluene solution. This compound is commercially available from a number of sources as a 20% toluene solution. The preparation of a solution of phosgene in toluene has been described.[2] Phosgene decomposes in the presence of moisture to form hydrochloric acid and carbon monoxide. This compound is a severe skin, eye, and mucous membrane irritant and may cause rapidly developing pulmonary edema with little warning.[3] Phosgene is used industrially as an intermediate and in the laboratory in organic synthesis.

Triphosgene [bis(trichloromethyl)carbonate, ($(CCl_3O)_2CO$)] [32315-10-9] is a white crystalline solid (mp 79-83°C) that undergoes the same reactions as phosgene. Thus triphosgene can be used as a convenient substitute for phosgene[4,5] although the compound is still very toxic. Triphosgene can produce pressure on storage[6] and trace amounts of moisture will produce phosgene.[7] Thus triphosgene should be handled with the same precautions used for phosgene itself.[7]

Principle of Destruction

Under controlled conditions phosgene is hydrolyzed with sodium hydroxide (NaOH) solution. Destruction of phosgene in toluene is greater than 99.995% and destruction of triphosgene is greater than 99.7%.[8]

Destruction of Phosgene in Toluene Solution[8]

Cautiously add 50 mL of a 20% solution of phosgene in toluene to 100 mL of 20% NaOH solution and stir the mixture in a 500 mL flask overnight. Cover the reaction vessel with aluminum foil to prevent uptake of atmospheric carbon dioxide (CO_2). Analyze for completeness of destruction, separate the aqueous and organic layers and discard them.

Destruction of Gaseous Phosgene[9]

Gaseous phosgene may be degraded by allowing it to pass through a 20% (w/v) NaOH solution.

Destruction of Triphosgene[8]

Cautiously add 20 mL of 20% NaOH solution to 2 g of triphosgene and stir the mixture overnight. Cover the reaction vessel with aluminum foil to prevent uptake of atmospheric CO_2. Analyze for completeness of destruction and discard it.

Analytical Procedure[8]

Analyze toluene layers directly. Analyze aqueous layers by adding 1 mL of the aqueous layer to 9 mL of toluene. Add anhydrous magnesium sulfate, shake, allow to settle, and analyze this mixture. Add 100 μL of the mixture to be tested to 2 mL of 2-methoxyethanol and then add 1 mL of a 5% (w/v) solution of 4-(4-nitrobenzyl)pyridine in 2-methoxyethanol. Determine the absorbance of the solution at 475 nm using an appropriate blank. The limit of detection was about 0.01 g/L.

Related Compounds

This procedure is specific for phosgene but similar procedures can be used to degrade related compounds as described in the monographs on Acid Halides and Anhydrides and Chlorosulfonic Acid.

References

1. Other names for this compound are carbon oxychloride, chloroformyl chloride, carbonyl chloride, carbonyl dichloride, and NCI-C60219.
2. Carter, H.E.; Frank, R.L.; Johnston, H.W. Carbobenzoxy chloride and derivatives. In *Organic Syntheses*; Horning, E.C., Ed., Wiley: New York, 1955; Coll. Vol. 3, pp. 167–169.
3. Lewis, R.J., Sr. *Sax's Dangerous Properties of Industrial Materials*, 8th ed.; Van Nostrand-Reinhold: New York, 1992; pp. 2782–2783.
4. Eckert, H.; Forster, B. Triphosgene, a crystalline phosgene substitute. *Angew. Chem. Int. Ed. Engl.* **1987**, *26*, 894–895.
5. Anonymous. Triphosgene. *Aldrichimica Acta* **1988**, *21*, 47.
6. Hollingsworth, M.D. Triphosgene warning. *Chem. Eng. News* **1992**, *July 13*, 4.
7. Damle, S.B. Safe handling of diphosgene, triphosgene. *Chem. Eng. News* **1993**, *Feb. 8*, 4.
8. Lunn, G. Unpublished observations.
9. Shriner, R.L.; Horne, W.H.; Cox, R.F.B. *p*-Nitrophenyl isocyanate. In *Organic Syntheses*; Blatt, A.H., Ed., Wiley: New York, 1943; Coll. Vol. 2, pp. 453–455.

PHOSPHORUS AND PHOSPHORUS PENTOXIDE

> **CAUTION!** Refer to safety considerations section on page 7 before starting any of these procedures.

Phosphorus (P) [7723-14-0] is used in the chemical laboratory. White phosphorus (sometimes called yellow phosphorus) is pyrophoric and can explode when it reacts with a variety of chemicals.[1] This compound is a poison and has a number of severe health effects including necrosis of the jaw.[1] Red phosphorus is not pyrophoric, but it is flammable and can explode when mixed with a variety of compounds.[2] Phosphorus pentoxide (P_2O_5) [1314-56-3][3] is corrosive and reacts violently with H_2O.[4] This compound is used as a drying or dehydrating agent in the laboratory.

Principles of Destruction

White phosphorus is oxidized by copper(II) sulfate to phosphoric acid[5] and red phosphorus is oxidized to phosphoric acid by potassium bromate.[6,7] Despite reports to the contrary,[8–12] including the first edition of this book, red phosphorus is not degraded by potassium chlorate.[6] Phosphorus pentoxide is hydrolyzed to phosphoric acid.[6]

Destruction Procedures

White Phosphorus

Cut 5 g of white phosphorus under H_2O into pellets that are no more than 5 mm across and add these pellets to 800 mL of 1 *M* cupric sulfate solution. Allow the reaction mixture to stand in a 2-L beaker in a hood for about a week. Stir occasionally. If one of the larger black pellets is cut under H_2O and no waxy white phosphorus is observed, the reaction is complete. Filter off the precipitate and, while keeping it wet, add it to 500 mL of 5.25% sodium hypochlorite (NaOCl) solution. Stir this mixture for 1 h to oxidize any copper phosphide to copper phosphate. Dispose of the final reaction mixture in an appropriate fashion. Use fresh NaOCl solution (see assay procedure below).

Red Phosphorus

Add 1 g of red phosphorus to 500 mL of 0.5 *M* sulfuric acid and add 12 g of potassium bromate while stirring. Stir the reaction mixture until all the phosphorus has dissolved. The reaction time depends on the physical characteristics of the phosphorus. Phosphorus in lumps takes about 24 h, whereas powdered phosphorus reacts in less than 1 h. If all the phosphorus has not dissolved in 24 h, add more potassium bromate. When the reaction is over add 16 g of sodium metabisulfite (more if necessary) to discharge the bromine color and discard the reaction mixture.

Phosphorus Pentoxide

Gradually add P_2O_5 to a stirred mixture of H_2O and crushed ice. Before discarding it ensure that no chunks of unreacted P_2O_5 are left.

Assay of Sodium Hypochlorite Solution

Sodium hypochlorite solutions tend to deteriorate with time so they should be periodically checked for the amount of active chlorine they contain. Pipette 10 mL of NaOCl solution into a 100-mL volumetric flask and fill to the mark with distilled H_2O. Pipette 10 mL of this solution into a conical flask containing 50 mL of distilled H_2O, 1 g of potassium iodide, and 12.5 mL of 2 *M* acetic acid. Titrate this solution against 0.1 *N* sodium thiosulfate solution using starch as an indicator. Each 1 mL of the sodium thiosulfate solution corresponds to 3.545 mg of active chlorine. The NaOCl solution used in these degradation reactions should contain 45–50 g of active chlorine per liter.

References

1. Lewis, R.J., Sr. *Sax's Dangerous Properties of Industrial Materials*, 8th ed.; Van Nostrand-Reinhold: New York, 1992; pp. 2791–2792.
2. Reference 1, p. 2791.
3. Other names for this compound are phosphoric anhydride, diphosphorus pentoxide, phosphorus(V) oxide, and POX.
4. Reference 1, p. 2795.
5. National Research Council, Committee on Hazardous Substances in the Laboratory. *Prudent Practices for Disposal of Chemicals from Laboratories;* National Academy Press: Washington, DC, 1983; pp. 92–93.
6. Lunn, G. Unpublished observations.
7. Deshmukh, G.S.; Sant, B.R. Determination of elementary (red) phosphorus by potassium bromate. *Anal. Chem.* **1952**, *24*, 901–902.
8. Slater, J.W. Einwirkung von Phosphor, Schwefel, Arsen und Antimon auf gewisse Arten Salze. *J. Prakt. Chem. (1)* **1853**, *60*, 247–248.
9. Slater, J.W. Action of phosphorus, sulphur, arsenic, and antimony upon certain classes of salts. *Chem. Gazz.* **1853**, *11*, 329–331.
10. Mellor, J.W. *A Comprehensive Treatise on Inorganic and Theoretical Chemistry*, Vol. VIII; Longmans, Green and Co.: London, 1947; p. 786.
11. Deshmukh, G.S.; Venugopalan, M. A note on the oxidation of red phosphorus by potassium chlorate. *J. Indian Chem. Soc.* **1956**, *33*, 355–356.
12. Reference 5, p. 93.

PICRIC ACID

CAUTION! Refer to safety considerations section on page 7 before starting any of these procedures.

Picric acid [88-89-1][1] is used in the chemical laboratory for preparing picrates for the characterization of organic compounds. This acid is also used in histological stains. When wet it is quite stable but in the dry form it is an explosive. Picric acid can form explosive salts with many metals and can cause local and systemic reactions with a variety of symptoms.[2] Only wet picric acid should be degraded. Seek professional help for dry picric acid.

Principles of Destruction

The nitro groups of picric acid may be reduced using sodium sulfide or tin in hydrochloric acid (HCl). Although the product should in theory be 2,4,6-triaminophenol it is likely that reduction will not be complete. In addition, after reduction has ceased air oxidation will convert the product to a mixture of nitroamines, various dimers, and other hazardous compounds. Thus reduction of picric acid will probably convert it to a mixture of hazardous compounds but at least they will not be explosive. The products of these reactions should be disposed of as hazardous waste.

Destruction Procedures

Destruction of Bulk Quantities

1. Dissolve 0.13 g of sodium hydroxide (NaOH) in 25 mL of H_2O, then add 2.7 g of sodium sulfide. When the sodium sulfide has dissolved, add 1 g of picric acid.[3] When the reaction appears to be complete dispose of the mixture with the hazardous waste.

2. Stir 1 g of picric acid, 10 mL of H_2O, and 4 g of granular tin in a flask that is cooled in an ice bath.[4] Add concentrated HCl (15 mL), cautiously at first, and, when addition is complete, allow the reaction mixture to warm to room temperature. Reflux the reaction for 1 h, cool, and filter it. Wash the unreacted tin with 10 mL of 2 *M* HCl and neutralize the filtrate with 10% NaOH. Refilter to remove tin chloride and discard the filtrate as hazardous aqueous waste. Discard the unreacted tin and the tin chloride.

Decontamination of Dilute Aqueous Solutions[4]

Dilute the aqueous solution of picric acid with H_2O, if necessary, so that the concentration does not exceed 0.4%, then for each 100 mL of solution add 2 mL of concentrated HCl to bring the pH to 2. Add granular tin (30 mesh, 1 g) and allow the mixture to stand at room temperature. After about 14 days the picric acid is completely degraded. Dispose of the reaction mixture as hazardous aqueous waste.

Analytical Procedures[4]

Picric acid can be determined by thin-layer chromatography on silica gel. The eluant is methanol:toluene:glacial acetic acid (8:45:4) and the picric acid forms a bright yellow spot of $R_f \sim 0.3$. Iodine vapor will increase the sensitivity of the procedure.

References

1. Other names for this compound are 2,4,6-trinitrophenol, carbazotic acid, 2-hydroxy-1,3,5-trinitrobenzene, nitroxanthic acid, phenol trinitrate, picronitric acid, Melinite, and C.I. 10305

2. Lewis, R.J., Sr. *Sax's Dangerous Properties of Industrial Materials*, 8th ed.; Van Nostrand-Reinhold: New York, 1992; p. 2805.
3. Manufacturing Chemists Association. *Laboratory Waste Disposal Manual*; Manufacturing Chemists Association: Washington DC, 1973; p. 133.
4. Armour, M-.A.; Browne, L.M.; Weir, G.L., Eds., *Hazardous Chemicals. Information and Disposal Guide*, 3rd ed.; University of Alberta: Edmonton, Alberta, 1987; p. 317.

POLYCYCLIC AROMATIC HYDROCARBONS

CAUTION! Refer to safety considerations section on page 7 before starting any of these procedures.

In an international collaborative study the destruction of the following polycyclic aromatic hydrocarbons (PAH) was investigated:[1]

Compound Name	Reference	Abbreviation	Structure	Registry Number
Benz[a]anthracene	2	BA	**I**	[56-55-3]
Benzo[a]pyrene	3	BP	**II**	[50-32-8]
7-Bromomethyl-benz[a]anthracene		B-r-MBA	**III**	[24961-39-5]
Dibenz[*a,h*]-anthracene	4	DBA	**IV**	[53-70-3]
7,12-Dimethyl-benz[a]anthracene	5	DMBA	**V**	[57-97-6]
3-Methylchol-anthrene	6	3-MC	**VI**	[56-49-5]

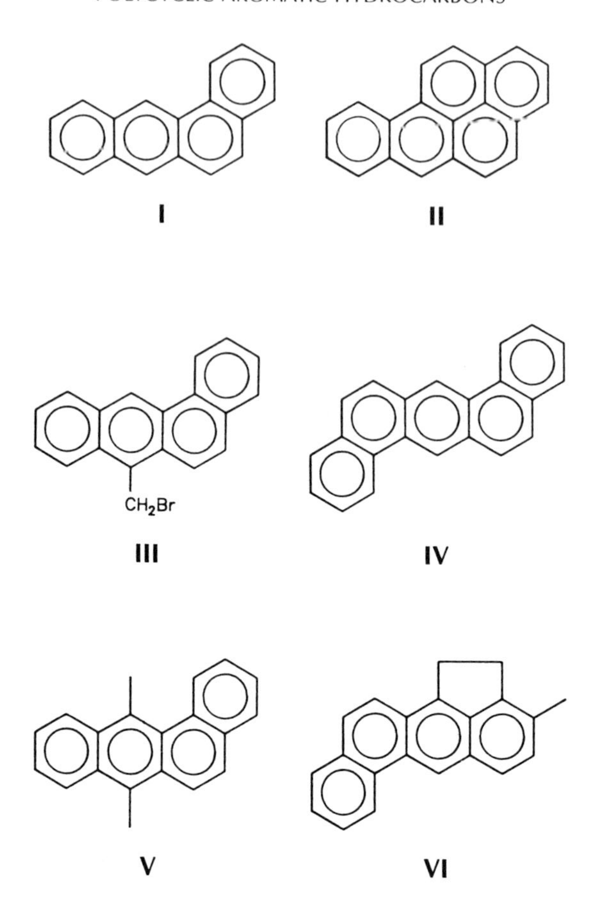

These compounds are all high-melting solids (mp >120°C) that are soluble in most organic solvents [e.g., benzene, toluene, cyclohexane, acetone, *N,N*-dimethylformamide (DMF), dimethyl sulfoxide (DMSO)], slightly soluble in alcohols, but only soluble in H_2O at the microgram per liter level. The compounds BA,[7,8] BP,[9,10] BrMBA,[11] DBA,[12,13] DMBA,[14,15] and 3-MC[16–18] cause cancer in experimental animals. While there is no direct evidence that they cause cancer in humans, coal tar and other materials known to be carcinogenic to humans may contain these PAH, so they should all be regarded as potential human carcinogens. The compounds BP,[19] DMBA,[20] and 3-MC[21] are teratogens. Polycyclic aromatic hydrocarbons are widely used in cancer research laboratories and these compounds are also found in the environment as products of combustion.

Principles of Destruction

Polycyclic aromatic hydrocarbons may be destroyed by oxidation with potassium permanganate in sulfuric acid ($KMnO_4$ in H_2SO_4) or by dissolution in concentrated H_2SO_4.[1] The solution of $KMnO_4$ in H_2SO_4 should be freshly prepared by stirring the calculated amount of $KMnO_4$ in H_2SO_4 for at least 15 min but no more than 1 h.[22] The products of these reactions have not been determined. Destruction efficiency was greater than 99% in each case.

Destruction Procedures

Destruction of Bulk Quantities of PAH

1. For every 5 mg of PAH add 2 mL of acetone and ensure that the PAH is completely dissolved, including any PAH that may be adhering to the walls of the container. For every 5 mg of PAH add 10 mL of a 0.3 *M* $KMnO_4$ solution in 3 *M* H_2SO_4 (freshly prepared) and swirl the mixture and allow to react for 1 h. The purple color should be maintained during this reaction time. If it is not, add more $KMnO_4$ solution until the reaction mixture remains purple for 1 h. At the end of the reaction decolorize with sodium metabisulfite, make strongly basic with 10 *M* potassium hydroxide (KOH) solution (**Caution!** Exothermic), dilute with H_2O, filter to remove manganese compounds,[22] neutralize the filtrate, check for completeness of destruction, and discard it.

Note: Use *at least* 1 mL of acetone and *at least* 10 mL of $KMnO_4$.

2. For every 5 mg of PAH add 2 mL of DMSO and ensure that the PAH is completely dissolved, including any PAH that may be adhering to the walls of the container. For every 5 mg of PAH add 10 mL of concentrated H_2SO_4 (**Caution!** Exothermic) and swirl the mixture and allow to react for at least 2 h. At the end of the reaction cautiously add the solution to at least three times the volume of cold H_2O (using an ice bath if desired as this is a very exothermic process), neutralize, test for completeness of destruction, and discard it.

Destruction of PAH in Organic Solvents (Except DMSO and DMF)

1. Remove the solvent by evaporation under reduced pressure using a rotary evaporator. For every 5 mg of PAH add 2 mL of acetone and ensure that the PAH is completely dissolved, including any PAH that may be

adhering to the walls of the container. For every 5 mg of PAH add 10 mL of a 0.3 *M* $KMnO_4$ solution in 3 *M* H_2SO_4 (freshly prepared) and swirl the mixture and allow to react for 1 h. The purple color should be maintained during this reaction time. If it is not, add more $KMnO_4$ solution until the reaction mixture remains purple for 1 h. At the end of the reaction decolorize the solution with sodium metabisulfite, make strongly basic with 10 *M* KOH (**Caution!** Exothermic), dilute with H_2O, filter to remove manganese compounds,[22] neutralize the filtrate, check for completeness of destruction, and discard it.
Note: Use *at least* 1 mL of acetone and *at least* 10 mL of $KMnO_4$.

2. Remove the solvent by evaporation under reduced pressure using a rotary evaporator. For every 5 mg of PAH add 2 mL of DMSO and ensure that the PAH is completely dissolved, including any PAH that may be adhering to the walls of the container. For every 5 mg of PAH add 10 mL of concentrated H_2SO_4 (**Caution!** Exothermic) and swirl the mixture and allow to react for at least 2 h. At the end of the reaction cautiously add the solution to at least three times the volume of cold H_2O (using an ice bath if desired as this is a very exothermic process), neutralize, test for completeness of destruction, and discard it.

Destruction of PAH in N,N-*Dimethylformamide*

For each 10 mL of DMF solution add 10 mL of H_2O and 20 mL of cyclohexane, shake and allow to separate. Extract the lower, aqueous layer twice more with 20-mL portions of cyclohexane, then combine the cyclohexane layers and remove the solvent by evaporation under reduced pressure using a rotary evaporator. For every 5 mg of PAH add 2 mL of acetone and ensure that the PAH is completely dissolved, including any PAH that may be adhering to the walls of the container. For every 5 mg of PAH add 10 mL of a 0.3 *M* $KMnO_4$ solution in 3 *M* H_2SO_4 (freshly prepared) and swirl the mixture and allow to react for 1 h. The purple color should be maintained during this reaction time. If it is not, add more $KMnO_4$ solution until the reaction mixture remains purple for 1 h. At the end of the reaction decolorize the solution with sodium metabisulfite, make strongly basic with 10 *M* KOH (**Caution!** Exothermic), dilute with H_2O, filter to remove manganese compounds,[22] neutralize the filtrate, check for completeness of destruction, and discard it.
Note: Use *at least* 1 mL of acetone and *at least* 10 mL of $KMnO_4$.

Destruction of PAH in Dimethyl Sulfoxide

1. For each 10 mL of DMSO solution add 5 mL of H_2O and 20 mL of cyclohexane, shake and allow to separate. Extract the lower, aqueous layer twice more with 20-mL portions of cyclohexane, then combine the cyclohexane layers and remove the solvent by evaporation under reduced pressure using a rotary evaporator. For every 5 mg of PAH add 2 mL of acetone and ensure that the PAH is completely dissolved, including any PAH that may be adhering to the walls of the container. For every 5 mg of PAH add 10 mL of a 0.3 *M* $KMnO_4$ solution in 3 *M* H_2SO_4 (freshly prepared) and swirl the mixture and allow to react for 1 h. The purple color should be maintained during this reaction time. If it is not, add more $KMnO_4$ solution until the reaction mixture remains purple for 1 h. At the end of the reaction decolorize the solution with sodium metabisulfite, make strongly basic with 10 *M* KOH (**Caution!** Exothermic), dilute with H_2O, filter to remove manganese compounds,[22] neutralize the filtrate, check for completeness of destruction, and discard it.
Note: Use *at least* 1 mL of acetone and *at least* 10 mL of $KMnO_4$.

2. Dilute the solution with more DMSO, if necessary, so that the PAH concentration does not exceed 2.5 mg/mL. For every 2 mL of DMSO add 10 mL of concentrated H_2SO_4 (**Caution!** Exothermic) and swirl the mixture and allow to react for at least 2 h. At the end of the reaction cautiously add the solution to at least three times the volume of cold H_2O (using an ice bath if desired as this is a very exothermic process), neutralize, test for completeness of destruction, and discard it.

Destruction of PAH in Water

Because these compounds are so insoluble, only trace amounts are likely to be present. Add enough $KMnO_4$ to make a 0.3 *M* solution and enough H_2SO_4 to make a 3 *M* solution and swirl the mixture and allow it to react for 1 h. The purple color should be maintained during this reaction time. If it is not, add more $KMnO_4$ solution until the reaction mixture remains purple for 1 h. At the end of the reaction decolorize the solution with sodium metabisulfite, make strongly basic with 10 *M* KOH (**Caution!** Exothermic), dilute with H_2O, filter to remove manganese compounds,[22] neutralize the filtrate, check for completeness of destruction, and discard it.

Destruction of PAH in Oil

For each 5 mL of oil solution add 20 mL of 2-methylbutane and 20 mL of acetonitrile, shake for at least 1 min (**Caution!** Pressure may develop) and allow the layers to separate. Extract the upper, hydrocarbon layer four more times with 20-mL portions of acetonitrile. If necessary, add a 10-mL portion of 2-methylbutane after the second extraction to avoid inversion of the layers caused by evaporation of the 2-methylbutane. Combine the acetonitrile layers and wash them with 20 mL of 2-methylbutane (discard the wash), then remove the solvent by evaporation under reduced pressure using a rotary evaporator (water bath temperature 25–30°C). For every 5 mg of PAH add 2 mL of acetone and ensure that the PAH is completely dissolved, including any PAH that may be adhering to the walls of the container. For every 5 mg of PAH add 10 mL of a 0.3 *M* $KMnO_4$ solution in 3 *M* H_2SO_4 (freshly prepared) and swirl the mixture and allow to react for 1 h. The purple color should be maintained during this reaction time. If it is not, add more $KMnO_4$ solution until the reaction mixture remains purple for 1 h. At the end of the reaction decolorize the solution with sodium metabisulfite, make strongly basic with 10 *M* KOH (**Caution!** Exothermic), dilute with H_2O, filter to remove manganese compounds,[22] neutralize the filtrate, check for completeness of destruction, and discard it. *Note:* Use *at least* 1 mL of acetone and *at least* 10 mL of $KMnO_4$.

Destruction of PAH in Agar

1. Cut the contents of the Petri dish into small pieces and homogenize with 30 mL of H_2O in a high speed blender. Extract the resulting solution twice with 30-mL portions of ethyl acetate (upper layer) and combine the extracts and dry them over anhydrous sodium sulfate. Remove the ethyl acetate by evaporation under reduced pressure using a rotary evaporator. For every 5 mg of PAH add 2 mL of acetone and ensure that the PAH is completely dissolved, including any PAH that may be adhering to the walls of the container. For every 5 mg of PAH add 10 mL of a 0.3 *M* $KMnO_4$ solution in 3 *M* H_2SO_4 (freshly prepared) and swirl the mixture and allow to react for 1 h. The purple color should be maintained during this reaction time. If it is not, add more $KMnO_4$ solution until the reaction mixture remains purple for 1 h. At the end of the reaction decolorize the solution with sodium metabisulfite, make strongly basic with 10 *M* KOH (**Caution!** Exothermic), dilute with H_2O, filter to remove manganese compounds,[22] neu-

tralize the filtrate, check for completeness of destruction, and discard it. *Note:* Use *at least* 1 mL of acetone and *at least* 10 mL of $KMnO_4$.

2. Cut the contents of the Petri dish into small pieces and homogenize with 30 mL of H_2O in a high speed blender. Extract the resulting solution twice with 30-mL portions of ethyl acetate (upper layer) and combine the extracts and dry them over anhydrous sodium sulfate. Remove the ethyl acetate by evaporation under reduced pressure using a rotary evaporator. For every 5 mg of PAH add 2 mL of DMSO and ensure that the PAH is completely dissolved, including any PAH that may be adhering to the walls of the container. For every 5 mg of PAH add 10 mL of concentrated H_2SO_4 (**Caution!** Exothermic) and swirl the mixture and allow to react for at least 2 h. At the end of the reaction cautiously add the solution to at least three times the volume of cold H_2O (using an ice bath if desired as this is a very exothermic process), neutralize, test for completeness of destruction, and discard it.

Decontamination of Glassware Contaminated with PAH

1. Rinse the glassware with four portions of acetone that are large enough to wet the glassware thoroughly. Analyze the fourth rinse for the absence of PAH. Combine the rinses and remove the acetone by evaporation under reduced pressure using a rotary evaporator. For every 5 mg of PAH add 2 mL of acetone and ensure that the PAH is completely dissolved, including any PAH that may be adhering to the walls of the container. For every 5 mg of PAH add 10 mL of a 0.3 *M* $KMnO_4$ solution in 3 *M* H_2SO_4 (freshly prepared) and swirl the mixture and allow to react for 1 h. The purple color should be maintained during this reaction time. If it is not, add more $KMnO_4$ solution until the reaction mixture remains purple for 1 h. At the end of the reaction decolorize the solution with sodium metabisulfite, make strongly basic with 10 *M* KOH (**Caution!** Exothermic), dilute with H_2O, filter to remove manganese compounds,[22] neutralize the filtrate, check for completeness of destruction, and discard it.
Note: Use *at least* 1 mL of acetone and *at least* 10 mL of $KMnO_4$.

2. Add sufficient DMSO to wet the surface of the glass and then add five times this volume of concentrated H_2SO_4 (**Caution!** Exothermic). Allow the mixture to react for 2 h with occasional swirling. At the end of the reaction cautiously add the solution to at least three times the volume of cold H_2O (using an ice bath if desired as this is a very exothermic process), neutralize, test for completeness of destruction, and discard it.

Treatment of Spills Involving PAH

Remove as much as possible of the spill by high efficiency particulate air (HEPA) vacuuming or using absorbents. Treat the removed material using one of the methods described above. Wet the surface with DMF, then add a 0.3 *M* solution of $KMnO_4$ in 3 *M* H_2SO_4 and allow to react for 1 h. Take up the residual solution with an absorbent, decolorize with sodium metabisulfite, make strongly basic with 10 *M* KOH (**Caution!** Exothermic), dilute with H_2O, filter to remove manganese compounds,[22] neutralize the filtrate, check for completeness of destruction, and discard it. Add more solvent to the spill area and absorb this with white paper. Examine the paper under long- and short-wavelength ultraviolet (UV) light for the presence of fluorescence attributable to PAH.

Analytical Procedures

High-performance liquid chromatography (HPLC) methods of analysis for PAH have been described.[23–27] Polycyclic aromatic hydrocarbons may also be analyzed by gas chromatography (GC) using a 1.8-m × 2-mm i.d. glass column packed with 3% OV-1 on 80/100 Supelcoport. The oven temperature is 260°C, the injection temperature is 300°C, and the temperature of the flame ionization detector is 300°C. Before analysis extract the neutralized reaction mixture three times with 10-mL portions of cyclohexane and dry the extracts over anhydrous sodium sulfate. Evaporate to dryness under reduced pressure using a rotary evaporator and take up the residue in 500 μL of acetone or other suitable solvent.

For BrMBA, extract the reaction mixture three times with 10-mL portions of cyclohexane and dry the extracts over anhydrous sodium sulfate and evaporate. Take up the residue in 250 μL of toluene and spot 10 μL of this solution on a silica gel 60 thin-layer chromatography (TLC) plate. Develop the TLC plate with cyclohexane : ether (60:40) and then examine **immediately** under long-wavelength UV light. 7-Bromomethylbenz-[*a*]anthracene has an R_f value of about 0.28. Use of a standard solution of BrMBA is helpful for quantitation.

Mutagenicity Assays[1]

The reaction mixtures from these procedures were tested using *Salmonella typhimurium* strains TA98 and TA100, with and without activation. In general, these mixtures were not mutagenic.

Related Compounds

Although these procedures have only been validated for the compounds listed above, they should be applicable to other PAH. When applied to other PAH the procedures should be thoroughly validated before being put into routine use. Similar procedures have been used for the analogous heterocyclic compounds, see the Polycyclic Heterocyclic Hydrocarbons monograph.

References

1. Castegnaro, M.; Grimmer, G.; Hutzinger, O.; Karcher, W.; Kunte, H.; Lafontaine, M.; Sansone, E.B.; Telling, G.; Tucker, S.P., Eds., *Laboratory Decontamination and Destruction of Carcinogens in Laboratory Wastes: Some Polycyclic Aromatic Hydrocarbons*; International Agency for Research on Cancer: Lyon, 1983 (IARC Scientific Publications No. 49).
2. Other names for this compound are 1,2-benzanthracene, 2,3-benzphenanthrene, benzanthrene, benzoanthracene, naphthanthracene, and tetraphene.
3. Other names for this compound are 3,4-benzpyrene, 1,2-benzpyrene, and benzo-(*d,e,f*)chrysene.
4. Another name for this compound is 1,2:5,6-dibenzanthracene.
5. Other names for this compound are 9,10-dimethyl-1,2-benzanthracene and 1,4-dimethyl-2,3-benzphenanthrene.
6. Other names for this compound are 20-methylcholanthrene, 3-MECA, and 1,2-dihydro-3-methylbenz(*j*)aceanthrylene.
7. International Agency for Research on Cancer. *IARC Monographs on the Evaluation of the Carcinogenic Risk of Chemicals to Man.* Volume 3, *Certain Polycyclic Aromatic Hydrocarbons and Heterocyclic Compounds*; International Agency for Research on Cancer: Lyon, 1973; pp. 45–68.
8. International Agency for Research on Cancer. *IARC Monographs on the Evaluation of the Carcinogenic Risk of Chemicals to Humans.* Volume 32, *Polynuclear Aromatic Compounds, Part 1, Chemical, Environmental and Experimental Data*; International Agency for Research on Cancer: Lyon, 1983; pp. 135–145.
9. Reference 7, pp. 91–136.
10. Reference 8, pp. 211–224.
11. Dipple, A.; Levy, L.S.; Lawley, P.D. Comparative carcinogenicity of alkylating agents: comparisons of a series of alkyl and aralkyl bromides of differing chemical reactivities as inducers of sarcoma at the site of a single injection in the rat. *Carcinogenesis* **1981**, *2*, 103–107.
12. Reference 7, pp. 178–196.
13. Reference 8, pp. 299–308.

14. Griswold, Jr., D.P.; Casey, A.E.; Weisburger, E.K.; Weisburger, J.H.; Schabel, F.M. On the carcinogenicity of a single intragastric dose of hydrocarbons, nitrosamines, aromatic amines, dyes, coumarins, and miscellaneous chemicals in female Sprague-Dawley rats. *Cancer Res.* **1966**, *26*, 619–625.

15. Griswold, Jr., D.P.; Casey, A.E.; Weisburger, E.K.; Weisburger, J.H. The carcinogenicity of multiple intragastric doses of aromatic and heterocyclic nitro or amino derivatives in young female Sprague-Dawley rats. *Cancer Res.* **1968**, *28*, 924–933.

16. Rigdon, R.H. Pulmonary neoplasms produced by methylcholanthrene in the white Pekin duck. *Cancer Res.* **1961**, *21*, 571–574.

17. Blumenthal, H.T.; Rogers, J.B. Studies of guinea pig tumors. II. The induction of malignant tumors in guinea pigs by methylcholanthrene. *Cancer Res.* **1962**, *22*, 1155–1162.

18. Homburger, F.; Hsueh, S-.S.; Kerr, C.S.; Russfield, A.B. Inherited susceptibility of inbred strains of Syrian hamsters to induction of subcutaneous sarcomas and mammary and gastrointestinal carcinomas by subcutaneous and gastric administration of polynuclear hydrocarbons. *Cancer Res.* **1972**, *32*, 360–366.

19. Lewis, R.J., Sr. *Sax's Dangerous Properties of Industrial Materials*, 8th ed.; Van Nostrand-Reinhold: New York, 1992; pp. 379–380.

20. Reference 19, p. 1350.

21. Reference 19, pp. 2289–2290.

22. Lunn, G.; Sansone, E.B.; De Méo, M.; Laget, M.; Castegnaro, M. Potassium permanganate can be used for degrading hazardous compounds. *Am. Ind. Hyg. Assoc. J.* **1994**, *55*, 167–171.

23. Lee, M.L.; Novotny, M.V.; Bartle, K. *Analytical Chemistry of Polycyclic Aromatic Compounds*; Academic Press: New York, 1981.

24. Pasquini, R.; Taningher, M.; Monarca, S.; Pala, M.; Angeli, G. Chemical composition and genotoxic activity of petroleum derivatives collected in two working environments. *J. Toxicol. Environ. Health* **1989**, *27*, 225–238.

25. Lafleur, A.L.; Plummer, E.F. Effect of column and mobile phase modifications on retention behavior in size exclusion chromatography of polycyclic aromatic hydrocarbons on poly(divinylbenzene). *J. Chromatogr. Sci.* **1991**, *29*, 532–537.

26. Pacakova, V.; Leclercq, P.A. Gas chromatography-mass spectrometry and high-performance liquid chromatographic analyses of thermal degradation products of common plastics. *J. Chromatogr.* **1991**, *555*, 229–237.

27. de Kok, T.M.C.M.; Levels, P.J.; van Faassen, A.; Hazen, M.; ten Hoor, F.; Kleinjans, J.C.S. Chromatographic methods for the determination of toxicants in faeces. *J. Chromatogr.* **1992**, *580*, 135–159.

POLYCYCLIC HETEROCYCLIC HYDROCARBONS

CAUTION! Refer to safety considerations section on page 7 before starting any of these procedures.

In a recent international collaborative study the destruction of the following polycyclic heterocyclic hydrocarbons (PHH) was investigated:[1]

Compound Name	Reference	Abbreviation	Structure	Registry Number
Dibenz[a,*j*]acridine	2	DB(a,*j*)AC	**I**	[224-42-0]
Dibenz[a,*h*]acridine	3	DB(a,*h*)AC	**II**	[226-36-8]
7*H*-Dibenzo[c,g]-carbazole	4	DB(c,g)C	**III**	[194-59-2]
13*H*-Dibenzo[a,*i*]-carbazole	5	DB(a,*i*)C	**IV**	[239-64-5]

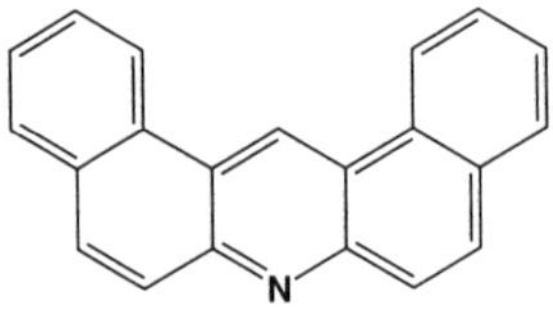

I

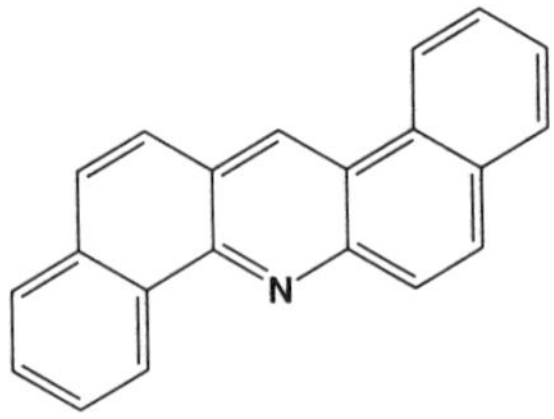

II

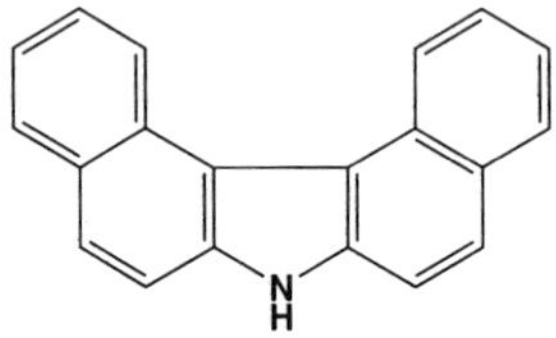

III

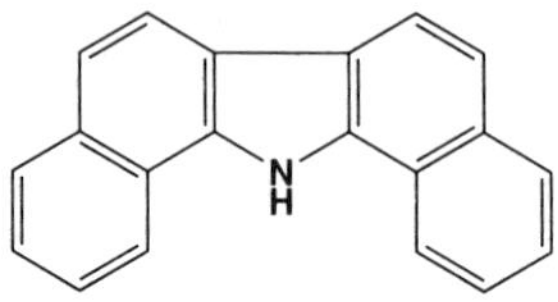

IV

These compounds are all high-melting solids (mp >150°C) that are soluble in organic solvents. The compounds DB(*a,j*)AC,[6] DB(*a,h*)AC,[7] and DB(*c,g*)C[8] cause cancer in experimental animals and have been classified as possibly carcinogenic to humans. The compound DB(*a,i*)C may be carcinogenic to rats.[9] These compounds are found in the environment as pollutants and are used as standards in analytical laboratories.

Principles of Destruction

All of the compounds considered can be degraded by oxidation with hydrogen peroxide (H_2O_2) and iron(II) chloride (Fenton's reagent)[1] as well as by oxidation with potassium permanganate in sulfuric acid ($KMnO_4$ in H_2SO_4).[8] The carbazoles can be degraded by dissolution in concentrated H_2SO_4[1] and DB(*c,g*)C, DB(*a,i*)C, and DB(*a,j*)AC can be degraded by oxidation with $KMnO_4$ in H_2O or sodium hydroxide (NaOH) solution.[1] The products of these reactions have not been determined. Destruction efficiency was greater than 99.5% in each case. Basification of the reaction mixtures containing manganese salts results in the precipitation of the manganese that is removed by filtration. Thus the filtrate contains less than 5 ppm of manganese and can be discarded with the nonhazardous aqueous waste.[9] When $KMnO_4$ dissolves in aqueous solutions it takes a finite time to dissolve, and then the permanganate starts to slowly decay. Thus it is important to observe the preparation times for the $KMnO_4$ solutions given below. These times should give $KMnO_4$ concentrations that are greater than 90% of the peak concentrations.[9]

Destruction Procedures

Destruction of Bulk Quantities of PHH

1. For every 5 mg of PHH add 3 mL of acetonitrile and ensure that the PHH is completely dissolved, including any PHH that may be adhering to the walls of the container. Prepare a 0.3 *M* solution of $KMnO_4$ in 3 *M* H_2SO_4 by stirring the calculated amount of $KMnO_4$ in 3 *M* H_2SO_4 for at least 15 min but no more than 1 h. For every 5 mg of PHH add 10 mL of a 0.3 *M* $KMnO_4$ solution in 3 *M* H_2SO_4 (freshly prepared) and swirl the mixture and allow to react for 1 h. The purple color should be maintained during this reaction time. If it is not, add more $KMnO_4$ solution until the reaction mixture remains purple for 1 h. At the end of the reaction decolorize the solution with 0.4 g of sodium metabisulfite for each 10 mL of $KMnO_4$ solution, basify by the addition of 4 mL of 10 *M* NaOH solution, dilute with 50 mL of H_2O, and filter. Test the filtrate for completeness of destruction and discard the filtrate and the precipitate appropriately.
Note: This reaction should be performed on at least this scale, even if less than 5 mg of PHH is present.

2. The reaction should be performed in a vessel that has a volume at least 20 times that of the H_2O_2 that is used. For every 5 mg of PHH add 5 mL of acetone and make sure that the PHH has dissolved completely. Surround the reaction vessel with an ice bath. For each 5 mg of PHH add 0.2 g of $FeCl_2$ or $FeCl_2.2H_2O$ or 0.3 g of $FeCl_2.4H_2O$, then add 10 mL of 30% H_2O_2 while stirring. The reaction has an induction period and the H_2O_2 should be added dropwise until the reaction mixture boils, then the addition should be completed over a 15-min period. (If the reaction mixture does not boil, the degradation will not be complete. In this case more iron chloride and more H_2O_2 should be added with more rapid addition of the H_2O_2.) When the addition of H_2O_2 is complete remove the ice bath and stir the reaction mixture for 30 min, neutralize, test for completeness of destruction, and discard it.

3. *Note:* This method works **only** for DB(*c,g*)C, DB(*a,i*)C, and DB(*a,j*)AC. For every 5 mg of DB(*c,g*)C, DB(*a,i*)C, or 1 mg of DB(*a,j*)AC add 2 mL of acetonitrile and ensure that the PHH is completely dissolved,

including any PHH that may be adhering to the walls of the container. Prepare a 0.3 *M* solution of $KMnO_4$ in 2 *M* NaOH by stirring the calculated amount of $KMnO_4$ in 2 *M* NaOH for at least 30 min but no more than 2 h. For every 2 mL of acetonitrile add 10 mL of a 0.3 *M* $KMnO_4$ solution in 2 *M* NaOH and swirl the mixture and allow to react for 3 h. The purple or green color should be maintained during this reaction time. If it is not, add more $KMnO_4$ solution until the reaction mixture remains purple for 1 h. (If splashing occurs, causing unreacted compound to be isolated on the wall of the reaction vessel, rinse the wall with the decontaminating solution.) At the end of the reaction decolorize the solution with 0.8 g of sodium metabisulfite for each 10 mL of $KMnO_4$ solution, dilute with an equal volume of H_2O, and check for completeness of destruction. Filter and discard the filtrate and the precipitate appropriately.

4. *Note:* This method works **only** for DB(*c,g*)C, DB(*a,i*)C, and DB(*a,j*)AC. For every 5 mg of DB(*c,g*)C, DB(*a,i*)C, or 1 mg of DB(*a,j*)AC add 2 mL of acetonitrile and ensure that the PHH is completely dissolved, including any PHH that may be adhering to the walls of the container. Prepare a 0.3 *M* solution of $KMnO_4$ in H_2O by stirring the calculated amount of $KMnO_4$ in H_2O for at least 30 min but no more than 2 h. For every 2 mL of acetonitrile add 10 mL of a 0.3 *M* $KMnO_4$ solution in H_2O and stir for at least 6 h. The purple color should be maintained during this reaction time. If it is not, add more $KMnO_4$ solution until the reaction mixture remains purple for 1 h. (If splashing occurs, causing unreacted compound to be isolated on the wall of the reaction vessel, rinse the wall with the decontaminating solution.) At the end of the reaction decolorize the solution with 0.8 g of sodium metabisulfite for each 10 mL of $KMnO_4$ solution (more if necessary for complete decolorization), basify with 1 mL of 10 *M* NaOH solution, check that the mixture is strongly basic, and check for completeness of destruction. Filter and discard the filtrate and the precipitate appropriately.

5. *Note:* This method works **only** for DB(*c,g*)C and DB(*a,i*)C. For every 5 mg of PHH add 2 mL of dimethyl sulfoxide (DMSO) and ensure that the PHH is completely dissolved, including any PHH that may be adhering to the walls of the container. For every 5 mg of PHH add 10 mL of concentrated H_2SO_4 (**Caution!** Exothermic) and swirl the mixture and allow to react for at least 2 h. At the end of the reaction cautiously add the solution to at least three times the volume of cold H_2O (using an ice bath if desired as this is a very exothermic process), neutralize, test for completeness of destruction, and discard it.

Destruction of PHH in Organic Solvents (Except DMSO and DMF)

1. Remove the solvent by evaporation under reduced pressure using a rotary evaporator (temperature of water bath ~40°C). For every 5 mg of PHH add 3 mL of acetonitrile and ensure that the PHH is completely dissolved, including any PHH that may be adhering to the walls of the container. Prepare a 0.3 *M* solution of $KMnO_4$ in 3 *M* H_2SO_4 by stirring the calculated amount of $KMnO_4$ in 3 *M* H_2SO_4 for at least 15 min but no more than 1 h. For every 5 mg of PHH add 10 mL of a 0.3 *M* $KMnO_4$ solution in 3 *M* H_2SO_4 (freshly prepared) and swirl the mixture and allow to react for 1 h. The purple color should be maintained during this reaction time. If it is not, add more $KMnO_4$ solution until the reaction mixture remains purple for 1 h. At the end of the reaction decolorize the solution with 0.4 g of sodium metabisulfite for each 10 mL of $KMnO_4$ solution, basify by the addition of 4 mL of 10 *M* NaOH solution, dilute with 50 mL of H_2O, and filter. Test the filtrate for completeness of destruction and discard the filtrate and the precipitate appropriately.
Note: This reaction should be performed on at least this scale, even if less than 5 mg of PHH is present.

2. Remove the solvent by evaporation under reduced pressure using a rotary evaporator (temperature of water bath ~40°C). The reaction should be performed in a vessel that has a volume at least 20 times that of the H_2O_2 that is used. For every 5 mg of PHH add 5 mL of acetone and make sure that the PHH has dissolved completely. Surround the reaction vessel with an ice bath. For each 5 mg of PHH add 0.2 g of $FeCl_2$ or $FeCl_2.2H_2O$ or 0.3 g of $FeCl_2.4H_2O$, then add 10 mL of 30% H_2O_2 while stirring. The reaction has an induction period and the H_2O_2 should be added dropwise until the reaction mixture boils then the addition should be completed over a 15-min period. (If the reaction mixture does not boil, the degradation will not be complete. In this case more iron chloride and more H_2O_2 should be added with more rapid addition of the H_2O_2.) When the addition of H_2O_2 is complete remove the ice bath and stir the reaction mixture for 30 min, neutralize, test for completeness of destruction, and discard it.

3. *Note:* This method works **only** for DB(*c,g*)C, DB(*a,i*)C, and DB(*a,j*)AC. Remove the solvent by evaporation under reduced pressure using a rotary evaporator (temperature of water bath ~40°C). For every 5 mg of DB(*c,g*)C, DB(*a,i*)C, or 1 mg of DB(*a,j*)AC add 2 mL of acetonitrile and ensure that the PHH is completely dissolved, including any PHH that may be adhering to the walls of the container. Prepare a 0.3 *M*

solution of $KMnO_4$ in 2 *M* NaOH by stirring the calculated amount of $KMnO_4$ in 2 *M* NaOH for at least 30 min but no more than 2 h. For every 2 mL of acetonitrile add 10 mL of a 0.3 *M* $KMnO_4$ solution in 2 *M* NaOH and swirl the mixture and allow to react for 3 h. The purple or green color should be maintained during this reaction time. If it is not, add more $KMnO_4$ solution until the reaction mixture remains purple for 1 h. (If splashing occurs, causing unreacted compound to be isolated on the wall of the reaction vessel, rinse the wall with the decontaminating solution.) At the end of the reaction decolorize the solution with 0.8 g of sodium metabisulfite for each 10 mL of $KMnO_4$ solution, dilute with an equal volume of H_2O, and check for completeness of destruction. Filter and discard the filtrate and the precipitate appropriately.

4. *Note:* This method works **only** for DB(*c,g*)C, DB(*a,i*)C, and DB(*a,j*)AC. Remove the solvent by evaporation under reduced pressure using a rotary evaporator (temperature of water bath ~40°C). For every 5 mg of DB(*c,g*)C, DB(*a,i*)C, or 1 mg of DB(*a,j*)AC add 2 mL of acetonitrile and ensure that the PHH is completely dissolved, including any PHH that may be adhering to the walls of the container. Prepare a 0.3 *M* solution of $KMnO_4$ in H_2O by stirring the calculated amount of $KMnO_4$ in H_2O for at least 30 min but no more than 2 h. For every 2 mL of acetonitrile add 10 mL of a 0.3 *M* $KMnO_4$ solution in H_2O and stir for at least 6 h. The purple color should be maintained during this reaction time. If it is not, add more $KMnO_4$ solution until the reaction mixture remains purple for 1 h. (If splashing occurs, causing unreacted compound to be isolated on the wall of the reaction vessel, rinse the wall with the decontaminating solution.) At the end of the reaction decolorize the solution with 0.8 g of sodium metabisulfite for each 10 mL of $KMnO_4$ solution (more if necessary for complete decolorization), basify with 1 mL of 10 *M* NaOH solution, check that the mixture is strongly basic, and check for completeness of destruction. Filter and discard the filtrate and the precipitate appropriately.

Destruction of PHH in DMSO

1. For each volume of DMSO solution add two volumes of H_2O. Extract the resulting mixture three times with an equal volume of cyclohexane, combine the cyclohexane layers, and remove the solvent by evaporation under reduced pressure using a rotary evaporator (temperature of water bath ~40°C). For every 5 mg of PHH add 3 mL of acetonitrile and ensure

that the PHH is completely dissolved, including any PHH that may be adhering to the walls of the container. Prepare a 0.3 *M* solution of $KMnO_4$ in 3 *M* H_2SO_4 by stirring the calculated amount of $KMnO_4$ in 3 *M* H_2SO_4 for at least 15 min but no more than 1 h. For every 5 mg of PHH add 10 mL of a 0.3 *M* $KMnO_4$ solution in 3 *M* H_2SO_4 (freshly prepared) and swirl the mixture and allow to react for 1 h. The purple color should be maintained during this reaction time. If it is not, add more $KMnO_4$ solution until the reaction mixture remains purple for 1 h. At the end of the reaction decolorize the solution with 0.4 g of sodium metabisulfite for each 10 mL of $KMnO_4$ solution, basify by the addition of 4 mL of 10 *M* NaOH solution, dilute with 50 mL of H_2O, and filter. Test the filtrate for completeness of destruction and discard the filtrate and the precipitate appropriately.
Note: This reaction should be performed on at least this scale, even if less than 5 mg of PHH is present.

2. *Note:* This method works **only** for DB(*c,g*)C, DB(*a,i*)C, and DB(*a,j*)AC. For each volume of DMSO solution add two volumes of H_2O. Extract the resulting mixture three times with an equal volume of cyclohexane, combine the cyclohexane layers and remove the solvent by evaporation under reduced pressure using a rotary evaporator (temperature of water bath ~40°C). For every 5 mg of DB(*c,g*)C, DB(*a,i*)C, or 1 mg of DB(*a,j*)AC add 2 mL of acetonitrile and ensure that the PHH is completely dissolved, including any PHH that may be adhering to the walls of the container. Prepare a 0.3 *M* solution of $KMnO_4$ in 2 *M* NaOH by stirring the calculated amount of $KMnO_4$ in 2 *M* NaOH for at least 30 min but no more than 2 h. For every 2 mL of acetonitrile add 10 mL of a 0.3 *M* $KMnO_4$ solution in 2 *M* NaOH and swirl the mixture and allow to react for 3 h. The purple or green color should be maintained during this reaction time. If it is not, add more $KMnO_4$ solution until the reaction mixture remains purple for 1 h. (If splashing occurs, causing unreacted compound to be isolated on the wall of the reaction vessel, rinse the wall with the decontaminating solution.) At the end of the reaction decolorize the solution with 0.8 g of sodium metabisulfite for each 10 mL of $KMnO_4$ solution, dilute with an equal volume of H_2O, and check for completeness of destruction. Filter and discard the filtrate and the precipitate appropriately.

3. *Note:* This method works **only** for DB(*c,g*)C, DB(*a,i*)C, and DB(*a,j*)AC. For each volume of DMSO solution add two volumes of H_2O. Extract the resulting mixture three times with an equal volume of cyclohexane, combine the cyclohexane layers and remove the solvent by evaporation under reduced pressure using a rotary evaporator (temperature of

water bath ~40°C). For every 5 mg of DB(*c,g*)C, DB(*a,i*)C, or 1 mg of DB(*a,j*)AC add 2 mL of acetonitrile and ensure that the PHH is completely dissolved, including any PHH that may be adhering to the walls of the container. Prepare a 0.3 *M* solution of $KMnO_4$ in H_2O by stirring the calculated amount of $KMnO_4$ in H_2O for at least 30 min but no more than 2 h. For every 2 mL of acetonitrile add 10 mL of a 0.3 *M* $KMnO_4$ solution in H_2O and stir for at least 6 h. The purple color should be maintained during this reaction time. If it is not, add more $KMnO_4$ solution until the reaction mixture remains purple for 1 h. (If splashing occurs, causing unreacted compound to be isolated on the wall of the reaction vessel, rinse the wall with the decontaminating solution.) At the end of the reaction decolorize the solution with 0.8 g of sodium metabisulfite for each 10 mL of $KMnO_4$ solution (more if necessary for complete decolorization), basify with 1 mL of 10 *M* NaOH solution, check that the mixture is strongly basic, and check for completeness of destruction. Filter and discard the filtrate and the precipitate appropriately.

4. *Note:* This method works **only** for DB(*c,g*)C and DB(*a,i*)C. If necessary, dilute so that the PHH concentration does not exceed 2.5 mg/mL. For every 2 mL of DMSO add 10 mL of concentrated H_2SO_4 (**Caution!** Exothermic) and swirl the mixture and allow to react for at least 2 h. At the end of the reaction cautiously add the solution to at least three times the volume of cold H_2O (using an ice bath if desired as this is a very exothermic process), neutralize, test for completeness of destruction, and discard it.

Destruction of PHH in DMF

1. For each volume of *N,N*-dimethylformamide (DMF) solution add two volumes of H_2O. Extract the resulting mixture three times with an equal volume of cyclohexane, combine the cyclohexane layers and remove the solvent by evaporation under reduced pressure using a rotary evaporator (temperature of water bath ~40°C). For every 5 mg of PHH add 3 mL of acetonitrile and ensure that the PHH is completely dissolved, including any PHH that may be adhering to the walls of the container. Prepare a 0.3 *M* solution of $KMnO_4$ in 3 *M* H_2SO_4 by stirring the calculated amount of $KMnO_4$ in 3 *M* H_2SO_4 for at least 15 min but no more than 1 h. For every 5 mg of PHH add 10 mL of a 0.3 *M* $KMnO_4$ solution in 3 *M* H_2SO_4 (freshly prepared) and swirl the mixture and allow to react for 1 h. The purple color should be maintained during this reaction time. If it is not,

add more $KMnO_4$ solution until the reaction mixture remains purple for 1 h. At the end of the reaction decolorize the solution with 0.4 g of sodium metabisulfite for each 10 mL of $KMnO_4$ solution, basify by the addition of 4 mL of 10 *M* NaOH solution, dilute with 50 mL of H_2O, and filter. Test the filtrate for completeness of destruction and discard the filtrate and the precipitate appropriately.
Note: This reaction should be performed on at least this scale, even if less than 5 mg of PHH is present.

2. *Note:* This method works **only** for DB(*c,g*)C, DB(*a,i*)C, and DB(*a,j*)AC. For each volume of DMF solution add two volumes of H_2O. Extract the resulting mixture three times with an equal volume of cyclohexane, combine the cyclohexane layers and remove the solvent by evaporation under reduced pressure using a rotary evaporator (temperature of water bath ~40°C). For every 5 mg of DB(*c,g*)C, DB(*a,i*)C, or 1 mg of DB(*a,j*)AC add 2 mL of acetonitrile and ensure that the PHH is completely dissolved, including any PHH that may be adhering to the walls of the container. Prepare a 0.3 *M* solution of $KMnO_4$ in 2 *M* NaOH by stirring the calculated amount of $KMnO_4$ in 2 *M* NaOH for at least 30 min but no more than 2 h. For every 2 mL of acetonitrile add 10 mL of a 0.3 *M* $KMnO_4$ solution in 2 *M* NaOH and swirl the mixture and allow to react for 3 h. The purple or green color should be maintained during this reaction time. If it is not, add more $KMnO_4$ solution until the reaction mixture remains purple for 1 h. (If splashing occurs, causing unreacted compound to be isolated on the wall of the reaction vessel, rinse the wall with the decontaminating solution.) At the end of the reaction decolorize the solution with 0.8 g of sodium metabisulfite for each 10 mL of $KMnO_4$ solution, dilute with an equal volume of H_2O, and check for completeness of destruction. Filter and discard the filtrate and the precipitate appropriately.

3. *Note:* This method works **only** for DB(*c,g*)C, DB(*a,i*)C, and DB(*a,j*)AC. For each volume of DMF solution add two volumes of H_2O. Extract the resulting mixture three times with an equal volume of cyclohexane, combine the cyclohexane layers and remove the solvent by evaporation under reduced pressure using a rotary evaporator (temperature of water bath ~40°C). For every 5 mg of DB(*c,g*)C, DB(*a,i*)C, or 1 mg of DB(*a,j*)AC add 2 mL of acetonitrile and ensure that the PHH is completely dissolved, including any PHH that may be adhering to the walls of the container. Prepare a 0.3 *M* solution of $KMnO_4$ in H_2O by stirring the calculated amount of $KMnO_4$ in H_2O for at least 30 min but no more than 2 h. For every 2 mL of acetonitrile add 10 mL of a 0.3 *M* $KMnO_4$ solution

in H_2O and stir for at least 6 h. The purple color should be maintained during this reaction time. If it is not, add more $KMnO_4$ solution until the reaction mixture remains purple for 1 h. (If splashing occurs, causing unreacted compound to be isolated on the wall of the reaction vessel, rinse the wall with the decontaminating solution.) At the end of the reaction decolorize the solution with 0.8 g of sodium metabisulfite for each 10 mL of $KMnO_4$ solution (more if necessary for complete decolorization), basify with 1 mL of 10 *M* NaOH solution, check that the mixture is strongly basic, and check for completeness of destruction. Filter and discard the filtrate and the precipitate appropriately.

Destruction of PHH in Water

1. Because these compounds are so insoluble, only trace amounts are likely to be present. Add enough $KMnO_4$ to make a 0.3 *M* solution and enough H_2SO_4 to make a 3 *M* solution and swirl the mixture and allow it to react for 1 h. The purple color should be maintained during this reaction time. If it is not, add more $KMnO_4$ solution until the reaction mixture remains purple for 1 h. At the end of the reaction (for each 10 mL of $KMnO_4$ solution) decolorize the solution with 0.4 g of sodium metabisulfite, basify by the addition of 4 mL of 10 *M* NaOH solution, dilute with 50 mL of H_2O, and filter. Test the filtrate for completeness of destruction and discard the filtrate and the precipitate appropriately.

2. *Note:* This method works **only** for DB(*c*,*g*)C, DB(*a*,*i*)C, and DB(*a*,*j*)AC. Because these compounds are so insoluble, only trace amounts are likely to be present. Add enough $KMnO_4$ to make a 0.3 *M* solution and enough NaOH to make a 2 *M* solution and swirl the mixture and allow it to react for 3 h. The purple or green color should be maintained during this reaction time. If it is not, add more $KMnO_4$ solution until the reaction mixture remains purple or green for 1 h. At the end of the reaction decolorize the solution with 0.8 g of sodium metabisulfite for each 10 mL of $KMnO_4$ solution, dilute with an equal volume of H_2O, and check for completeness of destruction. Filter and discard the filtrate and the precipitate appropriately.

3. *Note:* This method works **only** for DB(*c*,*g*)C, DB(*a*,*i*)C, and DB(*a*,*j*)AC. Because these compounds are so insoluble, only trace amounts are likely to be present. Add enough $KMnO_4$ to make a 0.3 *M* solution and allow it to react with stirring for 6 h. The purple color should be maintained during this reaction time. If it is not, add more $KMnO_4$ solution

until the reaction mixture remains purple for 1 h. At the end of the reaction decolorize the solution with 0.8 g of sodium metabisulfite for each 10 mL of $KMnO_4$ solution (more if necessary for complete decolorization), basify with 1 mL of 10 *M* NaOH solution, check that the mixture is strongly basic, and check for completeness of destruction. Filter and discard the filtrate and the precipitate appropriately.

Decontamination of Glassware Contaminated with PHH

1. Rinse the glassware with four portions of acetone that are large enough to wet the glassware thoroughly. Combine the rinses and remove the acetone by evaporation under reduced pressure using a rotary evaporator (temperature of water bath ~40°C). For every 5 mg of PHH add 3 mL of acetonitrile and ensure that the PHH is completely dissolved, including any PHH that may be adhering to the walls of the container. Prepare a 0.3 *M* solution of $KMnO_4$ in 3 *M* H_2SO_4 by stirring the calculated amount of $KMnO_4$ in 3 *M* H_2SO_4 for at least 15 min but no more than 1 h. For every 5 mg of PHH add 10 mL of a 0.3 *M* $KMnO_4$ solution in 3 *M* H_2SO_4 (freshly prepared) and swirl the mixture and allow to react for 1 h. The purple color should be maintained during this reaction time. If it is not, add more $KMnO_4$ solution until the reaction mixture remains purple for 1 h. In addition immerse the rinsed glassware in 0.3 *M* $KMnO_4$ in 3 *M* H_2SO_4 for 1 h. At the end of the reaction decolorize the solution with 0.4 g of sodium metabisulfite for each 10 mL of $KMnO_4$ solution, basify by the addition of 4 mL of 10 *M* NaOH solution, dilute with 50 mL of H_2O, and filter. Test the filtrate for completeness of destruction and discard the filtrate and the precipitate appropriately.
Note: This reaction should be performed on at least this scale, even if less than 5 mg of PHH is present.

2. Rinse the glassware with four portions of acetone that are large enough to wet the glassware thoroughly. Rinse with one volume of hexane and check the glassware for the absence of fluorescence under ultraviolet (UV) light. If fluorescence remains, repeat the washes. Combine the organic solvent rinses and remove the solvents by evaporation under reduced pressure using a rotary evaporator (temperature of water bath ~40°C). The reaction should be performed in a vessel that has a volume at least 20 times that of the H_2O_2 that is used. For every 5 mg of PHH add 5 mL of acetone and make sure that the PHH has dissolved completely. Surround the reaction vessel with an ice bath. For each 5 mg of PHH add 0.2 g of

$FeCl_2$ or $FeCl_2.2H_2O$ or 0.3 g of $FeCl_2.4H_2O$, then add 10 mL of 30% H_2O_2 while stirring. The reaction has an induction period and the H_2O_2 should be added dropwise until the reaction mixture boils then the addition should be completed over a 15-min period. (If the reaction mixture does not boil, the degradation will not be complete. In this case more iron chloride and more H_2O_2 should be added with more rapid addition of the H_2O_2.) When the addition of H_2O_2 is complete remove the ice bath and stir the reaction mixture for 30 min, neutralize, test for completeness of destruction, and discard it.

3. *Note:* This method works **only** for DB(*c,g*)C, DB(*a,i*)C, and DB(*a,j*)AC. Rinse the glassware with four portions of acetone that are large enough to wet the glassware thoroughly. Combine the rinses and remove the acetone by evaporation under reduced pressure using a rotary evaporator (temperature of water bath ~40°C). For every 5 mg of DB(*c,g*)C, DB(*a,i*)C, or 1 mg of DB(*a,j*)AC add 2 mL of acetonitrile and ensure that the PHH is completely dissolved, including any PHH that may be adhering to the walls of the container. Prepare a 0.3 *M* solution of $KMnO_4$ in 2 *M* NaOH by stirring the calculated amount of $KMnO_4$ in 2 *M* NaOH for at least 30 min but no more than 2 h. For every 2 mL of acetonitrile add 10 mL of a 0.3 *M* $KMnO_4$ solution in 2 *M* NaOH and swirl the mixture and allow to react for 3 h. The purple or green color should be maintained during this reaction time. If it is not, add more $KMnO_4$ solution until the reaction mixture remains purple for 1 h. (If splashing occurs, causing unreacted compound to be isolated on the wall of the reaction vessel, rinse the wall with the decontaminating solution.) In addition immerse the rinsed glassware in 0.3 *M* $KMnO_4$ in 2 *M* NaOH for 3 h. At the end of the reaction decolorize the solution with 0.8 g of sodium metabisulfite for each 10 mL of $KMnO_4$ solution, dilute with an equal volume of H_2O, and check for completeness of destruction. Filter and discard the filtrate and the precipitate appropriately.

4. *Note:* This method works **only** for DB(*c,g*)C, DB(*a,i*)C, and DB(*a,j*)AC. Rinse the glassware with four portions of acetone that are large enough to wet the glassware thoroughly. Combine the rinses and remove the acetone by evaporation under reduced pressure using a rotary evaporator (temperature of water bath ~40°C). For every 5 mg of DB(*c,g*)C, DB(*a,i*)C, or 1 mg of DB(*a,j*)AC add 2 mL of acetonitrile and ensure that the PHH is completely dissolved, including any PHH that may be adhering to the walls of the container. Prepare a 0.3 *M* solution of $KMnO_4$ in H_2O by stirring the calculated amount of $KMnO_4$ in H_2O for

at least 30 min but no more than 2 h. For every 2 mL of acetonitrile add 10 mL of a 0.3 *M* $KMnO_4$ solution in H_2O and stir for at least 6 h. The purple color should be maintained during this reaction time. If it is not, add more $KMnO_4$ solution until the reaction mixture remains purple for 1 h. (If splashing occurs, causing unreacted compound to be isolated on the wall of the reaction vessel, rinse the wall with the decontaminating solution.) In addition immerse the rinsed glassware in 0.3 *M* $KMnO_4$ solution for 6 h. At the end of the reaction decolorize the solution with 0.8 g of sodium metabisulfite for each 10 mL of $KMnO_4$ solution (more if necessary for complete decolorization), basify with 1 mL of 10 *M* NaOH solution, check that the mixture is strongly basic, and check for completeness of destruction. Filter and discard the filtrate and the precipitate appropriately.

5. *Note:* This method works **only** for DB(*c,g*)C and DB(*a,i*)C. Add enough DMSO to wet the glassware. For every 2 mL of DMSO used add 10 mL of concentrated H_2SO_4 (**Caution!** Exothermic) and swirl the mixture and allow to react for at least 2 h. At the end of the reaction cautiously add the solution to at least three times the volume of cold H_2O (using an ice bath if desired as this is a very exothermic process), neutralize, test for completeness of destruction, and discard it.

Treatment of Spills Involving PHH

1. Remove as much as possible of the spill by high efficiency particulate air (HEPA) vacuuming or using absorbents. Prepare a 0.3 *M* solution of $KMnO_4$ in 3 *M* H_2SO_4 by stirring the calculated amount of $KMnO_4$ in 3 *M* H_2SO_4 for at least 15 min but no more than 1 h. Collect solid on a cloth wetted with acetonitrile and absorb liquid on a dry cloth. Add enough acetonitrile to completely wet the area and add an excess of 0.3 *M* $KMnO_4$ in 3 *M* H_2SO_4. Immerse the cloth in 0.3 *M* $KMnO_4$ in 3 *M* H_2SO_4. Allow to react for 1 h. The purple color should be maintained during this reaction time. If it is not, add more $KMnO_4$ solution until the reaction mixture remains purple for 1 h. Remove the residual solution from the spill area with absorbents. Then, for each 10 mL of $KMnO_4$ solution, decolorize all solutions with 0.4 g of sodium metabisulfite, basify by the addition of 4 mL of 10 *M* NaOH solution, dilute with 50 mL of H_2O, and filter. Test the filtrate for completeness of destruction and discard the filtrate and the precipitate appropriately. Examine the area of spillage under long-wave and short-wave UV light for the absence of PHH fluorescence.

2. *Note:* This method works **only** for DB(*c,g*)C, DB(*a,i*)C, and DB(*a,j*)AC. Remove as much as possible of the spill by HEPA vacuuming or using absorbents. Prepare a 0.3 *M* solution of $KMnO_4$ in 2 *M* NaOH by stirring the calculated amount of $KMnO_4$ in 2 *M* NaOH for at least 30 min but no more than 2 h. Collect solid on a cloth wetted with acetonitrile and absorb liquid on a dry cloth. Add enough acetonitrile to completely wet the area and add an excess of 0.3 *M* $KMnO_4$ in 2 *M* NaOH. Immerse the cloth in 0.3 *M* $KMnO_4$ in 2 *M* NaOH. Allow to react for 3 h. The purple or green color should be maintained during this reaction time. If it is not, add more $KMnO_4$ solution until the reaction mixture remains purple for 1 h. Remove the residual solution from the spill area with absorbents. Then, for each 10 mL of $KMnO_4$ solution, decolorize all solutions with 0.8 g of sodium metabisulfite, dilute with an equal volume of H_2O, and check for completeness of destruction. Filter and discard the filtrate and the precipitate appropriately. Examine the area of spillage under long-wave and short-wave UV light for the absence of PHH fluorescence.

3. *Note:* This method works **only** for DB(*c,g*)C, DB(*a,i*)C, and DB(*a,j*)AC. Remove as much as possible of the spill by HEPA vacuuming or using absorbents. Prepare a 0.3 *M* solution of $KMnO_4$ in H_2O by stirring the calculated amount of $KMnO_4$ in H_2O for at least 30 min but no more than 2 h. Collect solid on a cloth wetted with acetonitrile and absorb liquid on a dry cloth. Add enough acetonitrile to completely wet the area and add an excess of 0.3 *M* $KMnO_4$ in H_2O. Immerse the cloth in 0.3 *M* $KMnO_4$ in H_2O. Allow to react for at least 6 h. The purple color should be maintained during this reaction time. If it is not, add more $KMnO_4$ solution until the reaction mixture remains purple for 1 h. Remove the residual solution from the spill area with absorbents. Then, for each 10 mL of $KMnO_4$ solution, decolorize all solutions with 0.8 g of sodium metabisulfite (more if necessary for complete decolorization), basify with 1 mL of 10 *M* NaOH solution, check that the mixture is strongly basic, and check for completeness of destruction. Filter and discard the filtrate and the precipitate appropriately. Examine the area of spillage under long-wave and short-wave UV light for the absence of PHH fluorescence.

Analytical Procedures

Polycyclic heterocyclic hydrocarbons may be determined by high-performance liquid chromatography (HPLC) as follows.[1,10] An aliquot of the reaction mixture was basified to pH 10–12 with NaOH solution and the PHH were extracted three times with a volume of cyclohexane equivalent to the volume of the aqueous layer. The pooled cyclohexane extracts were evaporated to dryness under reduced pressure and the residue was taken up in 0.5 mL of acetonitrile. The PHH were then determined by HPLC using a 250 × 3.6-mm i.d. column of Partisil ODS2 (10 μm) and a 50 × 4.6-mm guard column of pellicular ODS. The mobile phase was acetonitrile:water 80:20 flowing at 1.5 mL/min and the injection volume was 50 μL. The retention times were in the range 5.1–8.4 min. The detector was a Kontron spectrofluorometer operating at 366 nm excitation and 425 nm emission for the acridine compounds and 292 nm excitation and 389 nm emission for the carbazole compounds. If a spectrofluorometer is not available, UV detection at 254 nm can be used. The PHH may also be analyzed by gas chromatography (GC) using a 1.8-m × 2-mm i.d. glass column packed with 3% OV-1 on 80/100 Supelcoport.[1] The oven temperature is 260°C, the injection temperature is 300°C, and the temperature of the flame ionization detector is 300°C. The level of manganese in the filtrates was determined by atomic absorption spectroscopy at 279.5 nm.[10]

Mutagenicity Assays[1]

The reaction mixtures from these procedures were tested using *Salmonella typhimurium* strains TA97, TA98, TA100, and TA102 (TA98, TA100, TA1530, and TA1535 for the procedure involving concentrated H_2SO_4), with and without activation, and they were not mutagenic.

Related Compounds

Although these procedures have only been validated for the compounds listed above, they should be applicable to other PHH. However, it should be noted from the data above that not all procedures worked for all compounds. Therefore, when another PHH is to be degraded, the procedure should be thoroughly validated before use (see also the Polycyclic Aromatic Hydrocarbons monograph).

References

1. Castegnaro, M.; Barek, J.; Jacob, J.; Kirso, U.; Lafontaine, M.; Sansone, E.B.; Telling, G.; Vu Duc, T., Eds., *Laboratory Decontamination and Destruction of Carcinogens in Laboratory Wastes: Some Polycyclic Heterocyclic Hydrocarbons*; International Agency for Research on Cancer: Lyon, 1991 (IARC Scientific Publications No. 114).

2. Other names for this compound are 7-azadibenz(*a,j*)anthracene, 1,2:7,8-dibenzacridine, 3,4,5,6-dibenzacridine, dibenz(*a,f*)acridine, dibenzo(*a,j*)acridine, 3:4-6:7-dinaphthacridine, β-naphthacridine, and 1.2.1′.2′-dinaphthacridine.

3. Other names for this compound are 7-azadibenz(*a,h*)anthracene, 1,2:5,6-dibenzacridine, dibenz(*a,d*)acridine, 1,2,5,6-dibenzoacridine, 1,2,5,6-dinaphthacridine, 1.2.2′.1′-dinaphthacridine, and 3,4,7,8-dibenzacridine.

4. Other names for this compound are 7-aza-7*H*-dibenzo(*c,g*)fluorene, 3,4,5,6-dibenzocarbazole, 1,1′-dinaphtho-2,2′-carbazol, and *S*-1:2-dinaphthocarbazole.

5. Other names for this compound are 13-aza-13*H*-dibenzo(*a,i*)fluorene, 1,2,7,8-dibenzocarbazole, 1,2′-dinaphtho-2,1′-carbazole, 1:1′-imino-2:2′-dinaphthyl, dinaphthylcarbazole, and dinaphthyleneimine.

6. International Agency for Research on Cancer. *IARC Monographs on the Evaluation of the Carcinogenic Risk of Chemicals to Humans.* Volume 32, *Polynuclear Aromatic Compounds, Part 1, Chemical, Environmental and Experimental Data*; International Agency for Research on Cancer: Lyon, 1983; pp. 283–288.

7. Reference 6, pp. 277–281.

8. Reference 6, pp. 315–319.

9. Boyland, E.; Brues, A.M. The carcinogenic action of dibenzocarbazoles. *Proc. R. Soc. London.* **1937**, *122*, 429–441.

10. Lunn, G.; Sansone, E.B.; De Méo, M.; Laget, M.; Castegnaro, M. Potassium permanganate can be used for degrading hazardous compounds. *Am. Ind. Hyg. Assoc. J.* **1994**, *55*, 167–171.

POTASSIUM PERMANGANATE

CAUTION! Refer to safety considerations section on page 7 before starting any of these procedures.

Potassium permanganate ($KMnO_4$) [7722-64-7][1] is a purple solid that is toxic by ingestion and a strong irritant.[2] This compound is a powerful oxidizer and fires and explosions may occur when it comes in contact with organic or readily oxidizable materials, either when dry or in solution.[2,3] Potassium permanganate is also incompatible with many organic and inorganic compounds.[2,3] Potassium permanganate has many uses as an oxidant, particularly in organic synthesis and for the destruction of hazardous compounds.

Principle of Destruction

Potassium permanganate can be reduced with sodium metabisulfite and precipitated as manganese hydroxide at high pH. The brown precipitate of manganese hydroxide rapidly darkens to black as air oxidation to manganese dioxide occurs. Less than 1 ppm of mutagenic[4] manganese remains in solution.[5]

Destruction Procedure

Take up 5 g of $KMnO_4$ in 200 mL of 1 *M* sodium hydroxide solution and add 10 g of sodium metabisulfite. The purple color of the mixture should disappear, if it does not add more sodium metabisulfite. After stirring for 30 min dilute with 200 mL of H_2O, filter, and discard.[4]

Analytical Procedures

The filtrate can be tested for residual oxidant by adding a few drops to an equal volume of 10% (w/v) potassium iodide solution. Acidify with 1 *M* hydrochloric acid and add one drop of starch as an indicator. A deep blue color indicates the presence of oxidant and the procedure should be repeated. Manganese in the filtrate can be determined by atomic absorption spectroscopy at 279.5 nm.

References

1. Other names for this compound are Cairox, chameleon mineral, C.I. 77755, Condy's crystals, and permanganic acid potassium salt.
2. Lewis, R.J., Sr. *Sax's Dangerous Properties of Industrial Materials*, 8th ed.; Van Nostrand-Reinhold: New York, 1992; pp. 2877–2878.
3. Bretherick, L. *Bretherick's Handbook of Reactive Chemical Hazards*, 4th ed.; Butterworths: London, 1990; pp. 1295–1301.
4. Pagano, D.A.; Zeiger, E. Conditions for detecting the mutagenicity of divalent metals in *Salmonella typhimurium*. *Environ. Mol. Mutagen.* **1992**, *19*, 139–146.
5. Lunn, G.; Sansone, E.B.; De Méo, M.; Laget, M.; Castegnaro, M. Potassium permanganate can be used for degrading hazardous compounds. *Am. Ind. Hyg. Assoc. J.* **1994**, *55*, 167–171.

β-PROPIOLACTONE

CAUTION! Refer to safety considerations section on page 7 before starting any of these procedures.

β-Propiolactone [BPL (**I**) bp 162°C] [57-57-8] is a volatile liquid that is soluble to the extent of 37% in H_2O. This compound is used as a sterilant and industrially as an intermediate. β-Propiolactone is an animal carcinogen.[1] Other names for this compound are 2-oxetanone, hydracrylic acid β-lactone, β-propionolactone, propanolide, Betaprone, 3-propiolactone, β-propolactone, and 3-hydroxypropionic acid lactone.

I

Principle of Destruction

β-Propiolactone is oxidized by potassium permanganate in the presence of 3 *M* sulfuric acid ($KMnO_4$ in H_2SO_4). The products have not been determined. Destruction was complete and less than 0.21% BPL remained.[2]

Destruction Procedure

Take up 0.5 mL (573 mg) of β-propiolactone in 100 mL of 3 *M* H_2SO_4 and add 4.8 g of $KMnO_4$ in portions with stirring. Stir the reaction mixture at room temperature for 18 h, then decolorize it with sodium metabisulfite, make strongly basic with 10 *M* potassium hydroxide solution (**Caution!** Exothermic), dilute with H_2O, filter to remove manganese compounds,[3] neutralize the filtrate, check for completeness of destruction, and discard it.

Analytical Procedures[2]

Add 100 μL of the solution to be analyzed to 1 mL of a solution of 2 mL of acetic acid in 98 mL of 2-methoxyethanol. Swirl this mixture and add 1 mL of a solution of 5 g of 4-(4-nitrobenzyl)pyridine (4-NBP) in 100 mL of 2-methoxyethanol. Heat the solution at 100°C for 10 min, then cool in ice for 5 min. Add piperidine (0.5 mL) and 2-methoxyethanol (2 mL) and determine the violet color at 560 nm.

To check the efficacy of the analytical procedure, a small quantity of BPL can be added to the solution to be analyzed after the acetic acid-2-methoxyethanol has been added but before the 4-NBP is added. A positive response will indicate that the analytical technique is satisfactory. Using the analytical procedure described above with 10-mm disposable plastic cuvettes in a Gilford 240 spectrophotometer the limit of detection was 12 mg/L, but this can easily be reduced by using more than 100 μL of the reaction mixture.

Mutagenicity Assays

The mutagenicity assays were carried out as described on page 4 using tester strains TA98, TA100, TA1530, and TA1535. The final reaction mixture (tested at a level corresponding to 0.57 mg of undegraded BPL per plate) was not mutagenic. The pure compound was highly mutagenic.[2]

Related Compounds

This general oxidative procedure should work for other lactones although it would have to be validated for each compound.

References

1. International Agency for Research on Cancer. *IARC Monographs on the Evaluation of Carcinogenic Risk of Chemicals to Man.* Volume 4, *Some Aromatic Amines, Hydrazine and Related Substances,* N-*Nitroso Compounds and Miscellaneous Alkylating Agents*; International Agency for Research on Cancer: Lyon, 1974; pp. 259–269.
2. Lunn, G.; Sansone, E.B. Validated methods for degrading hazardous chemicals: Some alkylating agents and other compounds. *J. Chem. Educ.* **1990**, *67*, A249–A251.
3. Lunn, G.; Sansone, E.B.; De Méo, M.; Laget, M.; Castegnaro, M. Potassium permanganate can be used for degrading hazardous compounds. *Am. Ind. Hyg. Assoc. J.* **1994**, *55*, 167–171.

PROTEASE INHIBITORS

CAUTION! Refer to safety considerations section on page 7 before starting any of these procedures.

*N*α-*p*-Tosyl-L-lysine chloromethyl ketone [TLCK, *p*-$CH_3C_6H_4SO_2NHCH((CH_2)_4NH_2)C(O)CH_2Cl$] [4238-41-9][1] and *N*-*p*-tosyl-L-phenylalanine chloromethyl ketone [TPCK, *p*-$CH_3C_6H_4SO_2NHCH(CH_2C_6H_5)C(O)CH_2Cl$] [402-71-1][2] are widely used as enzyme inhibitors.[3–5] Both TLCK and TPCK are toxic and exhibit reproductive effects.[6,7] The compound TLCK is generally supplied as the hydrochloride (MW 369) which is a white solid, mp 160–161°C(decomp) and the compound TPCK is a white solid (MW 352), mp 106–108°C.

Principle of Destruction[8]

The stabilities of TLCK and TPCK decrease as the pH increases (Table 1). It was found that TLCK and TPCK could be degraded in 18 h by making the solution strongly alkaline (pH ≥ 12) by adding 1 *M* sodium hydroxide (NaOH) solution. With higher concentrations and shorter re-

Table 1 Stability of Protease Inhibitors in Aqueous Solution

Buffer[a] (pH)	Half-Life	
	TLCK	TPCK
3.0	22.5 days	113 h
5.0	41 h	103 h
6.4 Hanks'	>28 days[b]	74 h
7.0	106 min	33 h
7.3 Dulbecco's	93 min	53 h
7.4 PBS	98 min[c]	45 h
7.5 HEPES	84 min[d]	29 h
8.0 TBE	8 min	12 h
9.0	2 min	110 min
11.0	3 min[e]	26 min

[a] The pH of the reaction mixture was ± 0.2 units of that shown except where indicated.

[b] Actual pH 3.8.

[c] Actual pH 7.0.

[d] Actual pH 7.2.

[e] Actual pH 10.7.

action times degradation was incomplete. Either TLCK or TPCK at a concentration of 1 m*M* in the following buffers: pH 5.0 50 m*M* phthalate, pH 6.4 Hanks' balanced salts, pH 7.0 50 m*M* phosphate, pH 7.3 Dulbecco's phosphate buffered saline, pH 7.4 10 m*M* phosphate buffered saline, pH 7.5 50 m*M* 4-(2-hydroxyethyl)-1-piperazineethanesulfonic acid (HEPES), and pH 8.0 50 m*M* TRIS–borate–ethylenediaminetetraacetic acid (EDTA) (TBE), could be degraded under these conditions. Destruction of TLCK was greater than 99.8% and the final concentration was less than 0.37 ppm. Destruction of TPCK was greater than 99% (except in pH 7 and TBE buffers when it was >98.3%) and the final concentration was less than 3 ppm. Stock solutions of TLCK (5 m*M*, 1.85 mg/mL) in H_2O or dimethyl sulfoxide (DMSO) and TPCK (1 m*M*, 352 μg/mL) in isopropanol or DMSO could be degraded by adding 1 *M* NaOH solution. Destruction of TLCK was greater than 99.8% and the final concentration was less than 2 ppm. Destruction of TPCK was greater than 99% and the final concentration was less than 3 ppm. It was found that TPCK in isopropanol or DMSO solution was stable over 3 weeks. The stock solution of TLCK was freshly prepared with H_2O before each use and it was not stored. Stock solutions of TLCK in DMSO were reasonably stable; after 22 days the concentration of TLCK was 79% of the original concentration.

Destruction Procedures[8]

Destruction of TLCK and TPCK in Buffer

For each 10 mL of buffer containing up to 1 m*M* of TLCK or TPCK add 1 mL of 1 *M* NaOH solution. Ensure that the pH is greater than or equal to 12. After 18 h neutralize with 1 mL of acetic acid, test for completeness of destruction, and discard the reaction mixture.

Destruction of Stock Solutions of TLCK in Water

Dilute the solution, if necessary, so that the concentration of TLCK does not exceed 5 m*M* (1.85 mg/mL). For each 10 mL of solution add 1 mL of 1 *M* NaOH solution, shake to ensure complete mixing, and allow to stand for 18 h. Neutralize with 1 mL of acetic acid, test for completeness of destruction, and discard the reaction mixture.

Destruction of Stock Solutions of TLCK in DMSO

Dilute the solution, if necessary, so that the concentration of TLCK does not exceed 5 m*M* (1.85 mg/mL). For each 1 mL of solution add 5 mL of 1 *M* NaOH solution, shake to ensure complete mixing, and allow to stand for 18 h. Neutralize with 0.5 mL of acetic acid, test for completeness of destruction, and discard the reaction mixture.

Destruction of Bulk Quantities of TLCK

Dissolve the TLCK in H_2O so that the concentration does not exceed 5 m*M* (1.85 mg/mL). For each 10 mL of solution add 1 mL of 1 *M* NaOH solution, shake to ensure complete mixing, and allow to stand for 18 h. Neutralize with 1 mL of acetic acid, test for completeness of destruction, and discard the reaction mixture.

Destruction of Stock Solutions of TPCK in Isopropanol or DMSO

Dilute the solution, if necessary, so that the concentration of TPCK does not exceed 1 m*M* (352 μg/mL). For each 10 mL of solution add 1 mL of 1 *M* NaOH solution, shake to ensure complete mixing, and allow to stand for 18 h. Neutralize with 1 mL of acetic acid, test for completeness of destruction, and discard the reaction mixture.

Destruction of Bulk Quantities of TPCK

Dissolve the TPCK in isopropanol so that the concentration does not exceed 1 m*M* (352 μg/mL). For each 10 mL of solution add 1 mL of 1 *M* NaOH solution, shake to ensure complete mixing, and allow to stand for 18 h. Neutralize with 1 mL of acetic acid, test for completeness of destruction, and discard the reaction mixture.

Analytical Procedures[8]

The reaction mixtures should be neutralized as described above before analysis. Analyze TLCK by reverse phase high-performance liquid chromatography (HPLC) (ion-pair chromatography) using acetonitrile:0.1% trifluoroacetic acid 40:60 flowing at 1 mL/min and an ultraviolet (UV) detector set at 228 nm. The retention time was about 9.5 min and the limit of detection was about 0.37 μg/mL. Analyze TPCK by reverse phase HPLC using acetonitrile:10 m*M* phosphate buffer (pH 7) 48:52 flowing at 1 mL/min and a UV detector set at 228 nm. The retention time was about 10.5 min and the limit of detection was about 2 μg/mL. This analysis was complicated by the elution of numerous small unknown peaks. The analytical procedures should be checked by spiking an aliquot of the acidified reaction mixture with a small quantity of a dilute solution of the compound of interest.

Mutagenicity Assays[8]

The mutagenicity assays were carried out as described on page 4. The final reaction mixtures were not mutagenic. When dissolved in DMSO TLCK itself was not mutagenic. When dissolved in DMSO TPCK itself was mutagenic to tester strain TA98 with activation.

Related Compounds

The above techniques were investigated for TLCK and TPCK but they may also be applicable to other related enzyme inhibitors. However, they should be thoroughly investigated before being applied to other compounds. See also the monographs on Sulfonyl Fluoride Enzyme Inhibitors and on Diisopropyl Fluorophosphate.

References

1. Other names for this compound are 1-chloro-3-tosylamido-7-amino-L-2-heptanone, *N*-(5-amino-1-(chloroacetyl)pentyl)-4-methylbenzenesulfonamide, and tosyllysyl chloromethyl ketone.
2. Other names for this compound are L-1-*p*-tosylamino-2-phenylethyl chloromethyl ketone, tosyl-l-phenylalanylchloromethylketone, *N*-(chloroacetyl)-3-phenyl-*N*-(*p*-tolylsulfonyl)-alanine, and L-1-chloro-3-tosylamido-4-phenyl-2-butanone.
3. Carpenter, F.H. Treatment of trypsin with TPCK. In *Methods in Enzymology*, Vol. 11; Hirs, C.H.W., Ed., Academic Press: New York, 1967; p. 237.
4. Shaw, E. Site-specific reagents for chymotrypsin and trypsin. In *Methods in Enzymology*, Vol. 11; Hirs, C.H.W., Ed., Academic Press: New York, 1967; pp. 677–686.
5. Kezdy, F.J.; Thomson, A.; Bender, M.L. Studies on the reaction of chymotrypsin and L-1-chloro-3-tosylamido-4-phenyl-2-butanone. *J. Am. Chem. Soc.* **1967**, *89*, 1004–1009.
6. Lewis, R.J., Sr. *Sax's Dangerous Properties of Industrial Materials*, 8th ed.; Van Nostrand-Reinhold: New York, 1992; p. 3326.
7. Reference 6, p. 3327.
8. Lunn, G.; Sansone, E.B. Degradation and disposal of some enzyme inhibitors. *Appl. Biochem. Biotechnol.* (in press).

SODIUM AMIDE

CAUTION! Refer to safety considerations section on page 7 before starting any of these procedures.

Sodium amide ($NaNH_2$, sodamide) [7782-92-5] is a crystalline solid that can react explosively with H_2O, with heat, or on grinding. This compound may become explosive on storage.[1] Sodium amide is used in organic synthesis.

Destruction Procedure[2]

Immerse 5 g of $NaNH_2$ in 25 mL of toluene and slowly and cautiously add 30 mL of 95% ethanol with stirring. The $NaNH_2$ is converted to sodium ethoxide and ammonia. When the reaction is complete dilute the reaction mixture with 50 mL of H_2O, separate the layers, and discard them. Wash out contaminated apparatus with ethanol before cleaning.

Related Compounds

This procedure is also applicable to potassium amide [17242-52-3].

References

1. Lewis, R.J., Sr. *Sax's Dangerous Properties of Industrial Materials*, 8th ed.; Van Nostrand-Reinhold: New York, 1992; p. 3059.
2. Bergstrom, F.W. Sodium amide. In *Organic Syntheses*; Horning, E.C., Ed., Wiley: New York, 1955; Coll. Vol. 3, pp. 778–783.

STERIGMATOCYSTIN

CAUTION! Refer to safety considerations section on page 7 before starting any of these procedures.

Sterigmatocystin (I) [10048-13-2][1] is a fungal metabolite from *Aspergillus versicolor*. Sterigmatocystin is a solid (mp 242–244°C) which is insoluble in H_2O but soluble in organic solvents.[2] This compound is carcinogenic in laboratory animals.[3] The solid compound may become electrostatically charged and cling to glassware or protective clothing. In a recent collaborative study organized by the International Agency for Research on Cancer (IARC) the safe disposal of this compound was investigated.[2]

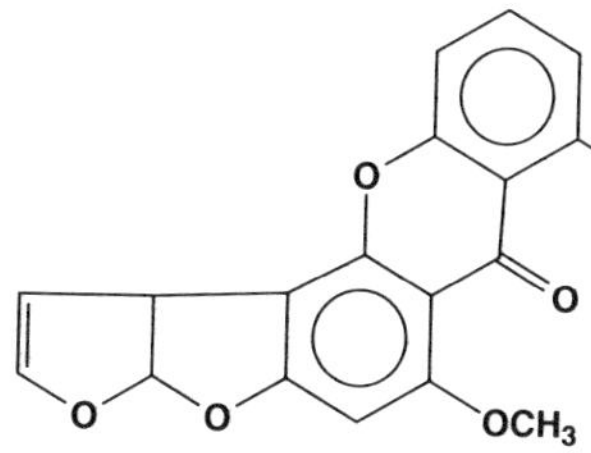

OH O O O O OCH_3

I

Principles of Destruction

Sterigmatocystin may be degraded using potassium permanganate in sodium hydroxide solution ($KMnO_4$ in NaOH) or dilute sodium hypochlorite (NaOCl) solution followed by acetone. The NaOCl solution is diluted to avoid the haloform reaction when the acetone is added. The acetone is added to destroy possibly hazardous dichloro compounds. The degradation efficiency was greater than 99.8% using NaOCl and greater than 99.9% using $KMnO_4$.

Destruction Procedures[2]

Destruction of Bulk Quantities of Sterigmatocystin

1. Prepare a dilute solution of NaOCl by adding 100 mL of commercial 5.25% NaOCl solution (Clorox bleach) to 200 mL of H_2O. Use fresh NaOCl solution (see assay procedure below). Dissolve each 1 mg of sterigmatocystin in 20 mL of methanol (sonication and heating to 50°C may be required). For each 20 mL of solution add 25 mL of the dilute NaOCl solution. After 1 h add a volume of acetone equal to 5% of the total reaction volume. After 5 min check for completeness of destruction and discard the reaction mixture.

2. Prepare a 0.3 *M* solution of $KMnO_4$ in 2 *M* NaOH solution by stirring the mixture for at least 30 min but no more than 2 h. Dissolve 300 μg of sterigmatocystin in 5 mL of acetonitrile and add 10 mL of $KMnO_4$ in NaOH. Stir for at least 3 h. The color should be either green or purple. If it is not, add more $KMnO_4$ in NaOH until the green or purple color persists for at least 1 h. For each 10 mL of $KMnO_4$ in NaOH add 0.8 g of sodium metabisulfite (more if necessary for complete decolorization), dilute with an equal volume of H_2O, filter to remove the manganese salts,[4] check for completeness of destruction, and discard the solid and filtrate appropriately.

Destruction of Sterigmatocystin in Aqueous Solution

1. Prepare a dilute solution of NaOCl by adding 100 mL of commercial 5.25% NaOCl solution (Clorox bleach) to 200 mL of H_2O. Use fresh NaOCl solution (see assay procedure below). For each volume of aqueous solution add 2 volumes of the dilute NaOCl solution. After 1 h add a volume of acetone equal to 5% of the total reaction volume. After 5 min check for completeness of destruction and discard the reaction mixture.

2. Dilute with H_2O, if necessary, so that the concentration of sterigmatocystin does not exceed 200 μg/mL. Add sufficient NaOH, with stirring, to make the concentration 2 *M*, then add sufficient solid $KMnO_4$ to make the concentration 0.3 *M*. Stir for at least 3 h. The color should be either green or purple. If it is not, add more $KMnO_4$ in NaOH until the green or purple color persists for at least 1 h. For each 10 mL of $KMnO_4$ in NaOH add 0.8 g of sodium metabisulfite (more if necessary for complete decolorization), dilute with an equal volume of H_2O, filter to remove the manganese salts,[4] check for completeness of destruction, and discard the solid and filtrate appropriately.

Destruction of Sterigmatocystin in Volatile Organic Solvents

1. Prepare a dilute solution of NaOCl by adding 100 mL of commercial 5.25% NaOCl solution (Clorox bleach) to 200 mL of H_2O. Use fresh NaOCl solution (see assay procedure below). Remove the organic solvent under reduced pressure using a rotary evaporator. Dissolve each 1 mg of sterigmatocystin in 20 mL of methanol (sonication and heating to 50°C may be required). For each 20 mL of solution add 25 mL of the dilute NaOCl solution. After 1 h add a volume of acetone equal to 5% of the total reaction volume. After 5 min check for completeness of destruction and discard the reaction mixture.

2. Prepare a 0.3 *M* solution of $KMnO_4$ in 2 *M* NaOH solution by stirring the mixture for at least 30 min but no more than 2 h. Remove the organic solvent under reduced pressure using a rotary evaporator. Dissolve 300 μg of sterigmatocystin in 5 mL of acetonitrile and add 10 mL of $KMnO_4$ in NaOH. Stir for at least 3 h. The color should be either green or purple. If it is not, add more $KMnO_4$ in NaOH until the green or purple color persists for at least 1 h. For each 10 mL of $KMnO_4$ in NaOH add 0.8 g of sodium metabisulfite (more if necessary for complete decolorization), dilute with an equal volume of H_2O, filter to remove the manganese salts,[4] check for completeness of destruction, and discard the solid and filtrate appropriately.

Destruction of Sterigmatocystin in Dimethyl Sulfoxide or N,N-Dimethylformamide

Prepare a dilute solution of NaOCl by adding 100 mL of commercial 5.25% NaOCl solution (Clorox bleach) to 200 mL of H_2O. Use fresh NaOCl solution (see assay procedure below). Dilute the dimethyl sulfoxide or *N,N*-dimethylformamide solution with 2 volumes of H_2O and extract three times with equal volumes of dichloromethane, pool the extracts, and dry them over anhydrous sodium sulfate. Remove the sodium sulfate by filtration and wash it with one volume of dichloromethane. Evaporate to dryness and make sure that all the dichloromethane is removed under reduced pressure using a rotary evaporator. Dissolve each 1 mg of sterigmatocystin in 20 mL of methanol (sonication and heating to 50°C may be required). For each 20 mL of solution add 25 mL of the dilute NaOCl solution. After 1 h add a volume of acetone equal to 5% of the total reaction volume. After 5 min check for completeness of destruction and discard the reaction mixture.

Decontamination of Glassware

1. Prepare a dilute solution of NaOCl by adding 100 mL of commercial 5.25% NaOCl solution (Clorox bleach) to 200 mL of H_2O. Use fresh NaOCl solution (see assay procedure below). Rinse the glassware with a small portion of methanol and evaporate the methanol to dryness. Dissolve each 1 mg of sterigmatocystin in 20 mL of methanol (sonication and heating to 50°C may be required). For each 20 mL of solution add 25 mL of the dilute NaOCl solution. After 1 h add a volume of acetone equal to 5% of the total reaction volume. After 5 min check for completeness of destruction and discard the reaction mixture. Immerse the glassware in the dilute NaOCl solution. After 1 h add a volume of acetone equal to 5% of the total reaction volume. After 5 min check for completeness of destruction and discard the reaction mixture.

2. Prepare a 0.3 *M* solution of $KMnO_4$ in 2 *M* NaOH solution by stirring the mixture for at least 30 min but no more than 2 h. Rinse the glassware five times with small portions of dichloromethane. Combine the rinses and evaporate the dichloromethane under reduced pressure using a rotary evaporator. Dissolve 300 μg of sterigmatocystin in 5 mL of acetonitrile and add 10 mL of $KMnO_4$ in NaOH. Stir for at least 3 h. The color should be either green or purple. If it is not, add more $KMnO_4$ in NaOH until the green or purple color persists for at least 1 h. For each 10 mL of $KMnO_4$ in NaOH add 0.8 g of sodium metabisulfite (more if necessary for complete decolorization), dilute with an equal volume of H_2O, filter to remove the manganese salts,[4] check for completeness of destruction, and discard the solid and filtrate appropriately.

Decontamination of Protective Clothing

Prepare a dilute solution of NaOCl by adding 100 mL of commercial 5.25% NaOCl solution (Clorox bleach) to 200 mL of H_2O. Use fresh NaOCl solution (see assay procedure below). Immerse the protective clothing in the dilute NaOCl solution. After 1 h add a volume of acetone equal to 5% of the total reaction volume. After 5 min check for completeness of destruction and discard the reaction mixture.

Decontamination of Spills

1. Prepare a dilute solution of NaOCl by adding 100 mL of commercial 5.25% NaOCl solution (Clorox bleach) to 200 mL of H_2O. Use fresh NaOCl solution (see assay procedure below). Collect spills of liquid with a dry tissue and spills of solid with a tissue wetted with methanol. Immerse all tissues in the dilute NaOCl solution. After 1 h add a volume of acetone equal to 5% of the total reaction volume. After 5 min check for completeness of destruction and discard the reaction mixture. Cover the spill area with the dilute NaOCl solution. After 1 h absorb the liquid onto tissues and immerse the tissues in 5% aqueous acetone After 5 min check for completeness of destruction and discard the reaction mixture. Check the surface for completeness of decontamination by using a wipe moistened with methanol and analyzing the wipe for the presence of sterigmatocystin.

2. Prepare a 0.3 *M* solution of $KMnO_4$ in 2 *M* NaOH solution by stirring the mixture for at least 30 min but no more than 2 h. Collect spills of liquid with a dry tissue and spills of solid with a tissue wetted with dichloromethane. Immerse all tissues in the $KMnO_4$ in NaOH solution. Allow to react for at least 3 h. The color should be either green or purple. If it is not, add more $KMnO_4$ in NaOH until the green or purple color persists for at least 1 h. For each 10 mL of $KMnO_4$ in NaOH add 0.8 g of sodium metabisulfite (more if necessary for complete decolorization), dilute with an equal volume of H_2O, filter to remove the manganese salts,[4] check for completeness of destruction, and discard the solid and filtrate appropriately. Cover the spill area with an excess of the $KMnO_4$ in NaOH solution and allow to react for 3 h. Collect the solution on a tissue and immerse the tissue in 2 *M* sodium metabisulfite solution. If the pH of this solution is acidic make it alkaline with sodium hydroxide. Rinse the spill area with a 2 *M* solution of sodium metabisulfite. Check the surface for completeness of decontamination by using a wipe moistened with methanol and analyzing the wipe for the presence of sterigmatocystin.

Decontamination of Thin-Layer Chromatography Plates

Prepare a dilute solution of NaOCl by adding 100 mL of commercial 5.25% NaOCl solution (Clorox bleach) to 200 mL of H_2O. Use fresh NaOCl solution (see assay procedure below). Spray the plate with the dilute NaOCl solution and allow to stand for 1 h. Then spray with 5% aqueous acetone and allow to react for 5 min. Check for completeness of destruction by scraping the plate and eluting any remaining sterigmatocystin with a suitable solvent.

Analytical Procedures[2]

1. For reaction mixtures obtained using the NaOCl procedures, acidify an aliquot of the final reaction mixture to pH 5–6 with glacial acetic acid and pass nitrogen through the mixture for at least 1 min. Analyze by reverse phase high-performance liquid chromatography (HPLC) using methanol:water 80:20 flowing at 1.5 mL/min and an ultraviolet (UV) detector set at 243 nm.

2. For reaction mixtures obtained using the $KMnO_4$ procedures, acidify an aliquot to pH 2–3 using concentrated hydrochloric acid. Extract this mixture three times with an equal volume of dichloromethane, pool the extracts, and dry them over anhydrous sodium sulfate. Remove the sodium sulfate by filtration, evaporate to dryness and take up the residue in 0.5 mL of methanol:water 80:20. Analyze by HPLC as above.

Mutagenicity Assays[2]

The residues from these degradation procedures were tested using tester strains TA97, TA98, TA100, and TA102 of *Salmonella typhimurium* with and without metabolic activation. No mutagenic activity was found.

Related Compounds

The above techniques were investigated for sterigmatocystin but they may also be applicable to some other mycotoxins. However, these techniques should be thoroughly investigated before being applied to other compounds. See also monographs for Aflatoxins, Citrinin, Patulin, and Ochratoxin A.

Assay of Sodium Hypochlorite Solution

Sodium hypochlorite solutions tend to deteriorate with time, so they should be periodically checked for the amount of active chlorine they contain. Pipette 10 mL of the NaOCl solution into a 100-mL volumetric flask and fill it to the mark with distilled H_2O. Pipette 10 mL of this solution into a conical flask containing 50 mL of distilled H_2O, 1 g of potassium iodide, and 12.5 mL of 2 *M* acetic acid. Titrate this solution against a 0.1 *N* sodium thiosulfate solution using starch as an indicator. Each 1 mL of the sodium thiosulfate solution corresponds to 3.545 mg of active chlorine. Commercially available NaOCl solution (Clorox bleach) contains 5.25% NaOCl and should contain about 45–50 g of active chlorine per liter.

References

1. The Chemical Abstracts name is 3*a*,12*c*-dihydro-8-hydroxy-6-methoxy-7*H*-furo-[3′,2′:4,5]furo[2,3-*c*]xanthen-7-one.
2. Castegnaro, M.; Barek, J.; Frémy, J-.M.; Lafontaine, M.; Miraglia, M.; Sansone, E.B.; Telling, G.M., Eds., *Laboratory Decontamination and Destruction of Carcinogens in Laboratory Wastes: Some Mycotoxins*; International Agency for Research on Cancer: Lyon, 1991 (IARC Scientific Publications No. 113).
3. International Agency for Research on Cancer. *IARC Monographs on the Evaluation of Carcinogenic Risk of Chemicals to Man*. Volume 10, *Some Naturally Occurring Substances*; International Agency for Research on Cancer: Lyon, 1975; pp. 245–251.
4. Lunn, G.; Sansone, E.B.; De Méo, M.; Laget, M.; Castegnaro, M. Potassium permanganate can be used for degrading hazardous compounds. *Am. Ind. Hyg. Assoc. J.* **1994**, *55*, 167–171.

SULFONYL FLUORIDE ENZYME INHIBITORS

> **CAUTION!** Refer to safety considerations section on page 7 before starting any of these procedures.

Phenylmethylsulfonyl fluoride (PMSF, $PhCH_2SO_2F$) [329-98-6],[1,2] 4-(2-aminoethyl)benzenesulfonyl fluoride (AEBSF, $H_2NCH_2CH_2C_6H_4SO_2F$) [34284-75-8],[3,4] and *p*-amidinophenylmethanesulfonyl fluoride (APMSF, *p*-APMSF, $H_2NC(=NH)C_6H_4CH_2SO_2F$) [74938-88-8][5,6] are widely used as enzyme inhibitors. Phenylmethylsulfonyl fluoride is a white crystalline solid (mp 92–94°C) that is soluble in organic solvents and H_2O. This compound is generally kept as a stock solution in isopropanol.[7] 4-(2-Aminoethyl)benzenesulfonyl fluoride is a white crystalline solid (mp 185–187°C)[8] and APMSF hydrochloride is a white crystalline solid (mp 190–191°C).[6]

Principle of Destruction[9]

PMSF and AEBSF are stable in isopropanol and dimethyl sulfoxide (DMSO) for at least 3 weeks. When dissolved in DMSO APMSF is stable

Table 1 Stability of Protease Inhibitors in Aqueous Solution

Buffer[a]	Half-Life		
	AEBSF	PMSF	APMSF
3.0	stable[b]	stable[c]	8 days
5.0	6 days	6 days	285 min
6.4 Hanks'	11 days[d]	6 h	90 h[e]
7.0	35 h	62 min	2.6 min
7.3 Dulbecco's	34 h	23 min	3.5 min[f]
7.4 PBS	33 h	23 min	3.5 min[f]
7.5 HEPES	17 h	11 min	<1 min[g]
8.0 TBE	136 min	7 min	<1 min
9.0	39 min	<1 min	<1 min
11.0	1 min	<1 min	<1 min

[a] The pH of the reaction mixtures was within ± 0.2 units of that shown except as indicated.

[b] After 39 days 73% remained.

[c] After 10.8 days 94% remained.

[d] In Hanks' balanced salts AEBSF demonstrates unusual behavior. Initially it degrades slowly with a half-life of approximately 11 days but between 5 and 7 days there is a sudden drop in concentration from 74 to 1.5% and another peak in the chromatogram becomes prominent suggesting that the AEBSF has been converted into another compound.

[e] Actual pH 3.7.

[f] Actual pH 6.9.

[g] Actual pH 7.2.

for at least 3 weeks and has a half-life of 26 days in 1:1 isopropanol:pH 3 buffer. The stability of these compounds decreases as the pH increases (Table 1). At each pH level AEBSF is more stable than PMSF, which is more stable than APMSF. All compounds may be degraded by making the aqueous solution strongly alkaline (pH $\geq$ 12) with 1 *M* sodium hydroxide (NaOH) solution. Although degradation should be complete at lower pH traces of some compounds were found, in some cases, when the mixtures were analyzed. At pH 12 degradation was complete in 1 h and less than 0.1% of the compounds remained. Complete degradation could be achieved for PMSF at 10 m*M* concentration, AEBSF at 1 m*M* concentration, and APMSF at 2.5 m*M* concentration in the following buffers: pH 5.0 50 m*M* phthalate, pH 6.4 Hanks' balanced salts, pH 7.0 50 m*M* phosphate, pH 7.3 Dulbecco's phosphate buffered saline, pH 7.4 10 m*M* phosphate buffered saline, pH 7.5 50 m*M* 4-(2-hydroxyethyl)-1-piperazine-ethanesulfonic acid (HEPES), and pH 8.0 50 m*M* TRIS-borate-ethylenediaminetetraacetic acid (EDTA) (TBE). Stock solutions of PMSF (100 m*M*, 17.4 mg/mL) and AEBSF (20 m*M*, 4.06 mg/mL) in isopropanol or DMSO and APMSF (25 m*M*, 6.25 mg/mL) in 1:1 isopropanol:pH 3 buffer or DMSO could be degraded by adding 1 *M* NaOH solution. The solubility of AEBSF in isopropanol is considerably less than that of PMSF. Although degradation of PMSF in isopropanol appeared to be instantaneous other peaks, perhaps including an ester formed from PMSF and isopropanol, were seen by high-performance liquid chromatography (HPLC). After 24 h these peaks had decayed to a very low level (~1 ppm). In each case degradation was greater than 99.9% and less than 2 ppm of the compound remained in solution.

Destruction Procedures[9]

Destruction of PMSF, AEBSF, or APMSF in Buffer

For each 10 mL of buffer containing up to 10 m*M* PMSF, 1 m*M* AEBSF, or 2.5 m*M* APMSF add 1 mL of 1 *M* NaOH solution. Ensure that the pH is greater than or equal to 12. After 1 h neutralize with 1 mL of acetic acid, test for completeness of destruction, and discard the reaction mixture.

Destruction of PMSF in Isopropanol or Dimethyl Sulfoxide

Dilute the solution, if necessary, so that the concentration of PMSF does not exceed 100 m*M* (17.4 mg/mL). For each 1 mL of solution add 5 mL of 1 *M* NaOH solution, shake to ensure complete mixing, and allow to stand for 24 h. Neutralize with 1 mL of acetic acid, test for completeness of destruction, and discard the reaction mixture.

Destruction of AEBSF in Isopropanol or Dimethyl Sulfoxide

Dilute the solution, if necessary, so that the concentration of AEBSF does not exceed 20 m*M* (4.06 mg/mL). For each 1 mL of solution add 10 mL of 1 *M* NaOH solution, shake to ensure complete mixing, and allow to stand for 24 h. Neutralize with 1 mL of acetic acid, test for completeness of destruction, and discard the reaction mixture.

Destruction of APMSF in 1:1 Isopropanol:pH 3 Buffer or Dimethyl Sulfoxide

Dilute the solution, if necessary, so that the concentration of APMSF does not exceed 25 m*M* (6.25 mg/mL). For each 1 mL of solution add 5 mL of 1 *M* NaOH solution, shake to ensure complete mixing, and allow to stand for 24 h. Neutralize with 1 mL of acetic acid, test for completeness of destruction, and discard the reaction mixture.

Destruction of Bulk Quantities of PMSF

Dissolve the PMSF in isopropanol so that the concentration does not exceed 100 m*M* (17.4 mg/mL). Stirring may be necessary. For each 1 mL of solution add 5 mL of 1 *M* NaOH solution, shake to ensure complete mixing, and allow to stand for 24 h. Neutralize with 1 mL of acetic acid, test for completeness of destruction, and discard the reaction mixture.

Destruction of Bulk Quantities of AEBSF

Dissolve the AEBSF in isopropanol so that the concentration does not exceed 20 m*M* (4.06 mg/mL). Stirring may be necessary. For each 1 mL of solution add 10 mL of 1 *M* NaOH solution, shake to ensure complete mixing, and allow to stand for 24 h. Neutralize with 1 mL of acetic acid, test for completeness of destruction, and discard the reaction mixture.

Destruction of Bulk Quantities of APMSF

Dissolve the APMSF in H_2O so that the concentration of APMSF does not exceed 100 m*M* (25 mg/mL). For each 1 mL of solution add 5 mL of 1 *M* NaOH solution, shake to ensure complete mixing, and allow to stand for 24 h. Neutralize with 1 mL of acetic acid, test for completeness of destruction, and discard the reaction mixture.

Analytical Procedures[9]

For reaction mixtures resulting from the destruction of PMSF, AEBSF, or APMSF in buffer acidify a 1 mL aliquot by adding 100 μL of acetic acid (pH 3–4). For reaction mixtures resulting from the destruction of PMSF or AEBSF in isopropanol or DMSO or APMSF in 1:1 isopropanol: pH 3 buffer or DMSO acidify a 1-mL aliquot by adding 200 μL of acetic acid (pH 3–4). Analyze PMSF by reverse phase high-performance liquid chromatography (HPLC) using acetonitrile:H_2O 50:50 flowing at 1 mL/min and an ultraviolet (UV) detector set at 220 nm. The retention time was about 8 min and the limit of detection was about 0.9 μg/mL. Analyze AEBSF and APMSF by reverse phase HPLC (ion-pair chromatography) using acetonitrile:0.1% trifluoroacetic acid 40:60 flowing at 1 mL/min and an ultraviolet (UV) detector set at 225 nm for AEBSF and 232 nm for APMSF. The retention time for AEBSF was about 9.5 min and the limit of detection was about 0.1 μg/mL and for APMSF the retention time was about 7.7 min and the limit of detection was about 0.5 μg/mL. The analytical procedures should be checked by spiking an aliquot of the acidified reaction mixture with a small quantity of a dilute solution of PMSF, AEBSF, or APMSF.

Mutagenicity Assays[9]

The mutagenicity assays were carried out as described on page 4. The inhibitors themselves were tested as solutions in DMSO (all compounds), H_2O (APMSF), and isopropanol (PMSF and AEBSF). The concentrations were such that amounts varying from 2000 to 50 μg were added to each plate. The mutagenicity varied somewhat depending on the solvent used. When dissolved in DMSO AEBSF was mutagenic to TA1530 and TA1535 with and without activation. When tested in isopropanol AEBSF was not mutagenic. When tested in any of the solvents PMSF and APMSF were not mutagenic. The final reaction mixtures obtained using the degradation procedures described above were tested and found not to be mutagenic. In a separate experiment PMSF in isopropanol was degraded with NaOH solution for 1 h. Although the PMSF appeared to be completely degraded large peaks of similar retention times were seen by HPLC. This reaction mixture was found to be mutagenic to TA1530 and TA1535 with and without activation. A reaction mixture that had been allowed to stand for 24 h (allowing the unknown peaks to degrade to very low levels) was tested and found not to be mutagenic. In a similar fashion degradation of stock solutions of AEBSF and APMSF that contained isopropanol for 1 h gave rise to unknown peaks in the HPLC chromatogram. Allowing the degradation procedure to proceed for 24 h gave a much cleaner result.

Related Compounds

The above techniques were investigated for PMSF, AEBSF, and APMSF, but they may also be applicable to other sulfonyl fluoride enzyme inhibitors.

However, they should be thoroughly investigated before being applied to other compounds.

References

1. Other names for this compound are phenylmethanesulfonyl fluoride, α-toluenesulfonyl fluoride.
2. Fahrney, D.E.; Gold, A.M. Sulfonyl fluorides as inhibitors of esterases. I. Rates of reaction with acetylcholinesterase, α-chymotrypsin, and trypsin. *J. Am. Chem. Soc.* **1963**, *85*, 997–1000.
3. Lawson, W.B.; Valenty, V.B.; Wos, J.D.; Lobo, A.P. Studies on the inhibition of human thrombin: Effects of plasma and plasma constituents. *Folia Haematol. (Leipzig)*, **1982**, *109*, 52–60.
4. Markwardt, F.; Hoffmann, J.; Körbs, E. Influence of synthetic thrombin inhibitors on the thrombin-antithrombin reaction. *Thromb. Res.* **1973**, *2*, 343–348.
5. Markwardt, F.; Drawert, J.; Walsmann, P. Synthetic low molecular weight inhibitors of serum kallikrein. *Biochem. Pharmacol.* **1974**, *23*, 2247–2256.
6. Laura, R.; Robinson, D.J.; Bing, D.H. (*p*-Amidinophenyl)methanesulfonyl fluoride, an irreversible inhibitor of serine proteases. *Biochemistry* **1980**, *19*, 4859–4864.
7. James, G.T. Inactivation of the protease inhibitor phenylmethylsulfonyl fluoride in buffers. *Anal. Biochem.* **1978**, *86*, 574–579.
8. Personal communication from Calbiochem, La Jolla, CA, 6/15/93.
9. Lunn, G.; Sansone, E.B. Degradation and disposal of some enzyme inhibitors. *Appl. Biochem. Biotechnol.* (in press).

SULFUR-CONTAINING COMPOUNDS

> **CAUTION!** Refer to safety considerations section on page 7 before starting any of these procedures.

The destruction of a variety of sulfur-containing compounds was investigated. Mercaptans are organic compounds of the general formula R—SH, where R is an alkyl or aryl group. The alkyl compounds are also known as thiols and the aromatic compounds are also called thiophenols. Ethanethiol [75-08-1] (bp 35°C),[1] 1-butanethiol [109-79-5] (bp 98°C),[2] and thiophenol [108-98-5] (bp 169°C)[3] are liquids and most other mercaptans are liquids, although some solids are known. Most liquid and some solid mercaptans are volatile and possess an overpowering objectionable odor. Ethanethiol is a skin and eye irritant and has effects on the central nervous system.[4] 1-Butanethiol is an eye irritant and a teratogen.[5] Thiophenol is an eye irritant and can cause dermatitis, headaches, and dizziness.[6] These compounds are also incompatible with various reagents, including oxidizing agents. Mercaptans should only be handled in a properly functioning chemical fume hood.

Disulfides are organic compounds of the general formula R—SS—R and they are readily formed by the mild oxidation of the corresponding thiols. These compounds are much less volatile than the corresponding thiols but still possess an objectionable odor. The destruction of methyl disulfide [624-92-0] (bp 109°C),[7] ethyl disulfide [110-81-6] (bp 151–153°C),[8] and butyl disulfide [629-45-8] (bp 229–233°C)[9] was investigated. Methyl disulfide is incompatible with oxidizing agents[10] and ethyl disulfide is a skin and eye irritant and incompatible with oxidizing agents.[11] Mercaptans and disulfides are used in synthetic organic chemistry. Lower molecular weight thiols are added to natural gas to give it a characteristic odor.[12]

Inorganic sulfides are solids that release hydrogen sulfide, an extremely toxic gas with an unpleasant odor,[13] on treatment with acid. Sodium sulfide (Na_2S) [1313-82-2][14] can explode on rapid heating or percussion[15] and is incompatible with a variety of inorganic and organic compounds.[15,16] These compounds are widely used in chemical laboratories.

Carbon disulfide (CS_2) [75-15-0][17] is widely used as a solvent. This compound is highly volatile (bp 46°C), a teratogen, highly toxic, flammable, and is incompatible with a variety of organic and inorganic compounds.[18,19]

Principle of Destruction

Ethanethiol, thiophenol, 1-butanethiol, methyl disulfide, ethyl disulfide, butyl disulfide, sodium sulfide, and carbon disulfide can be oxidized by NaOCl. The products depend on the starting materials. Ethanethiol and 1-butanethiol are initially oxidized to the corresponding disulfides and then, on prolonged reaction, are oxidized to the corresponding sulfonic acid. Addition of a detergent (Triton X-100) helps to solubilize butyl disulfide, and hence increase the rate of the reaction. Thiophenol is oxidized to phenyl disulfide, which is so insoluble in H_2O that the reaction stops at this compound. In a similar fashion, inorganic sulfide can be oxidized to sulfate[20] and CS_2 is oxidized to carbon dioxide and sulfuric acid.[21] These products are generally nonvolatile and odorless. Because the products of these reactions are acid the pH tends to fall as the reaction progresses. This fall in pH causes the hypochlorite to decompose and so the reaction will not go to completion. Adding sodium hydroxide (NaOH) solution stabilizes the pH and so the reaction goes to completion. Destruction was greater than 99% in each case (although an analytical procedure was not available for carbon disulfide). When thiols were degraded less than 1%

of the corresponding disulfides was found in the final reaction mixture although the degradation of thiophenol stopped with the production of phenyl disulfide.

Destruction Procedures[22]

Destruction of Methyl Disulfide, Carbon Disulfide, Thiophenol, and Sodium Sulfide

Stir 600 mL of a 5.25% NaOCl solution (see assay procedure below) and 200 mL of 1 *M* NaOH solution at room temperature and add 0.05 mol of methyl disulfide (4.7 g, 4.5 mL) or carbon disulfide (3.8 g, 3 mL) or 0.1 mol of thiophenol (11.0 g, 10.25 mL) or sodium sulfide (7.8 g) in portions over 1 h. Stir the reaction mixture for another hour and remove solid phenyl disulfide by filtration. Check that the aqueous layer is still oxidizing, check for completeness of destruction, and discard it.

Destruction of Ethyl Disulfide, Butyl Disulfide, Ethanethiol, and 1-Butanethiol

Stir 600 mL of a 5.25% NaOCl solution (see assay procedure below) and 200 mL of 1 *M* NaOH solution at room temperature and add 3.2 mL of Triton X-100 if 1-butanethiol or butyl disulfide are to be degraded. Add 0.05 mol of ethyl disulfide (6.1 g, 6.1 mL) or butyl disulfide (8.9 g, 9.5 mL) or 0.1 mol of ethanethiol (6.2 g, 7.4 mL) or 1-butanethiol (9.0 g, 10.7 mL) in portions over 1 h. Stir the reaction mixture for 18 h, check that it is still oxidizing, check for completeness of destruction, and discard it.

Assay of Sodium Hypochlorite Solution

Sodium hypochlorite solutions deteriorate with time so they should be used fresh or periodically checked for the amount of active chlorine they contain. Pipette 10 mL of NaOCl solution into a 100-mL volumetric flask and fill to the mark with distilled H_2O. Pipette 10 mL of this solution into a conical flask containing 50 mL of distilled H_2O, 1 g of potassium iodide (KI), and 12.5 mL of 2 *M* acetic acid. Titrate this solution against 0.1 *N* sodium thiosulfate solution using starch as an indicator. Each 1 mL of the sodium thiosulfate solution corresponds to 3.545 mg of active chlorine. The NaOCl solution used in these degradation reactions should contain 45–50 g of active chlorine per liter.

Analytical Procedures[22]

To determine that the reaction mixture is still oxidizing add a few drops to an equal volume of 10% (w/v) KI solution, then acidify with a drop of 1 *M* hydrochloric acid (HCl) and add a drop of starch solution as an indicator. The deep blue color of the starch–iodine complex indicates that excess oxidant is present.

The thiols and disulfides are conveniently analyzed by gas chromatography (GC) using a 1.8-m × 2-mm i.d. column packed with 5% Carbowax 20 M on 80/100 Supelcoport. The column was fitted with a precolumn that was changed when it became contaminated. The carrier gas was nitrogen flowing at 30 mL/min, the injection temperature was 200°C and the flame ionization detector temperature was 300°C. Oven temperatures (°C) and approximate retention times (min) were as follows: ethanethiol (60; 0.56), 1-butanethiol (60; 1.5), methyl disulfide (60; 5.0), ethyl disulfide (100; 3.0), thiophenol (140; 3.3), and butyl disulfide (150; 3.1). Phenyl disulfide can be determined using a column packed with 5% Carbowax 20 M on 80/100 Chromosorb W HP with an oven temperature of 225°C and an injection temperature of 250°C. The GC conditions given above are only a guide; the exact conditions would have to be determined experimentally. Using 5-μL injections the detection limits were about 50 μg/mL. Before analysis the oxidizing power of the reaction mixture should be removed by adding an aliquot to sodium metabisulfite. Ethanethiol can be determined by direct analysis of this mixture but the other compounds should be determined after extraction into ether (1-butanethiol) or *tert*-butyl methyl ether. The analytical procedures should be periodically validated by spiking the reaction mixture, after addition of sodium metabisulfite, with a small amount of the compound to be determined.

Residual sodium sulfide was determined by adding 20 mL of the reaction mixture to 10 mL of 10% (w/v) sodium metabisulfite solution and acidifying this mixture by the addition of 10 mL of 2 *M* HCl. A 2-mL aliquot of this solution was added to 200 μL of a 10% (w/v) solution of $CuSO_4.5H_2O$. A dark precipitate of copper sulfide indicated the presence of residual sulfide. The absorbance of the solution was determined at 500 nm against a suitable blank. No sulfide was detected but the addition of 200 μL of a dilute solution of sodium sulfide produced a significant increase in absorbance. The limit of detection for sodium sulfide was 26 μg/mL.

Related Compounds

These procedures should be generally applicable to other thiols, disulfides, and inorganic sulfides provided that these compounds have some H_2O solubility. In some cases, as was found for thiophenol, insoluble disulfides may be formed which cause the reaction to halt at this stage.

References

1. Other names for this compound are ethyl mercaptan, ethyl hydrosulfide, ethyl sulfhydrate, ethyl thioalcohol, LPG ethyl mercaptan 1010, mercapto ethane, thioethanol, and thioethyl alcohol.
2. Other names for this compound are phenyl mercaptan, benzenethiol, and USAF XR-19.
3. Other names for this compound are *n*-butyl mercaptan, NCI-C60866, normal butyl thioalcohol, and thiobutyl alcohol.
4. Lewis, R.J., Sr. *Sax's Dangerous Properties of Industrial Materials*, 8th ed.; Van Nostrand-Reinhold: New York, 1992; pp. 1631–1632.
5. Reference 4, p. 623.
6. Reference 4, p. 2762.
7. Another name for this compound is dimethyl disulfide.
8. Another name for this compound is diethyl disulfide.
9. Another name for this compound is dibutyl disulfide.
10. Reference 4, p. 1371.
11. Reference 4, p. 1218.
12. Norell, J; Louthan, R.P. Thiols. In *Kirk-Othmer Encyclopedia of Chemical Technology*, 3rd ed., Vol 22; Wiley: New York, 1983; pp. 946–964.
13. Reference 4, pp. 1903–1904.
14. Other names for this compound are sodium monosulfide and sodium sulfuret.
15. Reference 4, p. 3114.
16. Bretherick, L. *Bretherick's Handbook of Reactive Chemical Hazards*, 4th ed.; Butterworths: London, 1990; pp. 1386–1387.
17. Other names for this compound are carbon bisulfide, carbon sulfide, dithiocarbonic anhydride, sulphocarbonic anhydride, Weeviltox, and NCI-C04591.
18. Reference 4, pp. 698–699.
19. Reference 15, pp. 196–198.

20. National Research Council, Committee on Hazardous Substances in the Laboratory. *Prudent Practices for Disposal of Chemicals from Laboratories*; National Academy Press: Washington, DC, 1983; p. 86.

21. Reference 20, p. 66.

22. Lunn, G. Unpublished observations.

6-THIOGUANINE AND 6-MERCAPTOPURINE

CAUTION! Refer to safety considerations section on page 7 before starting any of these procedures.

The degradation of a number of antineoplastic drugs, including 6-thioguanine and 6-mercaptopurine, was investigated by the International Agency for Research on Cancer (IARC).[1] 6-Thioguanine (mp >360°C) (**I**) [154-42-7][2] and 6-mercaptopurine (mp 313–314°C) (**II**) [50-44-2][3] are solids; they are insoluble in H_2O and organic solvents but soluble in dilute acid or base. 6-Mercaptopurine is mutagenic.[4,5] 6-Thioguanine[6] and 6-mercaptopurine[4] are teratogens. These compounds are employed as antineoplastic drugs.

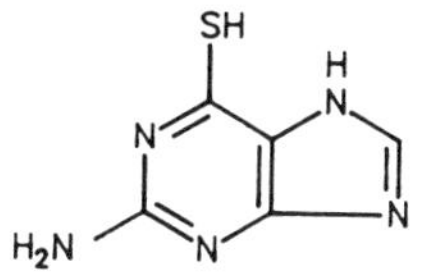

I

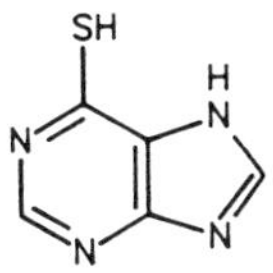

II

Principle of Destruction

6-Thioguanine and 6-mercaptopurine are destroyed by oxidation with potassium permanganate in sulfuric acid ($KMnO_4$ in H_2SO_4).[1] Destruction is greater than 99.5%.

Destruction Procedures

Destruction of Bulk Quantities of 6-Thioguanine and 6-Mercaptopurine

Dissolve in 3 *M* H_2SO_4 so that the concentration does not exceed 0.9 mg/mL, then add 0.5 g of $KMnO_4$ for each 80 mL of solution and stir overnight. Decolorize with sodium metabisulfite, make strongly basic with 10 *M* potassium hydroxide (KOH) solution (**Caution!** Exothermic), dilute with H_2O, filter to remove manganese compounds,[7] neutralize the filtrate, check for completeness of destruction, and discard it.

Destruction of Aqueous Solutions of 6-Thioguanine and 6-Mercaptopurine

Dilute with H_2O, if necessary, so that the concentration does not exceed 0.9 mg/mL, then add enough concentrated H_2SO_4 to obtain a 3 *M* solution and allow it to cool to room temperature. For each 80 mL of solution add 0.5 g of $KMnO_4$ and stir overnight. Decolorize with sodium metabisulfite, make strongly basic with 10 *M* KOH (**Caution!** Exothermic), dilute with H_2O, filter to remove manganese compounds,[7] neutralize the filtrate, check for completeness of destruction, and discard it.

Destruction of Pharmaceutical Preparations of 6-Thioguanine and 6-Mercaptopurine

To solutions of 7.5 mg of 6-thioguanine in 50 mL of 5% dextrose solution or 10 mg of 6-mercaptopurine in 10 mL of 5% dextrose solution add enough concentrated H_2SO_4 to obtain a 3 *M* solution. Allow the solution to cool to room temperature. Dissolve oral preparations of these drugs in 3 *M* H_2SO_4. For each 80 mL of any of these solutions add 4 g of $KMnO_4$, in small portions to avoid frothing, and stir the mixture overnight. Decolorize with sodium metabisulfite, make strongly basic with 10 *M* KOH (**Caution!** Exothermic), dilute with H_2O, filter to remove manganese compounds,[7] neutralize the filtrate, check for completeness of destruction, and discard it.

Destruction of Solutions of 6-Thioguanine and 6-Mercaptopurine in Volatile Organic Solvents

Remove the solvent under reduced pressure using a rotary evaporator and take up the residue in 3 *M* H_2SO_4, so that the concentration of the drug does not exceed 0.9 mg/mL. For each 80 mL of solution add 0.5 g of $KMnO_4$ and stir overnight. Decolorize with sodium metabisulfite, make strongly basic with 10 *M* KOH (**Caution!** Exothermic), dilute with H_2O, filter to remove manganese compounds,[7] neutralize the filtrate, check for completeness of destruction, and discard it.

Destruction of Dimethyl Sulfoxide or N,N-Dimethylformamide Solutions of 6-Thioguanine and 6-Mercaptopurine

Dilute with H_2O so that the concentration of dimethyl sulfoxide (DMSO) or *N,N*-dimethylformamide (DMF) does not exceed 20% and the concentration of the drug does not exceed 0.9 mg/mL, then add enough concentrated H_2SO_4 to obtain a 3 *M* solution and allow it to cool to room temperature. For each 80 mL of solution add 4 g of $KMnO_4$, in small portions to avoid frothing, and stir overnight. Decolorize with sodium metabisulfite, make strongly basic with 10 *M* KOH (**Caution!** Exothermic), dilute with H_2O, filter to remove manganese compounds,[7] neutralize the filtrate, check for completeness of destruction, and discard it.

Decontamination of Glassware Contaminated with 6-Thioguanine and 6-Mercaptopurine

Immerse the glassware in a 0.3 *M* solution of $KMnO_4$ in 3 *M* H_2SO_4 for 10–12 h, then decolorize it by the addition of sodium metabisulfite, make strongly basic with 10 *M* KOH (**Caution!** Exothermic), dilute with H_2O, filter to remove manganese compounds,[7] neutralize the filtrate, check for completeness of destruction, and discard it.

Decontamination of Spills of 6-Thioguanine and 6-Mercaptopurine

Allow any organic solvent to evaporate and remove as much of the spill as possible by high efficiency particulate air (HEPA) vacuuming (not sweeping), then rinse the area with 0.1 *M* H_2SO_4. Take up the rinse with absorbents and allow the rinse and absorbents to react with 0.3 *M* $KMnO_4$ solution in 3 *M* H_2SO_4 overnight. If the color fades, add more solution. Decolorize with sodium metabisulfite, make strongly basic with 10 *M* KOH

(**Caution!** Exothermic), dilute with H_2O, filter to remove manganese compounds,[7] neutralize the filtrate, check for completeness of destruction, and discard it. Check for completeness of decontamination by using a wipe moistened with 0.1 *M* sodium hydroxide solution. Analyze the wipe for the presence of the drug.

Analytical Procedures

These drugs can be analyzed by high-performance liquid chromatography (HPLC) using a 25-cm reverse phase column and ultraviolet (UV) detection at 340 nm. The mobile phases that have been recommended are as follows:

6-Thioguanine 0.02 *M* KH_2PO_4 : acetonitrile (98:2) flowing at 1.5 mL/min
or
0.1 m*M* KH_2PO_4 : methanol (92.5:7.5) flowing at 1 mL/min

6-Mercaptopurine 0.02 *M* KH_2PO_4 : acetonitrile (98:2) flowing at 1.5 mL/min
or
0.1 m*M* KH_2PO_4 flowing at 1 mL/min

Mutagenicity Assays

In the IARC study[1] tester strains TA98, TA100, and TA1535 of *Salmonella typhimurium* were used with and without mutagenic activation. The reaction mixtures were not mutagenic.

Related Compounds

Potassium permanganate in H_2SO_4 is a general oxidative method and should, in principle, be applicable to many drugs. However, any new application should be thoroughly validated both for complete destruction of the compound and for the production of nonmutagenic reaction mixtures.

References

1. Castegnaro, M.; Adams, J.; Armour, M-.A.; Barek, J.; Benvenuto, J.; Confalonieri, C.; Goff, U.; Ludeman, S.; Reed, D.; Sansone, E.B.; Telling, G., Eds., *Laboratory De-*

contamination and Destruction of Carcinogens in Laboratory Wastes: Some Antineoplastic Agents; International Agency for Research on Cancer: Lyon, 1985 (IARC Scientific Publications No. 73).

2. Other names for this compound are 2-amino-1,7-dihydro-6*H*-purine-6-thione, Lanvis, 2-aminopurine-6-thiol, 2-aminopurine-6(1*H*)-thione, 2-amino-6-mercaptopurine, Tabloid, 2-amino-6-MP, BW 5071, 6-mercapto-2-aminopurine, 6-mercaptoguanine, NSC-752, TG, ThG, thioguanine, tioguanin, and Wellcome U3B.
3. Other names for this compound are purine-6-thiol, 7-mercapto-1,3,4,6-tetrazaindene, 6MP, Leukerin, Mercaleukin, Purinethol, 6-purinethiol, 1,7-dihydro-6*H*-purine-6-thione, Ismipur, Leukeran, Leupurin, Mercapurin, Mern, MP, NCI-C04886, NSC-755, Purimethol, thiohypoxanthine, 6-thioxopurine, U-4748, and Puri-Nethol.
4. Lewis, R.J., Sr. *Sax's Dangerous Properties of Industrial Materials*, 8th ed.; Van Nostrand-Reinhold: New York, 1992; p. 2941.
5. Mosesso, P.; Palitti, F. The genetic toxicology of 6-mercaptopurine. *Mutat. Res.* **1993**, *296*, 279–294.
6. Reference 4, p. 209.
7. Lunn, G.; Sansone, E.B.; De Méo, M.; Laget, M.; Castegnaro, M. Potassium permanganate can be used for degrading hazardous compounds. *Am. Ind. Hyg. Assoc. J.* **1994**, 55, 167–171.

URANYL COMPOUNDS

> **CAUTION!** Refer to safety considerations section on page 7 before starting any of these procedures.

Compounds containing the uranyl ion (UO_2^{2+}) are used in biological stains, in photography, in pottery glazes, and as reagents in analytical chemistry. These compounds are mildly radioactive but are not regulated by the Nuclear Regulatory Commission (NRC). Uranyl acetate [541-09-3][1] is a poison[2] and the dihydrate is a teratogen.[3]

Principle of Decontamination

Uranyl ions can be removed from solution by using Amberlite IR-120(plus) a strongly acid gel-type resin with a sulfonic acid functionality.[4] On a small scale it is most convenient to stir the resin in the solution to be decontaminated but on a larger scale or for routine use it might be most convenient to pass the solution through a column packed with the resin. Two examples for decontaminating solutions of uranyl acetate containing 1000 and 60 ppm uranium are given. The final concentration of uranium was less than

0.5 ppm in each case. Solutions containing other concentrations can be decontaminated by adjusting the solution volume:weight of resin ratio. Although the volume of waste that must be disposed of is greatly reduced by using this technique a small amount (i.e., the resin contaminated with the uranium) remains and it should be discarded appropriately. The resin can be regenerated by washing with acid but the concentrated uranium containing solution generated by this technique must also be disposed of appropriately.

Decontamination Procedures

1. For each 200 mL of solution containing no more than 1000 ppm of uranium add 1 g of Amberlite IR-120(plus) resin. Stir the mixture for 24 h, filter, check the filtrate for completeness of decontamination, and discard it. The speed and efficiency of decontamination will depend on factors such as the size and shape of the flask and the rate of stirring. The beads now contain the uranium and should be discarded appropriately.

2. For each liter of solution containing no more than 60 ppm of uranium add 1 g of Amberlite IR-120(plus) resin. Stir the mixture for 4 h, filter, check for the filtrate for completeness of decontamination, and discard it. The speed and efficiency of decontamination will depend on factors such as size and shape of flask and the rate of stirring. The beads now contain the uranium and should be discarded appropriately.

Analytical Procedures

Although atomic absorption spectroscopy is generally used to determine metal ions in solution it is a relatively insensitive technique for uranium. The following colorimetric procedure[5] gave good results. Dissolve 0.30 g of Arsenazo III in 200 mL of H_2O and 25 mL of glacial acetic acid. Stir for 30 min, filter, wash the filter paper with glacial acetic acid, and make up the filtrate to 1 L with glacial acetic acid. Mix 1 mL of the reaction mixture, 2 mL of 0.3 *M* hydrochloric acid, and 1 mL of the Arsenazo reagent and determine the absorbance at 653 nm against an appropriate blank. The limit of detection was 0.5 ppm uranium; the response was linear from 25 to 0.5 ppm. The reagent solution is a deep red color but the blue color of the uranium complex is readily detected at 653 nm. Waste solutions containing uranium also frequently contain lead. It was found that this procedure was insensitive to lead up to a concentration of at least 100 ppm.

Related Compounds

This procedure is specific for uranium but ion-exchange resins can be used to decontaminate solutions containing ions of other metals (see the monographs on Heavy Metals and Mercury). Waste solutions containing uranium also frequently contain lead and both metals can be removed simultaneously using ion-exchange resin. For example, about 5 L of an aqueous solution containing 38.6 ppm uranium and 163 ppm lead was decontaminated with 100 g of Amberlite IR-120(plus) ion-exchange resin. After stirring for 72 h the concentration of uranium was less than 0.5 ppm and the concentration of lead was less than 0.25 ppm.

References

1. Other names for this compound are uranium oxyacetate, uranium acetate, bis(aceto)-dioxouranium, bis(aceto-o)dioxouranium, and bis(acetato)dioxouranium.
2. Lewis, R.J.,Sr. *Sax's Dangerous Properties of Industrial Materials*, 8th ed.; Van Nostrand-Reinhold: New York, 1992; p. 3468
3. Reference 2, p. 3469.
4. Lunn, G. Unpublished observations.
5. Kressin, I.K. Spectrophotometric method for the determination of uranium in urine. *Anal. Chem.* **1984**, *56*, 2269–2271.

APPENDIXES

APPENDIX I

Recommendations for Wipe Solvents for Use After Spill Cleanup

After a chemical spill has been cleaned up and the area decontaminated it is frequently helpful to use a moistened wipe to take a sample of the decontaminated surface to ensure that decontamination is complete. The wipe should be moistened with a reagent that will dissolve the spilled compound and will not interfere with the subsequent analysis. This list has been compiled on the basis of information in the published literature and a few tests. The use of a wipe sample is generally not appropriate with highly reactive compounds and these cases are indicated by NA. In some cases a different reagent may be required depending on the circumstances of the spill and the cleanup. These instances are detailed in the footnotes.

We list here the names of compounds as used in this book. Please consult the Molecular Formula Index, the CAS Registry Number Index, or the Name Index for compounds that may be known by alternate names.

Acetic anhydride	NA
Acetonitrile	NA
Acetyl chloride	NA
Acid halides	NA
Acid chlorides	NA
Acid anhydrides	NA
Acridine orange	Water

Acyl chlorides	NA
Acyl halides	NA
Adriamycin	Water
Aflatoxin B_1	Methanol
Aflatoxin B_2	Methanol
Aflatoxin G_1	Methanol
Aflatoxin G_2	Methanol
Aflatoxin M_1	Methanol
Aflatoxins	Methanol
Alcian blue 8GX	Water
Alizarin red S	Water
Alkali metal alkoxides	NA
Alkali metals	NA
Alkaline earth metals	NA
Alkylating agents	See individual compounds
p-Amidinophenylmethanesulfonyl fluoride (hydrochloride)	Water
2-Aminoanthracene	Methanol[1]
3-Aminobenzotrifluoride	Methanol
4-Aminobiphenyl	Methanol[1]
4-(2-Aminoethyl)benzenesulfonyl fluoride	Isopropanol
N-Aminomorpholine	Methanol[1]
N-Aminopiperidine	Methanol[1]
N-Aminopyrrolidine	Methanol[1]
Ammonium hydrogen difluoride	Water
Ammonium dichromate	Water
Amoxicillin	Water
Ampicillin	Water
Anhydrides	NA
Antineoplastic drugs	See individual compounds
Aromatic amines	See individual compounds
Azides	See individual compounds
Azobenzene	Methanol[1]
Azocompounds	See individual compounds
Azoxyanisole	Methanol
Azoxybenzene	Methanol
Azoxycompounds	See individual compounds
Azoxymethane	Methanol
Azure A	Water

Azure B	Water
Barium	NA
BCME	Methanol
BCNU	Methanol
Benz[*a*]anthracene	Acetone
Benzal chloride	Methanol
Benzenesulfonyl chloride	Methanol
Benzidine	Methanol[1]
Benzo[*a*]pyrene	Acetone
Benzonitrile	Methanol
Benzoyl chloride	NA
Benzoyl peroxide	Acetone
Benzyl azide	Methanol
Benzyl bromide	Methanol
Benzyl chloride	Methanol
Benzyl cyanide	Methanol
Biological stains	See individual compounds
Bis(chloromethyl)ether	Methanol
Bleomycin sulfate	Water
Borane. THF complex	NA
Boron trifluoride	NA
Brilliant blue R	Water
Bromine	Methanol
Bromobenzene	Methanol
4-Bromobenzoic acid	Methanol[2]
1-Bromobutane	Methanol
2-Bromobutane	Methanol
1-Bromodecane	Methanol
2-Bromoethanol	Water
2-Bromoethylamine	Water
7-Bromomethylbenz[*a*]anthracene	Acetone
2-Bromo-2-methylpropane	Methanol
1-Bromononane	Methanol
4-Bromophenylacetic acid	Methanol[2]
Butadiene diepoxide	Water
1-Butanethiol	Methanol
n-Butyl bromide	Methanol
sec-Butyl bromide	Methanol
tert-Butyl bromide	Methanol

n-Butyl chloride	Methanol
Butyl disulfide	Methanol
tert-Butyl hydroperoxide	Methanol
tert-Butyl hypochlorite	Cyclohexane
n-Butyl iodide	Methanol
sec-Butyl iodide	Methanol
tert-Butyl iodide	Methanol
tert-Butyl peroxide	Methanol
Butyllithium	NA
Cadmium acetate	Water
Calcium	NA
Calcium carbide	NA
Calcium hydride	NA
Calcium hypochlorite	Water
Carbamic acid esters	Water
Carbon disulfide	NA
Carmustine	Methanol
CCNU	Methanol
Cefadroxil	Water
Cefmenoxime	pH6 buffer
Cefsulodin	pH6 buffer
Cephalothin, sodium salt	Water
Chlorambucil	Methanol
Chloroacetic acid	Water
2-, 3-, and 4-Chloroaniline	Methanol[1]
Chlorobenzene	Methanol
2-Chlorobenzoic acid	Methanol[2]
1-Chlorobutane	Methanol
1-Chlorodecane	Methanol
2-Chloroethanol	Water
2-Chloroethylamine	Water
2-Chloro-5-fluorobenzoic acid	Methanol[2]
2-Chlorohydroquinone	Methanol
2-Chloroisophthalic acid	Methanol[2]
Chloromethylmethylether	Methanol
4-(4-Chloro-2-methylphenyl)butyric acid	Methanol[2]
4-(4-Chloro-3-methylphenyl)butyric acid	Methanol[2]
Chloromethylsilanes	NA
2-, 3-, and 4-Chloronitrobenzene	Methanol
m-Chloroperbenzoic acid	Water

4-Chlorophenol	Methanol
4-Chlorophenoxyacetic acid	Methanol[2]
2-Chlorophenylacetic acid	Methanol[2]
3-Chloropyridine	Methanol
N-Chlorosuccinimide	Water
Chlorosulfonic acid	NA
Chlorotrimethylsilane	NA
Chlorozotocin	Methanol
Chromerge	Water
Chromic acid	Water
Chromium(VI)	Water
Chromium trioxide	Water
Cisplatin	Water
Citrinin	Ethanol
CMME	Methanol
Cobalt(II) sulfate	Water
Complex metal hydrides	NA
Congo red	Water
Coomassie brilliant blue G	Water
Copper(II) sulfate	Water
Cresyl violet acetate	Water
Crystal violet	Water
Cyanogen bromide	Water
Cyanogen chloride	Water
Cyanogen iodide	Water
Cyclophosphamide	Water
Cycloserine	Water
Dacarbazine	0.1 M HCl[3]
Daunomycin	Water
Daunorubicin	Water
Dialkyl sulfates	See individual compounds
Diaminobenzidine	Methanol[1]
Di-(4-amino-3-chlorophenyl)methane	Methanol[1]
N,N'-Diaminopiperazine	Methanol[1]
2,4-Diaminotoluene	Methanol[1]
Dibenz[*a,h*]acridine	Acetonitrile
Dibenz[*a,j*]acridine	Acetonitrile
Dibenz[*a,h*]anthracene	Acetone
13*H*-Dibenzo[*a,i*]carbazole	Acetonitrile
7*H*-Dibenzo[*c,g*]carbazole	Acetonitrile

1,1-Dibutylhydrazine	Methanol[1]
3,3′-Dichlorobenzidine	Methanol[1]
2,4-Dichlorobenzoic acid	Methanol[2]
3,4-Dichlorobenzoic acid	Methanol[2]
cis-Dichloro-*trans*-dihydroxy-bis(isopropylamine)platinum(IV)	Water
Dichlorodimethylsilane	NA
Dichloromethotrexate	0.1 M HCl[3]
2,4-Dichlorophenylacetic acid	Methanol[2]
3,4-Dichlorophenylacetic acid	Methanol[2]
α,α-Dichlorotoluene	Methanol
1,1-Diethylhydrazine	Methanol[1]
Diethyl sulfate	Methanol
Diisopropyl fluorophosphate	DMF
1,1-Diisopropylhydrazine	Methanol[1]
Diisopropylnitramine	Methanol
3,3′-Dimethoxybenzidine	Methanol[1]
N,*N*′-Dimethyl-4-amino-4′-hydroxyazobenzene	Methanol[1]
3,3′-Dimethylbenzidine	Methanol[1]
7,12-Dimethylbenz[*a*]anthracene	Acetone
Dimethylcarbamoyl chloride	NA
Dimethyl disulfide	Methanol
1,2-Dimethylhydrazine	Water
1,1-Dimethylhydrazine	Water
Dimethyl sulfate	Methanol
N,*N*′-Dinitrosopiperazine	Methanol
1,5-Diphenylcarbazide	Methanol
Diphenyl disulfide	Acetone
1,2-Diphenylhydrazine	Methanol[1]
Disulfides	See individual compounds
Doxorubicin	Water
Drugs	See individual compounds
ENNG	Methanol
ENU	Methanol
ENUT	Methanol
Enzyme inhibitors	See individual compounds
Eosin B	Water
Eosin Y	Water

Erythritol anhydride	Water
Erythrosin B	Water
Ethanethiol	Water
Ethidium bromide	Water
3-Ethoxypropionitrile	Water
Ethyl carbamate	Water
Ethyl disulfide	Methanol
Ethyl mercaptan	Water
Ethyl methanesulfonate	Water
N-Ethyl-*N'*-nitro-*N*-nitrosoguanidine	Methanol
N-Ethyl-*N*-nitrosourea	Methanol
N-Ethyl-*N*-nitrosourethane	Methanol
N-Ethylurethane	Water
Etoposide	Ethanol
Fast garnet	Methanol
4-Fluoroaniline	Methanol[1]
Fluorobenzene	Methanol
Fluorobenzoic acid	Methanol[2]
2-Fluoroethanol	Water
4-Fluoronitrobenzene	Methanol
Giemsa stain	Water
Haloethers	See individual compounds
Heavy metal salts	Water[5]
Hexamethylphosphoramide	Water
Homidium bromide	Water
Hydrazine	Water
Hydrazobenzene	Methanol[1]
Hydrides, complex metal	NA
Hydrogen cyanide	NA
Hydrogen peroxide	Water
Hypochlorites	See individual compounds
Ifosfamide	Water
Inorganic cyanides	Water
Inorganic fluorides	Water
Iodine	Methanol
Iodobenzene	Methanol
1-Iodobutane	Methanol
2-Iodobutane	Methanol
Iodomethane	Methanol

2-Iodo-2-methylpropane	Methanol
Iproniazid phosphate	Water
Iron(II) sulfate	Water
Isoniazid	Water
Isophosphamide	Water
Janus green B	Water
LAH	NA
Lead(II) acetate	Water
Lithium	NA
Lithium aluminum hydride	NA
Lithium hydride	NA
Lomustine	Methanol
Magnesium	NA
Manganese(II) sulfate	Water
Mechlorethamine	Water
Melphalan	Methanol
Mercaptans	See individual compounds
6-Mercaptopurine	0.1 M KOH[3]
Mercuric acetate	Water
Mercury(II) chloride	Water
Mercury-containing compounds	Water[5]
Metal hydrides	NA
Methanesulfonates	See individual compounds
Methanesulfonyl chloride	NA
Methanethiol	NA
Methotrexate	0.1 M HCl[3]
2-Methylaziridine	Water
Methyl carbamate	Water
Methyl CCNU	Methanol
3-Methylcholanthrene	Acetone
Methyl disulfide	Methanol
Methylene blue	Water
Methylhydrazine	Methanol[1]
Methyl iodide	Methanol
Methyl methanesulfonate	Methanol
N-Methyl-*N'*-nitro-*N*-nitrosoguanidine	Methanol
N-Methyl-*N*-nitrosoacetamide	Methanol
N-Methyl-*N*-nitroso-*p*-toluenesulfonamide	Methanol
N-Methyl-*N*-nitrosourea	Methanol

N-Methyl-*N*-nitrosourethane	Methanol
1-Methyl-1-phenylhydrazine	Methanol[1]
1-Methyl-4-phenyl-1,2,3,6-tetrahydropyridine	Methanol[1]
3-Methyl-1-*p*-tolyltriazene	Methanol[1]
Methyltrichlorosilane	NA
N-Methylurethane	Water
Metronidazole	Water
Mitomycin C	Methanol
MNNG	Methanol
MNTS	Methanol
MNU	Methanol
MNUT	Methanol
MOCA	Methanol[1]
MPTP	Methanol[1]
1-Naphthylamine	Methanol[1]
2-Naphthylamine	Methanol[1]
Neutral red	Water
Nickel(II) sulfate	Water
Nigrosin	Water
4-Nitrobiphenyl	Methanol
Nitrogen mustards	See individual compounds
N-Nitromorpholine	Methanol
N-Nitrosamides	See individual compounds
N-Nitrosamines	See individual compounds
N-Nitrosodibutylamine	Methanol
N-Nitrosodiethylamine	Water
N-Nitrosodiisopropylamine	Methanol
N-Nitrosodimethylamine	Water
N-Nitrosodiphenylamine	Methanol
N-Nitrosodipropylamine	Methanol
N-Nitroso-*N*-methylaniline	Methanol
N-Nitrosomorpholine	Water
N-Nitrosopiperidine	Water
N-Nitrosopyrrolidine	Water
Nitrosoureas	See individual compounds
Norethindrone	Methanol
Ochratoxin A	Methanol
Orcein	Water

Organic azides	See individual compounds
Organic bromides	See individual compounds
Organic chlorides	See individual compounds
Organic cyanides	See individual compounds
Organic fluorides	See individual compounds
Organic halides	See individual compounds
Organic nitriles	See individual compounds
Osmium tetroxide	[4]
Patulin	Water
PCNU	Methanol
Peracetic acid	Water
Peracids	See individual compounds
Peroxides	See individual compounds
Phenyl azide	Methanol
4-Phenylazoaniline	Methanol[1]
4-Phenylazophenol	Methanol[1]
Phenyl cyanide	Methanol
Phenyl disulfide	Acetone
Phenylhydrazine	Methanol
Phenyl mercaptan	Methanol
Phenylmethylsulfonyl fluoride	Isopropanol
Phosgene	NA
Phosphomolybdic acid	Water
Phosphorus	NA
Phosphorus pentoxide	NA
Picric acid	Water
Polycyclic aromatic hydrocarbons	See individual compounds
Polycyclic heterocyclic hydrocarbons	See individual compounds
Potassium	NA
Potassium amide	NA
Potassium *tert*-butoxide	NA
Potassium cyanide	Water
Potassium dichromate	Water
Potassium fluoride	Water
Potassium hydride	NA
Potassium permanganate	Water
Procarbazine hydrochloride	Water
1,3-Propane sultone	Water
Propidium iodide	Water

β-Propiolactone	Water
Propionyl chloride	NA
Propyleneimine	Water
Protease inhibitors	See individual compounds
Rose Bengal	Water
Safranine O	Water
Silicon tetrachloride	NA
Silver nitrate	Water
Sodium	NA
Sodium amide	NA
Sodium azide	Water
Sodium borohydride	NA
Sodium cyanide	Water
Sodium cyanoborohydride	NA
Sodium dichromate	Water
Sodium ethoxide	NA
Sodium fluoride	Water
Sodium hexafluorosilicate	Water
Sodium hydride	NA
Sodium hypochlorite	Water
Sodium methoxide	NA
Sodium peroxide	Water
Sodium sulfide	Water
Spirohydantoin mustard	Methanol
Sterigmatocystin	Methanol
Streptozotocin	Water
Strontium	NA
Sulfamethoxazole	Methanol
Sulfides, inorganic	Water
Sulfonyl chlorides	NA
Sulfonyl fluoride enzyme inhibitors	See individual compounds
Sulfonyl halides	NA
Sulfur-containing compounds	See individual compounds
Sulfuryl chloride	NA
Teniposide	Ethanol
Tetramethyltetrazene	Methanol[1]
Tetrazenes	See individual compounds
Thallium(I) nitrate	Water
6-Thioguanine	0.1 M KOH[3]

Thiols	See individual compounds
Thionyl chloride	NA
Thiophenol	Methanol
Tin(II) chloride	Water
Tin(II) fluoride	Water
TLCK (hydrochloride)	Water
p-Toluenesulfonyl chloride	Methanol
Toluidine blue O	Water
p-Tolylhydrazine hydrochloride	Water
TPCK	Isopropanol
Triazenes	See individual compounds
2,4,6-Tribromophenol	Methanol
2,2,2-Trichloroacetic acid	Water[2]
2,4,6-Trichlorophenol	Methanol
Trimethoprim	Methanol
Trimethylsilyl chloride	NA
2,4,6-Trinitrophenol	Water
Triphosgene	NA
Trypan blue	Water
Uracil mustard	Methanol
Uranyl acetate	Water
Uranyl compounds	Water[5]
Urethane	Water
Verapamil hydrochloride	Water
Vinblastine sulfate	Water
Vincristine sulfate	Water
Zinc chloride	Water

References

1. If the compound is present as the hydrochloride or other salt or if the decontamination procedure produces the hydrochloride or other salt, water should be used instead of methanol. Neutralization of this solution may be necessary before analysis.
2. If the compound is present as the sodium or other salt or if the decontamination procedure produces the sodium or other salt, water should be used instead of methanol. Neutralization of this solution may be necessary before analysis.
3. Neutralization of this solution may be necessary before analysis.
4. Osmium tetroxide is soluble in water but a better method for determining completeness of decontamination may be to suspend either a glass cover slip coated in corn oil (Mazola) or a piece of filter paper soaked in corn oil over the area. Blackening indicates that osmium tetroxide is still present.
5. Although all of the salts mentioned in this book are water soluble this will not necessarily always be the case.

APPENDIX II

Procedures for Drying Organic Solvents

Dry organic solvents are frequently used in organic chemistry and it is not uncommon to find that highly reactive and hazardous reagents (e.g., complex metal hydrides) are recommended for their preparation. However, Burfield and Smithers[1-9] showed in a series of papers that comparatively nonhazardous reagents can be used effectively to dry a wide range of organic solvents. The reader is referred to the original papers for full details. The most versatile reagent is molecular sieve but barium oxide, boric anhydride, calcium chloride, and potassium hydroxide have also been recommended (Table 1). In most cases, these reagents are the most effective desiccants.

Molecular sieve in various pore sizes has been recommended; 3-Å molecular sieve is generally more effective than 4-Å molecular sieve. The molecular sieve should be activated before use by heating at 300°C overnight, and then allowing the sieve to cool in a desiccator. In general the ratio of desiccant to solvent is 5% (w/v) but greater drying efficiency can be achieved by using a larger amount of desiccant. The results in Table 1 were obtained using static loading, that is, no stirring. Stirring initially accelerates drying but prolonged stirring is less effective, perhaps because of physical breakdown of the sieve beads and desorption of water from the destroyed cavities. Solvents can also be dried using a column of molecular sieve. Distillation from molecular sieve is not recommended because

Table 1 Procedures for Drying Organic Solvents[a]

Compound	Reagent	Residual Water (ppm)	Time (h)	Notes[b]	Reference
Acetone	3-Å sieve	115	6		6
Acetone	Boric anhydride	18	24	A,B	6
Acetonitrile	3-Å sieve	27	168		7
Acetonitrile	3-Å sieve	0.6	192	C	1
Benzene	4-Å sieve	0.03	24		7
1,2-Butanediol	3-Å sieve	140	6	A,D	8
1,4-Butanediol	3-Å sieve	420	6	A,D	8
2-Butanol	3-Å sieve	14	6	A,D	8
tert-Butanol	3-Å sieve	13	6	A,D	8
1,3-Diaminopropane	3- or 4-Å sieve	<25	24		9
Dichloromethane	4-Å sieve	0.07	96		1
Diisopropylamine	3- or 4-Å sieve	<25	24		9
2,6-Dimethylpyridine	3-Å sieve	128	168		9
Dioxane	4-Å sieve	40	24		7
Dioxane	4-Å sieve	13	96		1
N,N-Dimethylformamide	4-Å sieve	138	96		1
N,N-Dimethylformamide	3-Å sieve	1.5	72	C	6
Dimethylsulfoxide	4-Å sieve	10	72	C	6
1,2-Ethanediol	3-Å sieve	360	6	A,D	8
Ethanol	3-Å sieve	18	6	A,D	8
Ether	4-Å sieve	0.29	6	E	2
Ether	4-Å sieve	0.095	6	A,E	2
Ether	$CaCl_2$	0.24	6	A,E	2
Ethyl acetate	4-Å sieve	128	96		1
HMPA $[((CH_3)_2N)_3PO]$	4-Å sieve	29	72	C,F	6
Methanol	3-Å sieve	95	24		8
2-Methylpyridine	Barium oxide	27	24		9
2-Methylpyridine	3-Å sieve	55	24		9
1,5-Pentanediol	3-Å sieve	220	6	A,D	8
Pyridine	4-Å sieve	0.3	24	C	9
Tetrahydrofuran	4-Å sieve	28	168		1
Toluene	4-Å sieve	0.01	96		1
Triethylamine	4-Å sieve	28	24	C	9

Table 1 (***Continued***)

Compound	Reagent	Residual Water (ppm)	Time (h)	Notes[b]	Reference
Triethylamine	KOH	23	168	A	9
2,4,6-Trimethylpyridine	3-Å sieve	47	24	G	9
2,4,6-Trimethylpyridine	KOH	27	24	A,G	9

[a] Except where shown static drying (i.e., no stirring) was used and the desiccant loading was 5% (w/v).

[b] Notes A Powdered desiccant used.

B The acetone was stirred with the desiccant, distilled, and sequentially dried with the desiccant.

C Sequential drying with two batches of desiccant was employed.

D The powdered desiccant was added to the solvent and the mixture shaken vigorously then allowed to settle for 6 h.

E Desiccant loading was 10% w/v.

F Distillation from P_2O_5 gave a H_2O content of 22 ppm, static drying with P_2O_5 was ineffective.

G Static drying with calcium carbide gave a final water content of 8 ppm.

water may desorb on heating. The best procedure is to decant the solvent from the beads and then distil.[1]

Molecular sieve can be recycled. For reasons of safety, the sieve is first placed in a fume hood to allow the organic solvent to evaporate. Solvents that are not volatile can be removed by washing with hexane or a similar hydrocarbon solvent. When all the solvent has been removed the sieve can be reactivated by heating at 300°C overnight.[1]

A frequent procedure in organic chemistry is the extraction of the desired product from aqueous solution using ether. The ether extract is then dried before proceeding further. Powdered 4-Å molecular sieve was the most efficient reagent but calcium chloride, magnesium sulfate, and potassium carbonate (in that order) were also effective. Sodium and calcium sulfate were relatively ineffective.[2] However, in some cases the desiccant may react with the compound of interest in the ether solution so efficiency is not the only criterion when selecting a drying agent.

Chloroform is widely used as a solvent in nuclear magnetic resonance (NMR) spectroscopy and it is generally stabilized by the addition of ethanol. On prolonged storage phosgene can form by oxidation of the chloroform. Ethanol[3,4] and phosgene[4] can be removed from chloroform by using a 5-Å molecular sieve. In addition, the presence of the sieve inhibits phosgene formation.[5]

References

1. Burfield, D.R.; Gan, G.-H.; Smithers, R.H. Molecular sieves—desiccants of choice. *J. Appl. Chem. Biotechnol.* **1978**, *28*, 23–30.
2. Burfield, D.R.; Smithers, R.H. Drying of grossly wet ether extracts. *J. Chem. Educ.* **1982**, *59*, 703–704.
3. Burfield, D.R. Purification of chloroform for NMR spectroscopy. *J. Chem. Educ.* **1979**, *56*, 486.
4. Burfield, D.R.; Smithers, R.H. Applications of molecular sieves to purification of chloroform and diethyl ether: selective absorption of ethanol and phosgene. *Chem. Ind. (London)* **1980**, 240–241.
5. Burfield, D.R.; Goh, E.H.; Ong, E.H.; Smithers, R.H. Storage and decomposition of chloroform. *Gazz. Chim. Ital.* **1983**, *113*, 841–843.
6. Burfield, D.R.; Smithers, R.H. Desiccant efficiency in solvent drying. 3. Dipolar aprotic solvents. *J. Org. Chem.* **1978**, *43*, 3966–3968.
7. Burfield, D.R.; Lee, K.-H.; Smithers, R.H. Desiccant efficiency in solvent drying. A reappraisal by application of a novel method for solvent water assay. *J. Org. Chem.* **1977**, *42*, 3060–3065.
8. Burfield, D.R.; Smithers, R.H. Desiccant efficiency in solvent and reagent drying. 7. Alcohols. *J. Org. Chem.* **1983**, *48*, 2420–2422.
9. Burfield, D.R.; Smithers, R.H.; Tan, A.S.C. Desiccant efficiency in solvent and reagent drying. 5. Amines. *J. Org. Chem.* **1981**, *46*, 629–631.

APPENDIX III

Safety Considerations with Potassium Permanganate

Manganese is known to be a carcinogen,[1,2] it is mutagenic[3–6] in yeast,[7–13] *Escherichia coli*,[14,15] and *Salmonella typhimurium*,[16] and can damage DNA.[17–20] In a recent paper De Méo et al.[16] pointed out that when potassium permanganate in sulfuric acid ($KMnO_4$ in H_2SO_4) was used to degrade hazardous compounds mutagenic reaction mixtures were produced because manganese was left in solution. The Ames test was used with tester strain *S. typhimurium* TA102, which was most sensitive to manganese, to assess mutagenic activity. Mutagenic activity was also detected with strain TA100 but at a lower level. Disposal of reaction mixtures that contain manganese is not desirable; however, by manipulating the workup conditions, $KMnO_4$ can be used to degrade hazardous reagents and the manganese can subsequently be removed from solution.[21]

The concentrations of manganese left in solution when $KMnO_4$ was used in acidic, basic, or neutral conditions were measured and the reaction conditions were modified to reduce the amount of manganese left in solution. The relative efficiencies of degradation of ethanol by $KMnO_4$ in acidic, neutral, or basic solution were also investigated.

Reactions

Each reaction mixture consisted of 10 mL of a 0.3 *M* solution of $KMnO_4$ in 3 *M* H_2SO_4, 1 *M* sodium hydroxide (NaOH) solution, or H_2O. The solutions were prepared by adding the requisite amount of $KMnO_4$ to 3 *M* H_2SO_4, 1 *M* NaOH solution, or H_2O and stirring this mixture for at least 15 min but no more than 1 h (3 *M* H_2SO_4) or at least 30 min but no more than 2 h (1 *M* NaOH solution or H_2O). In control experiments the calculated amount of solid $KMnO_4$ required to make a 0.3 *M* solution was stirred with 3 *M* H_2SO_4, 1 *M* NaOH solution, and H_2O. The level of $KMnO_4$ in solution was determined by periodically removing aliquots, diluting 1000-fold, and measuring the absorbance at 522 nm. It was found that there was a rapid increase in concentration followed by a slow decline. Stirring times of 15–60 min for 3 *M* H_2SO_4 and 30–120 min for 1 *M* NaOH solution and H_2O gave solutions that contained greater than 90% of the peak concentration. These stirring times gave good results in actual practice but the variation of concentration with time may change with the rate of stirring, shape and size of the container, and possible contaminants.

Each reaction mixture was stirred at room temperature for 18 h, then decolorized with a reducing agent that was either 0.5 g of ascorbic acid or 0.8 g of sodium metabisulfite. Various workup procedures were then employed as described below. In each case the final reaction mixture was filtered and the amount of manganese in the solution determined by atomic absorption spectroscopy at 279.5 nm (limit of detection 0.4 ppm) and the amount of ethanol determined by GC (1.8-m $\times$ 2-mm i.d. packed column of 10% Carbowax 20 M + 2% KOH on 80/100 Chromosorb W AW at 80°C, retention time 0.9 min, limit of detection 0.03 mg/mL) with flame ionization detection.

To test the reaction mixtures for mutagenicity, the plate incorporation technique of the *Salmonella*/mammalian microsome mutagenicity assay was performed essentially as recommended by Ames et al.[22] with the modifications of De Méo et al.[16,23] Liquid preincubation for 1 h was used. Tester strains TA98, TA100, TA102, TA1530, and TA1535 were used with and without S9 rat liver microsomal activation and 100 μL of solution was applied to each plate. The reaction mixtures were neutralized with acetic acid before testing if necessary.

When 50 μL of ethanol was used per 10 mL of oxidizing solution the reaction mixture remained purple or green. Larger amounts of ethanol exhausted the oxidizing power of the solutions and the reaction mixtures became brown or colorless. Control experiments in which no ethanol was added were also carried out and these mixtures remained purple or green. For all of these reactions, one of the steps of the technique was to decolorize the reaction mixture with sodium metabisulfite or ascorbic acid. This process reduces the manganese(VII) of the permanganate ion to manganese(II). When the reaction mixture is acidic, a clear solution containing a high concentration of Mn^{2+} ions is obtained. However, when the reaction mixture is basic or neutral, a brown precipitate of manganese hydroxide [$Mn(OH)_2$] is obtained. The brown precipitate rapidly darkens to black as the $Mn(OH)_2$ is oxidized to manganese dioxide (MnO_2) by air.[24] The MnO_2 is also insoluble in H_2O.

Reactions Involving 3 *M* Sulfuric Acid

When the solvent was 3 *M* H_2SO_4 initial experiments involved the addition of 8 mL of 10 *M* potassium hydroxide (KOH) to precipitate the manganese (produced after the reducing agent was added) as $Mn(OH)_2$. When 50 μL of ethanol was degraded and decolorization was effected with sodium metabisulfite the manganese concentration was only 1.6 ppm (Table 1); however, a large precipitate (5.7 g, Table 2) was removed by filtration. In addition to $Mn(OH)_2$ the precipitate contained potassium sulfate, which precipitated from the highly concentrated solution. Accordingly, the re-

Table 1 Amount of Manganese (ppm) Remaining in Solution after Degradation of Ethanol with 10 mL of a Potassium Permanganate Solution for 18 h

		Amount of Ethanol Degraded (μL)			
		0		50	
		Reducing Agent[a]			
Solvent	Workup	A	S	A	S
H_2SO_4	Add KOH	1200	3.8	120	1.6
H_2SO_4	Add KOH + 20 mL H_2O	102	1.9	81	8.1
H_2SO_4	Add KOH + 100 mL H_2O	135	1.0	8.4	1.0
H_2SO_4	Add $NaHCO_3$	3100	760	7200	350
H_2SO_4	Add $NaHCO_3$ + H_2O	58	15	111	11
H_2SO_4	Adjust pH to 7.3-7.8	104	4280	104	3950
NaOH		260	<0.4	830	<0.4
NaOH	Add H_2O	280	<0.4	380	<0.4
H_2O		230	4800	620	450
H_2O	Add NaOH	600	3.2	1400	3.6
H_2O	Add NaOH + H_2O	1150	0.7	1500	0.5

[a] A = ascorbic acid; S = sodium metabisulfite.

Table 2 Dry Weight of Precipitate (g) after Degradation of Ethanol with 10 mL of a Potassium Permanganate Solution for 18 h

		Amount of Ethanol Degraded (μL)			
		0		50	
		Reducing Agent[a]			
Solvent	Workup	A	S	A	S
H_2SO_4	Add KOH	6.4	5.9	5.8	5.7
H_2SO_4	Add KOH + 20 mL H_2O	3.1	2.8	3.3	2.4
H_2SO_4	Add KOH + 100 mL H_2O	0.8	0.7	0.9	0.7
H_2SO_4	Add $NaHCO_3$	Cent.[b]	Cent.	Cent.	Cent.
H_2SO_4	Add $NaHCO_3$ + H_2O	Cent.	Cent.	Cent.	Cent.
H_2SO_4	Adjust pH to 7.3–7.8	3.6	2.1	2.0	2.7
NaOH		Cent.	1.4	1.2	1.4
NaOH	Add H_2O	0.9	1.2	1.0	1.2
H_2O		0.6	0.9	0.6	0.8
H_2O	Add NaOH	1.0	1.3	Cent.	1.3
H_2O	Add NaOH + H_2O	0.7	1.0	0.7	1.1

[a] A = ascorbic acid, S = sodium metabisulfite.

[b] The reaction mixture was too viscous to be filtered and centrifugation of the solution was necessary. Thus the dry weight of precipitate could not be obtained.

action mixtures were diluted with H_2O before filtration. This procedure caused the soluble potassium sulfate to redissolve in the H_2O while the insoluble $Mn(OH)_2$ remained as a precipitate. When 20 mL of H_2O was used, the amount of precipitate was reduced to 2.4 g and when 100 mL of H_2O was used, the amount of precipitate was 0.7 g. When sodium metabisulfite was used as a reducing agent only low concentrations of manganese were found. The use of 100 mL of H_2O gave slightly better results. The results were not as good, however, when ascorbic acid was used as a reducing agent.

Sodium bicarbonate has been recommended as a neutralizing agent for reactions involving 3 *M* H_2SO_4[25] and it has the advantage that a neutral pH is produced and the end of the neutralization process is quite apparent because effervescence ceases. However, this procedure left high concentrations of manganese in solution. These reaction mixtures were very thick and viscous, which may have inhibited complete neutralization. Diluting the reaction mixture with 20 mL of H_2O gave better results but the results were still not as good as before.

Finally, the pH of reaction mixtures involving 3 *M* H_2SO_4 was adjusted to 7.3–7.8 with NaOH solution and hydrochloric acid as recommended by De Méo et al.[16] Moderate amounts of manganese still remained in solution when ascorbic acid was used and large amounts of manganese remained in solution when sodium metabisulfite was used.

Reactions Involving 1 *M* Sodium Hydroxide and Water

When the solvent was 1 *M* NaOH solution and the reducing agent was sodium metabisulfite, no manganese was found in solution, to the limit of detection, after filtration. However, when the reducing agent was ascorbic acid quite large amounts of manganese were found. Similar results were obtained when the reaction mixtures were diluted with H_2O before filtration, in which case, the dry weight of the precipitate was somewhat lower.

When the solvent was H_2O, relatively large amounts of manganese were found in all cases. Accordingly, this procedure was modified by adding 1 mL of 10 *M* NaOH solution to precipitate most of the manganese. In this case sodium metabisulfite gave much better results than ascorbic acid. Similar results were obtained when the reaction mixtures were diluted with H_2O before filtration, which reduced the weight of precipitate.

The efficiency of ethanol degradation was also investigated. It was found that 50 μL of ethanol was completely degraded in every case. The degradation efficiencies were greater than 98.6% when the solvent was 3 *M* H_2SO_4 and greater than 99.2% when the solvents were H_2O or 1 *M* NaOH solution. When 250 and 500 μL of ethanol were used only partial degradation was achieved. The rate of degradation, however, depended on the solvent. When the solvent was 3 *M* H_2SO_4 or 1 *M* NaOH solution, less than 1.5% (the limit of detection) of the ethanol remained after 2 min. However, when the solvent was H_2O, degradation was much slower and 8% remained after 6 h. The reaction required 18 h before the amount of ethanol was reduced below the limit of detection. The reaction demonstrated pseudo-first-order kinetics.

Conclusions

When the solvent was 3 *M* H_2SO_4, the best procedure was decolorization with sodium metabisulfite, basification with KOH, dilution with 100 mL of H_2O, and filtration. When the solvent was 1 *M* NaOH solution, the best procedure was decolorization with sodium metabisulfite, dilution with H_2O, and filtration; and when the solvent was H_2O, the best procedure was decolorization with sodium metabisulfite, basification, dilution with H_2O, and filtration. Each of these reaction mixtures was tested for mutagenicity and no mutagenic activity was found. In each of these cases the amount of manganese in the final solution was less than 2 ppm. The Environmental Protection Agency (EPA) does not specify an acceptable limit for manganese in solution in its definitions of hazardous waste. However, for comparison, the EPA limits for lead and chromium are 5 ppm.[26]

If a new method is to be developed it should be borne in mind that, in a number of cases, oxidation by permanganate is pH dependent. Oxidation of cyanide[27] and benzylamines[28] is more rapid at high pH and alcohols, aldehydes, and fluoral hydrate are oxidized more rapidly at high or low pH.[27] The way in which the rate varies with pH appears to depend on the substrate. Procedures that should be tested are $KMnO_4$ in 3 *M* H_2SO_4 with the work up described in the previous paragraph or $KMnO_4$ in 1 *M* NaOH solution followed by decolorization with sodium metabisulfite, dilution with an equal amount of H_2O, and filtration. Potassium permanganate in H_2O followed by decolorization with sodium metabisulfite, basification with 10 *M* NaOH solution, dilution with an equal amount of H_2O, and filtration can also be employed, but because of the slow rate of this reaction (at least for ethanol), it should only be considered if the other reactions are unsatisfactory. Details of these procedures are given in Recommended Degradation Procedures (below).

For spills it is probably more convenient to use 0.3 *M* $KMnO_4$ in 3 *M* H_2SO_4. When the decontamination is complete, the spill area can be decolorized with sodium metabisulfite then mopped up with paper towels. The solution can be squeezed out of the paper towels, basified, diluted, and filtered to remove manganese. The filtrate can be discarded with the aqueous waste and the precipitate and the paper towels discarded with the heavy metal waste. If NaOH solution is used as a solvent, then decolorization with sodium metabisulfite will produce a heavy brown precipitate over the spill area, which will be hard to clean up. Test for completeness of decontamination by wiping the surface with a wipe moistened with an appropriate solvent. Analyze the squeezings from the wipe for the presence of the spilled compound. Appropriate protective equipment should be worn. This equipment should include rubber gloves, eye protection, and a lab coat. Additional equipment (e.g., respirator, rubber boots) may be necessary depending on the nature of the compound spilled and the extent of the spill.

Recommended Degradation Procedures

The volume of $KMnO_4$ solution required to degrade a given amount of a hazardous compound depends on the nature of the compound. Some compounds require more oxidant than others. However, an initial starting point might be 10 mL of solution for each 40 mg of compound. Before the method is put into routine use experiments should be carried out to determine the ratio of oxidant : substrate that is required to obtain complete chemical destruction. The final reaction mixtures should also be checked for mutagenicity and, if possible, for the presence of manganese.

Potassium Permanganate in 3 M Sulfuric Acid

Prepare a 0.3 *M* solution of $KMnO_4$ in 3 *M* H_2SO_4 by stirring solid $KMnO_4$ in 3 *M* H_2SO_4 for at least 15 min, but no more than 1 h. Add a sufficient volume of this solution to the compound to be degraded to cause complete destruction and stir this mixture at room temperature overnight. The reaction mixture should remain purple. If it turns brown, add sufficient $KMnO_4$ in 3 *M* H_2SO_4 so that the purple color persists for at least 1 h. Decolorize the reaction mixture by adding 0.8 g of sodium metabisulfite (more if necessary for complete decolorization) for each 10 mL of reaction mixture and check for completeness of destruction. For each 10 mL of reaction mixture add 8 mL of 10 *M* KOH solution (**Caution!** Exothermic), check that the mixture is strongly basic, dilute with 100 mL of H_2O, and filter. Discard the precipitate with the heavy metal waste and neutralize the filtrate, check for the presence of manganese, and discard it with the nonhazardous aqueous waste.

Potassium Permanganate in 1 M Sodium Hydroxide Solution

Prepare a 0.3 *M* solution of $KMnO_4$ in 1 *M* NaOH solution by stirring solid $KMnO_4$ in 1 *M* NaOH solution for at least 30 min, but no more than 2 h. Cover the mixture during this procedure to prevent absorption of atmospheric carbon dioxide. Add a sufficient volume of this solution to the compound to be degraded to cause complete destruction, cover the mixture, and stir it overnight at room temperature. The reaction mixture should remain green or purple. If it does not, add sufficient $KMnO_4$ in 1 *M* NaOH solution to cause the green or purple color to persist for at least 1 h. Decolorize the reaction mixture by adding 0.8 g of sodium metabisulfite (more if necessary for complete decolorization) for each 10 mL of reaction mixture, dilute with an equal volume of H_2O, and check for completeness of destruction. Filter and discard the precipitate with the heavy metal waste. Neutralize the filtrate, check for the presence of manganese, and discard it with the nonhazardous aqueous waste.

Aqueous Solution of Potassium Permanganate

Prepare a 0.3 *M* aqueous solution of $KMnO_4$ by stirring solid $KMnO_4$ in H_2O for at least 30 min, but no more than 2 h. Add a sufficient volume of this solution to the compound to be degraded to cause complete destruction and stir this mixture at room temperature for 18 h. The reaction mixture should remain purple. If it does not, add sufficient aqueous $KMnO_4$ to cause the purple color to persist for at least 1 h. Decolorize the reaction mixture by adding 0.8 g of sodium metabisulfite (more if necessary for complete decolorization) for each 10 mL of reaction mixture, dilute with an equal volume of H_2O, and check for completeness of destruction. For each 10 mL of reaction mixture add 1 mL of 10 *M* NaOH solution, check that the mixture is strongly basic, and filter. Discard the precipitate with the heavy metal waste and neutralize the filtrate, check for the presence of manganese, and discard it with the nonhazardous aqueous waste.

References

1. Stoner, G.D.; Shimkin, M.B.; Troxell, M.C.; Thompson, T.L.; Terry, L.S. Test for carcinogenicity of metallic compounds by the pulmonary tumor response in Strain A mice. *Cancer Res.* **1976**, *36*, 1744–1747.
2. DiPaolo, J.A. The potentiation of lymphosarcomas in the mouse by manganous chloride. *Fed. Proc.* **1964**, *23*, 393 (Abstract).
3. Flessel, C.P. Metals as mutagens. In *Inorganic and Nutritional Aspects of Cancer (Advances in Experimental Medicine and Biology; 91)*; Schrauzer, G.N., Ed., Plenum Press: New York, 1978; pp. 117–128.
4. Goodman, M.F.; Keener, S.; Guidotti, S.; Branscomb, E.W. On the enzymatic basis for mutagenesis by manganese. *J. Biol. Chem.* **1983**, *258*, 3469–3475.
5. El-Deiry, W.S.; Downey, K.M.; So, A.G. Molecular mechanisms of manganese mutagenesis. *Proc. Natl. Acad. Sci. USA* **1984**, *81*, 7378–7382.
6. Beckman, R.A.; Mildvan, A.S.; Loeb, L.A. On the fidelity of DNA replication: Manganese mutagenesis in vitro. *Biochemistry* **1985**, *24*, 5810–5817.
7. Putrament, A.; Baranowska, H.; Ejchart, A.; Prazmo, W. Manganese mutagenesis in yeast. A practical application of manganese for the induction of mitochondrial antibiotic-resistant mutations. *J. Gen. Microbiol.* **1975**, *62*, 265–270.
8. Putrament, A.; Baranowska, H.; Ejchart, A.; Prazmo, W. Manganese mutagenesis in

yeast. IV. The effects of magnesium, protein synthesis inhibitors and hydroxyurea on ant^{R} induction in mitochondrial DNA. *Mol. Gen. Genet.* **1975**, *140*, 339–347.

9. Prazmo, W.; Balbin, E.; Baranowska, H.; Ejchart, A.; Putrament, A. Manganese mutagenesis in yeast. II. Conditions of induction and characteristics of mitochondrial respiratory deficient *Saccharomyces cerevisiae* mutants induced with manganese and cobalt. *Genet. Res.* **1975**, *26*, 21–29.
10. Baranowska, H.; Ejchart, A.; Putrament, A. Manganese mutagenesis in yeast. V. On mutation and conversion induction in nuclear DNA. *Mutat. Res.* **1977**, *42*, 343–348.
11. Putrament, A.; Baranowska, H.; Ejchart, A.; Jachymczyk, W. Manganese mutagenesis in yeast. VI. Mn(2+) uptake, mitDNA replication and E^{R} induction. Comparison with other divalent cations. *Mol. Gen. Genet.* **1977**, *151*, 69–76.
12. Singh, I. Induction of gene conversion and reverse mutation by manganese sulphate and nickel sulphate in *Saccharomyces cerevisiae*. *Mutat. Res.* **1984**, *137*, 47–49.
13. Putrament, A.; Baranowska, H.; Ejchart, A.; Prazmo, W. Manganese mutagenesis in yeast. In *Methods in Cell Biology*, Volume 20; Prescott, D.M., Ed., Academic Press: New York, 1978; pp. 25–34.
14. Zakour, R.A.; Glickman, B.W. Metal-induced mutagenesis in the *lacI* gene of *Escherichia coli*. *Mutat. Res.* **1984**, *126*, 9–18.
15. Rossman, T.G.; Molina, M. The genetic toxicology of metal compounds: II. Enhancement of ultraviolet-light induced mutagenesis in *Escherichia coli* WP2. *Environ. Mutagen.* **1986**, *8*, 263–271.
16. De Méo, M.; Laget, M.; Castegnaro, M.; Duménil, G. Genotoxic activity of potassium permanganate in acidic solutions. *Mutat. Res.* **1991**, *260*, 295–306.
17. Hamilton-Koch, W.; Snyder, R.D.; Lavelle, J.M. Metal-induced DNA damage and repair in human diploid fibroblasts and chinese hamster ovary cells. *Chem.-Biol. Inter.* **1986**, *59*, 17–28.
18. Snyder, R.D. Role of active oxygen species in metal-induced DNA strand breakage in human diploid fibroblasts. *Mutat. Res.* **1988**, *193*, 237–246.
19. Joardar, M.; Sharma, A. Comparison of clastogenicity of inorganic Mn administered in cationic and anionic forms in vivo. *Mutat. Res.* **1990**, *240*, 159–163.
20. Mukhopadhyay, M.J.; Sharma, A. Comparison of different plants in screening for Mn clastogenicity. *Mutat. Res.* **1990**, *242*, 157–161.
21. Lunn, G.; Sansone, E.B.; De Méo, M.; Laget, M.; Castegnaro, M. Potassium permanganate can be used for degrading hazardous compounds. *Am. Ind. Hyg. Assoc. J.* **1994**, *55*, 167–171.
22. Ames, B.N.; McCann, J.; Yamasaki, E. Methods for detecting carcinogens and mutagens with the *Salmonella*/mammalian-microsome mutagenicity test. *Mutat. Res.* **1975**, *31*, 347–364.
23. De Méo, M.; Miribel, V.; Botta, N.; Laget, M.; Duménil, G. Applicability of the SOS chromotest to detect urinary mutagenicity caused by smoking. *Mutagenesis* **1988**, *3*, 277–283.
24. Cotton, F.A.; Wilkinson, G. *Advanced Inorganic Chemistry*, 5th ed.; Wiley: New York, 1988; pp. 697–708.

25. Lunn, G.; Sansone, E.B. Validated methods for degrading hazardous chemicals: Some alkylating agents and other compounds. *J. Chem. Educ.* **1990**, *67*, A249–A251.

26. 40CFR Part 261.24; 1991.

27. Stewart, R. Oxidation by permanganate. In *Oxidation in Organic Chemistry*, Part A; Wiberg, K.B., Ed., Academic Press: New York, 1965; pp. 1–68.

28. Wei, M.-M.; Stewart, R. The mechanisms of permanganate oxidation. VIII. Substituted benzylamines. *J. Am. Chem. Soc.* **1966**, *88*, 1974–1979.

APPENDIX IV

Emerging Technologies Applicable to the Treatment of Hazardous Waste in Biomedical Research Institutions

by Steven W. Rhodes, Environmental Research Section, Environmental Control and Research Program, NCI-Frederick Cancer Research and Development Center, Frederick MD 21702

Currently, the primary methods for the treatment of hazardous waste streams include incineration, stripping, adsorption, evaporation, and bio-treatment.[1-3] Stripping processes may use either air or steam. Air stripping towers can remove from 80% to greater than 90% of volatile organics in aqueous streams.[3] Steam stripping is essentially a continuous fractional distillation process[2] and is more effective than air stripping for the removal of organics with low volatilities.[2-4] Stripping processes normally incorporate a vapor recovery system to control organic emissions.[4] Activated carbon is often used for the adsorption of organics in aqueous waste streams,[2,3,5] although certain polymeric resins have been reported to be effective in the adsorption of organic wastes.[6-10] Evaporation processes have also been used to treat wastes.[2,11] In the pervaporation process,[11] the waste stream is brought into contact with a membrane while the opposite side is under vacuum. Water and organics adsorbed onto one side of the membrane diffuse through and are drawn off by the vacuum into a recovery system.[11] A common drawback to stripping, adsorption, and evaporation is that even with 100% vapor recovery all that has occurred is that the organics of concern have been shifted from one matrix to another, still leaving a potentially hazardous waste. Incineration relies on high temperatures, optimal residence times, and complete turbulence for the destruction of hazardous components; however, there is the potential to produce other hazardous materials that could be released to the environment.[1] Biotreatment utilizes naturally occurring or genetically engineered microorganisms to degrade organics in waste streams.[12–15] A number of organic compounds have been found to be amenable to biotreatment; on the other hand, some organics have been found to be refractory or inhibitory to conventional biological processes.[13] One major obstacle to biotreatment is waste streams that vary in composition. The organisms evolve so that they can degrade new compounds by the adaptation of enzyme pathways.[15] However, this is a slow process so that if new compounds are suddenly introduced into the waste stream they will not be degraded. Thus, biotreatment cannot be viewed as a reasonable treatment method where the composition of the waste stream varies because new and varied waste components do not allow the acclimation of the organisms.

These waste treatment processes suffer from the drawbacks of moving the compounds from one matrix to another, the potential release of hazardous materials to the environment, or the inability to handle a varying waste stream. There are, however, a number of recently developed processes that show promise for the treatment of waste streams from biomedical research institutions. Some examples of compounds that are found in the waste streams from biomedical research institutions include trichloroacetic acid, chloroform, methanol, acetic acid, acetonitrile, phenol, pseudocumene, naphthalene, 2-methoxyethanol, xylenes, trifluoroacetic acid, and halogenated dibenzo-*p*-dioxins.

Hydrogenation

Hydrogenation is the addition of hydrogen to an organic moiety. The reaction proceeds at a negligible rate without the addition of a catalyst, thus the term catalytic hydrogenation is often employed. Catalytic hydrogenation has long been used in the refining of petroleum[16] and the hydrogenation of vegetable oils to convert them into solids for use in cooking oils.[17] The widespread use and reliability of this technology has led some to investigate its application to the treatment of hazardous materials. Catalytic hydrogenation has been used successfully to degrade polychlorinated biphenyls (PCBs), halogenated byproducts from a vinyl chloride manufacturing plant, halogenated distillation bottoms from the production and purification of epichlorhydrin, and residue from distillate oil H_2SO_4 treatment.[18] It has been suggested that the remaining liquid organic fractions may be distilled to yield high value petrochemicals or used as a clean fuel source. Hagenmaier et al.[19] have indicated that dechlorination-hydrogenation can be used for the decomposition of PCBs. Lunn and Sansone[20] reported the successful use of this technique for the degradation of 54 halogenated compounds. Gioia[16] reviewed the literature concerning the kinetics of hydrogenation of compounds containing heteroatoms and suggests that the data show concrete evidence of the potential for this technology in the detoxification of hazardous organic wastes.

Generally, this technique is best suited for certain well-defined waste streams having organic compounds with groups that are susceptible to hydrogenation, for example, double bonds and halogens. Although it clearly can be carried out on a large scale (e.g., in the hydrogenation of cooking oils) the process involves considerable engineering problems with the use of flammable hydrogen. On a small scale, nickel-aluminum alloy can be used to effect hydrogenation without the use of exogenous hydrogen. On balance, catalytic hydrogenation is probably not a useful process for the waste streams encountered at biomedical research institutions.

Fenton's Reagent

In 1894 Fenton reported the oxidation of malic acid in a mixture of hydrogen peroxide (H_2O_2) and ferrous ion.[21] Fenton's reagent has been noted to be an effective oxidant for many organics.[22] This reaction is thought to involve the hydroxyl radical,[22–24] however, this mechanism has recently been questioned.[25,26]

The Fenton reaction has been successfully applied to the degradation of 2-nitrophenol, 4-nitrophenol, 2,4-di-nitrophenol, and nitrobenzene, although only 4-nitrophenol and 2,4-di-nitrophenol went to completion after 7 h.[27] Total decomposition and mineralization has been reported for 2-chlorophenol, 3-chlorophenol, 4-chlorophenol, 3,4-dichlorophenol, and 2,4,5-trichlorophenol; the rates of decomposition were enhanced by increasing the concentration of the ferrous ion.[24] The oxidation of 1,4-dioxane has been reported to be completed in 10 h with a 10-fold molar excess of H_2O_2.[28] Watts et al.[29] reported the mineralization of pentachlorophenol in a silica sand system, but suggested that bench-scale studies will be needed to determine the effectiveness in natural matrices. The oxidation of *N*-nitrosodimethylamine has also been reported.[25]

Both light and dark Fenton reactions have been studied. In a study of the oxidation of 4-nitrophenol it was found that the best results were obtained when the reaction mixture was irradiated with polychromic ultraviolet (UV) light.[30] In kinetic studies of the photo-Fenton reaction, Zepp et al.[23] indicated that the process effectively degraded trace amounts of nitrobenzene and anisole and that light strongly accelerated the reaction. Zepp et al. suggested that the results had important implications for the removal of undesirable impurities from H_2O. Complete mineralization of the herbicides 2,4-dichlorophenoxyacetic acid and 2,4,5-trichlorophenoxyacetic acid has been reported; degradation was accelerated by irradiation with visible light that had a UV component.[31] Photochemical reactions involving the ferrous ion are also suggested to be important in atmospheric and surface waters.[32,33]

The Fenton reaction is a versatile oxidative process that shows great promise for treating a variety of waste products. However, the reaction is carried out as a batch process and our experience shows it to be an *extremely* vigorous reaction. Accordingly, the process may be best suited for treating small quantities of waste in the laboratory rather than large quantities. However, the photo-Fenton reaction may be adaptable to a flow-through process that could used to treat some components of the wastes at biomedical research institutions.

Advanced Oxidation Processes

Ultraviolet light, ozone, or H_2O_2 can each effect the oxidation of organics, although their individual use for degradation of organics has been shown to have limitations.[34–36] However, oxidation can be greatly enhanced when a combination of UV and ozone or UV and H_2O_2 is used.[35,37–39] When ozone or H_2O_2 is irradiated with UV light, hydroxyl radicals and other reactive species that can oxidize organics to carbon dioxide (CO_2) and H_2O are formed.[37,38]

Ultraviolet Light and Hydrogen Peroxide. A number of studies have shown the effectiveness of the combination of UV and H_2O_2 on low concentrations of organics in aqueous matrices. Destruction of greater than 98% of initial benzene (0.2 m*M*) in 90 min has been reported in bench top studies.[35] Similarly, the concentration of trichloroethylene was reported to decrease from 50 ppm to less than 1 ppm in 50 min and all of the reacted chlorine was converted to the chloride ion.[39] Other halogenated aliphatics (tetrachloroethylene, 1,1,2,2-tetrachloroethane, dichloromethane, chloroform, carbon tetrachloride, and ethylene dibromide) were included in this study although disappearance kinetics were not reported. Sundstrom et al.[40] and Mansour[41] investigated the oxidation of aromatic compounds with UV and H_2O_2 and reported that the process was successful for degrading benzene and some substituted aromatics. The photooxidation of 2,4-dinitrotoluene has been reported to be greater than 99% depending on the ratio of H_2O_2 to 2,4-dinitrotoluene and the length of time of the reaction.[42] Substantial degradation of phenol, 2,5-dimethylphenol, 2,5-dichlorophenol, 2-chlorophenol, and *m*-cresol also have been reported to be dependent on the reaction time and the ratio of H_2O_2 to the phenolics.[43] Concentrations of dimethyl phthalate and isophorone were reduced by more than 99% in 60 min using the UV/H_2O_2 technique.[36] Lipczynska-Kochany[30] investigated the combinations UV/H_2O_2 and UV/Fenton reagent for the destruction of 4-nitrophenol and reported approximately 80% degradation in 180 min with the UV/H_2O_2 process and 100% degradation in 100 min with the UV/Fenton reagent system. In a study of a mixture of benzene and trichloroethylene it was reported that benzene had a strong adverse effect on the destruction of trichloroethylene.[38] It was suggested that the adverse effect on the reaction rate may be due to intermediates formed during the oxidation of benzene. This result points out the importance of testing batches of actual waste as well as solutions of pure compound. When waste is treated the presence of one compound may interfere with the destruction of another compound. The combination of UV and H_2O_2 has also been investigated for the treatment of wastewaters from the chlorination and alkaline extraction stages of paper pulp bleaching. It was found that although the process produced a significant reduction in color, it was not economically competitive with other treatments including photochemical processes and ozonation.[44]

Ultraviolet Light and Ozone. Like the combination of UV and H_2O_2, the irradiation of ozone with UV produces hydroxyl radicals that ultimately can mineralize (mineralization is the complete oxidation of organics to carbon dioxide, water, and inorganic compounds) organic reactants.[37,45] Peyton et al.[43] reported the destruction of 63% of tetrachloroethylene in 7 min with UV and ozone; however, they reported a significant retardation of the oxidation in lake water and speculated that it may be due to radical intermediates formed. Significant destruction of trihalomethane precursors in surface water with UV irradiation and ozone has also been reported.[46] McShea et al.[47] reported significant removal efficiencies of pentachlophenol and phenol from wood preserving treatment wastewater.

Commercial applications of advanced oxidation processes[48] utilized UV in conjunction with both ozone and H_2O_2 (the Ultrox process)[49–51] or a combination of UV, ozone, and a proprietary reducing agent (the Rayox process).[52,53] The use of UV/ozone/H_2O_2 appears to be superior to UV/ozone or UV/H_2O_2. However, comparative studies do not appear to have been published. Evolving designs of the commercially available systems have reduced the initial capital costs and the energy costs of these systems. These cost factors were thought to be two of the main obstacles to their commercial success. Presently, these systems are being used in conjunction with a "polishing" step of carbon adsorption or air stripping, which has further lowered the costs involved in the treatment of waste streams.[48]

Advanced oxidation processes involving some combination of UV, H_2O_2, and ozone show great potential for treating some of the wastes stored at biomedical research institutions. It would be particularly valuable to carry out a comparative study on the efficacy of the various combinations of oxidants to determine which combination is best for a given component of the waste stream. A possible complication is the presence of strongly UV absorbing buffers that might limit the efficacy of the process. Also, the formation of colored products may tend to slow the reaction. It should be possible to overcome these problems, however, by varying the reactor design, perhaps by using a thin-film reactor in which a thin-film of liquid moves past the light source.

Photocatalytic Oxidation

Like advanced oxidation processes, photocatalytic oxidation makes use of UV radiation. However, instead of using an oxidizing agent, this process involves the use of semiconductors as catalysts. In this process, photons of sufficient energy are absorbed on the surface of the semiconductor and electron-hole pairs are generated, with free electrons produced in the conduction band and positive holes remaining in the valence band.[54,55] The hole reacts with the H_2O to produce a hydroxyl radical and a proton.[56,57] The common catalyst used is the anatase phase of titanium dioxide (TiO_2), which is an *n*-type semiconductor.[56]

Izumi et al.[58] used powdered TiO_2 and UV to oxidize a number of hydrocarbons (benzene, hexane, cyclohexane, nonane, decane, and kerosene), as well as benzoic acid and adipic acid.[59] The combination of UV, TiO_2, and Fenton's reagent was reported to be successful in the photolytic oxidation of various aromatic compounds (benzene, toluene, and acetophenone).[60] Similarly, the combination of UV, TiO_2, and Fenton's reagent was reported to oxidize aromatic compounds to CO_2.[61] In an investigation of the photocatalytic oxidation of liquid 4-*tert*-butyltoluene to 4-*tert*-butylbenzaldehyde, it was found that over time there was further oxidation to the corresponding acid.[62] The complete mineralization of chloroform has been demonstrated with near-UV and TiO_2 in aqueous slurries.[63] Matthews[64] investigated the effect of UV and TiO_2 on 21 organic compounds known to be water contaminants and concluded that, since some of the compounds, highly resistant to oxidation, were readily converted to CO_2, most organics would be completely oxidized under the same conditions. Complete mineralization of 2,4,5-trichlorophenoxyacetic acid and 2,4,5-trichlorophenol has also been reported, with half-lives of the compounds ranging from 30 to 90 min.[65] Photocatalytic oxidation has been used to study the first step of the degradation of acetic and chloroacetic acid and, while formation of CO_2 and the chloride ion was noted, data on disappearance kinetics and products were not given.[66] The degradation of tetrachloroethylene and trichloroethylene has been studied by Glaze et al.[67] and a variety of degradation products were observed, including dichloroacetic acid and trichloroacetic acid. Glaze concludes that TiO_2-mediated photocatalysis may not be of use for the degradation of tetra- and trichloroethylene. Complete oxidation of phenol, with the UV/TiO_2 procedure, can be accomplished within 4 h at temperatures below 90°C.[54] Pacheco and Holmes[55] achieved degradation of salicylic acid to below detectable limits in 15 s using solar photocatalytic reactors.[55] Recently, the feasibility of using TiO_2 attached to floating glass microbeads for the remediation of oil spills has been considered.[56]

In terms of the equipment required, the UV/TiO_2 procedure is very similar to the advanced oxidation processes discussed above. Similarly, the UV/TiO_2 procedure shows great promise for treating wastes generated at biomedical research institutions. The problems are also likely to be similar (e.g., UV absorbing buffers and the formation of colored products). Additionally, the work of Glaze et al.[67] suggests that the products of these reactions should be thoroughly investigated.

Wet Air Oxidation

Wet air oxidation is the process for aqueous phase oxidation of dissolved or suspended organics at elevated temperatures (175–340°C) and pressures

(300–3000 lb/in^2), but below the supercritical point of H_2O.[13,68–71] The aqueous phase allows the oxidation reactions to proceed at relatively low temperatures [13,68,70,71] and provides a heat transfer medium that allows the process to be thermally self-sustaining with low organic feed concentrations.[71] Air or oxygen is bubbled through the liquid phase of the reactor, which gives rise to the term "wet air oxidation."[68,71] The system is optimal when the waste stream contains 1–20 or 30% (w/w) oxidizable waste.[72,73] At organic concentrations above the high end of this range, so much heat may be evolved that the system may boil dry.[72] In catalyzed wet air oxidation the rate of the transfer of oxygen to the dissolved state is increased, allowing the reaction to occur at lower temperatures (165–200°C).[74] The catalyst used in this process consists of bromide, nitrate, and manganese ions in acidic solution.[74]

Wet air oxidation converts hazardous waste streams to less toxic compounds,[75] which may include partially oxidized species, such as alcohols, aldehydes, and organic acids.[72] Sulfur compounds are oxidized to inorganic sulfates; organic nitrogen is converted to ammonia; and chlorinated compounds yield hydrochloric acid and salts.[68,70,75] Wet air oxidation has been evaluated for a number of organic compounds and waste streams including *m*-xylene,[73] ammonium thiocyanate,[72] phenolics,[68,70,72,76] methanol,[72,76] propionaldehyde,[72] acetone,[72,76] acetic acid,[72] propionic acid,[72] isobutyric acid,[72] spent caustic,[68,70,72] cyanide waste,[68,70,72] chloroform,[68,70,72] carbon tetrachloride,[68,70,72] DDT, malathion, kepone, and pesticide residues.[68,70,72,76]

It has been reported that wet air oxidation is not very effective in the destruction of refractory chlorinated wastes,[71] and as such should not be considered a stand-alone process since the effluent may require further treatment.[68,72,75] An advantage of wet air oxidation is that the conditions of temperature and pressure required should be readily achievable. On the other hand, oxidation does not go to completion (i.e., to CO_2 and H_2O) and so it would be necessary to fully characterize the products. Wet air oxidation may be useful for degrading hazardous wastes at biomedical research institutions but it is probably not the first choice.

Supercritical Water Oxidation

When increasing temperature and pressure are applied to a liquid and a gas in equilibrium, the liquid becomes less dense due to the temperature and the gas becomes more dense due to the pressure.[77] As the system approaches the critical point the difference in the density of the gas and the liquid becomes indistinguishable; the substance is too thin to be a liquid and too dense to be a gas.[78] A substance that has reached its critical temperature and pressure is termed a supercritical fluid. The critical temperature and pressure for H_2O are 374°C and 3209 lb/in^2, respectively. When H_2O is subjected to temperatures and pressures above the critical point it exhibits properties that are quite different from H_2O at room temperature. The dielectric constant of H_2O drops from 80 to 2 indicating substantially decreased hydrogen bonding and polarity.[79,80] Thus, ionic substances are immiscible and many organics are miscible in all proportions.[78,79,81] Some woods will even fully dissolve in supercritical H_2O.[78] A recent patent[82] describes a reactor design that can minimize problems caused when salts come out of solution under supercritical conditions. Reaction inefficiencies due to mass transport are overcome when organics and oxygen are solubilized in supercritical H_2O because they are in a single phase.[83] As the result of the unique properties of supercritical H_2O, oxidation of organics occurs quickly and heteroatoms are converted to acids, salts, and oxides.[83]

In 1975 Modell et al.[79] determined that organics in aqueous solutions could be degraded to low molecular weight products by treating them at high temperatures and pressures. Supercritical H_2O oxidation was reported to have destruction efficiencies on 2,4-dinitrotoluene of 99.7%, 99.992%, and 99.9998% with temperatures of 457, 513, and 574°C, respectively, with a residence time of 30 s.[80] Destruction efficiencies of greater than 99.9% have also been reported for various halogenated aliphatics, halogenated aromatics, and oxygenated compounds.[84] Supercritical H_2O oxidation has also been reported to have degraded over 99% of cyclohexane, biphenyl, and methyl ethyl ketone.[85] The process has been investigated as a method for treating wastewater on spacecraft and, at temperatures above 650°C, urea was completely broken down to N_2 gas, CO_2, and H_2O.[86] Staszak et al.[87] reported the degradation of organic components of two waste streams (one containing polychlorinated biphenyls and the other a dilute isopropyl alcohol stream containing several priority pollutants) to below detectable limits and indicated that any halogen present was converted to a halo-acid form. Yang and Eckert[88] reported increased oxidation rates of *p*-chlorophenol when copper(II) or manganese(II) salt catalysts were added to the reaction mixture.[88] Destruction efficiencies greater than 99% have been reported for a number of components of pharmaceutic and biopharmaceutic wastes, including: spores, fluorescein, ethylene glycol, endotoxin, malaria antigen, and bacteria.[89] In studies of the oxidation of carbon monoxide, ammonia, and ethanol it was reported that supercritical H_2O is an excellent environment to destroy hazardous organics since there is essentially complete oxidation with no char formation.[90] Swallow et al.[91] reported destruction efficiencies greater than 99.5% for 2-chlorophenol, nitrobenzene, 1,1,2-trichloroethylene, chloroform, carbon tetrachloride, and PCBs. Shanableh and Gloyna[92] reported the efficient destruction of biological treatment plant sludges, acetic acid, 2-nitrophenol, 2,4-dimethylphenol, phenol, and 2,4-dinitrotoluene. It has been reported that partial oxidation of nitrogen-containing organics leads to the formation of ammonia.[93] Lee and Gloyna[94,95] reported 99.6% hydrolysis of acetamide at 400°C in 300 s and high destruction efficiencies for six other toxic organics. Recently, it was reported that the technology is capable of achieving a degradation efficiency of 99.99% in a "single step."[96] In the only reported case where oxidation of an organic compound was not achieved Webley and Tester[97] reported that at temperatures below 450°C methanol in an aqueous solution was essentially not oxidized.

Huppert et al.[98] investigated the reaction mechanism of guaiacol in supercritical H_2O and reported that the reaction is via parallel pyrolysis and hydrolysis pathways. In an investigation of the oxidation of 1,4-dichlorobenzene in supercritical H_2O, it was reported that the oxidation involves two reaction pathways: an oxidation pathway and a reaction with H_2O leading to the sequential removal of the chlorine from the ring.[87] In a study of reactions of model coal compounds in supercritical H_2O it was reported there were both pyrolytic and hydrolytic components.[99] In an investigation of the oxidation kinetics of carbon monoxide, Helling and Tester[100] suggested that during the oxidation, H_2O forms a "cage" around the reactant. An investigation of the incomplete oxidation of phenol in supercritical H_2O at temperatures below those proposed for commercial applications revealed a number of intermediate and breakdown products that appeared in the reaction including: mono- and dicarboxylic acids, dihydroxybenzene, phenoxyphenols, dibenzofuran, and dibenzo-*p*-dioxin.[101,102] This result suggests a series of complex reactions in the degradation process and underscores the necessity for determining the products of these reactions as well as monitoring the destruction of the starting material.

Supercritical H_2O oxidation shows great promise for treating the waste streams from biomedical research institutions.

Other Thermal Treatments

Molten Salt. The molten salt process can be applied to either solid or liquid waste.[71,103] It has been noted to be particularly applicable to grossly toxic and highly halogenated compounds.[71] In this process the waste stream, along with air or oxygen, is introduced into a pool of molten sodium carbonate (~900°C) causing the organics to be oxidized to CO_2 and H_2O.[69,71,103] Inorganic elements form oxygenated salts and remain in the molten salt.[71,103]

Molten Glass. Similar to molten salt, molten glass makes use of a pool of molten glass to destroy organics. The process would accept contaminated soils directly.[69] The organics would be degraded at temperatures of 1200°C.[69,71]

Plasma Arc. Plasma, consisting of charged and neutral particles with an overall charge near zero, is produced when air is passed through an electric arc.[69,71] As the activated parts of the plasma arc decay the energy is transferred to the waste stream, which is then destroyed.[69]

These processes require expensive and difficult to obtain equipment and are probably not suited for the treatment of wastes at biomedical research institutions.

References

1. Cheremisinoff, P.N. Thermal treatment technologies for hazardous wastes. *Pollut. Eng.* **1988**, *20(8)(Aug)*, 50–55.
2. Turner, R.J. Waste treatability tests of spent solvent and other organic wastewater. *Environ. Prog.* **1989**, *8*, 113–119.
3. Heilshorn, E.D. Removing VOCs from contaminated water. *Chem. Eng.* **1992**, *99(2)*, 40–44.
4. Fair, G.E.; Dryden, F.E. Comparison of air stripping versus steam stripping for treatment of volatile organic compounds in contaminated groundwater. *Hazard. Mater. Control* **1990**, *3(5), (Sep–Oct)*, 18–22.
5. Cheremisinoff, P.N. Treating wastewater. *Pollut. Eng.* **1990**, *22(9)*, 60–65.
6. Lunn, G.; Sansone, E.B. Decontamination of aqueous solutions of biological stains. *Biotech. Histochem.* **1991**, *66*, 307–315.
7. Rock,S.L.; Stevens,B.W. Polymeric adsorption-ion exchange process for decolorizing dye waste streams. *Textile Chem. Colorist* **1975**, *7*, 169–171.
8. Crook, E.H.; McDonnell, R.P.; McNulty, J.T. Removal and recovery of phenols from industrial waste effluents with Amberlite XAD polymeric adsorbents. *Ind. Eng. Chem., Prod. Res. Dev.* **1975**, *14*, 113–118.
9. Kennedy, D.C. Treatment of effluent from manufacture of chlorinated pesticides with a synthetic, polymeric adsorbent, Amberlite XAD-4. *Environ. Sci. Technol.* **1973**, *7*, 138–141.
10. Suffet, I.H.; Brenner, L.; Coyle, J.T.; Cairo, P.R. Evaluation of the capability of granular activated carbon and XAD-2 resin to remove trace organics from treated drinking water. *Environ. Sci. Technol.* **1978**, *12*, 1315–1322.

11. Lipski, C.; Cote, P. The use of pervaporation for the removal of organic contaminants from water. *Environ. Prog.* **1990**, *9(4) (Nov)*, 254–261.
12. Heilshorn, E.D. Removing VOCs from contaminated water.Part 2. *Chem. Eng.* **1991**, *98(3)*, 152–158.
13. Bowers, A.R.; Eckenfelder, W.W.; Gaddipati, P.; Monsen, R.M. Toxicity reduction in industrial wastewater discharges. *Pollut. Eng.* **1988**, *20(Feb)*, 68–72.
14. Goronszy, M.C.; Eckenfelder, W.W.; Froelich, E. Waste water: A guide to industrial pretreatment. *Chem. Eng.* **1992**, *99(6)*, 78–83.
15. Cheremisinoff, P.N. New strategies for haz waste treatment and disposal. *Pollut. Eng.* **1988**, *20(4)(Apr)*, 64–71.
16. Gioia, F. Detoxification of organic waste liquids by catalytic hydrogenation. *J. Hazard. Mater.* **1991**, *26*, 243–260.
17. Morrison, R.T.; Boyd, R.N. *Organic Chemistry*, 3rd ed.; Allyn and Bacon, Inc.: Boston, 1975; pp. 1062–1063.
18. Kalnes, T.N.; James, R.B. Hydrogenation and recycle of organic waste streams. *Environ. Prog.* **1988**, *7(3), (Aug)*, 185–191.
19. Hagenmaier, H.; Brunner, H.; Haag, R.; Kraft, M. Copper-catalyzed dechlorination/hydrogenation of polychlorinated dibenzo-*p*-dioxins, polychlorinated dibenzofurans, and other chlorinated aromatic compounds. *Environ. Sci. Technol.* **1987**, *21*, 1085–1088.
20. Lunn, G.; Sansone, E.B. Validated methods for degrading hazardous chemicals: Some halogenated compounds. *Am. Ind. Hyg. Assoc. J.* **1991**, *52*, 252–257.
21. Fenton, H.J.H. Oxidation of tartaric acid in presence of iron. *J. Chem. Soc.* **1894**, *65*, 899–910.
22. Walling, C. Fenton's reagent revisited. *Acc. Chem. Res.* **1975**, *8*, 125–131.
23. Zepp, R.G.; Faust, B.C.; Hoigné, J. Hydroxyl radical formation in aqueous reactions (pH 3-8) of iron(II) with hydrogen peroxide: The photo-Fenton reaction. *Environ. Sci. Technol.* **1992**, *26*, 313–319.
24. Barbeni, M.; Minero, C.; Pelizzetti, E.; Borgarello, E.; Serpone, N. Chemical degradation of chlorophenols with Fenton's reagent (Fe^{2+} + H_2O_2). *Chemosphere* **1987**, *16*, 2225–2237.
25. Wink, D.A.; Nims, R.W.; Desrosiers, M.F.; Ford, P.C.; Keefer, L.K. A kinetic investigation of intermediates formed during the Fenton reagent mediated degradation of *N*-nitrosodimethylamine: Evidence for an oxidative pathway not involving hydroxyl radical. *Chem. Res. Toxicol.* **1991**, *4*, 510–512.
26. Wink, D.A.; Nims, R.W.; Saavedra, J.E.; Desrosiers, M.F.; Ford, P.C. The oxidation of alkylnitrosamines via the Fenton reagent. The use of nitrosamines to probe oxidative intermediates in the Fenton reaction. In *The Chemistry and Biochemistry of Nitrosamines and Other N-Nitroso Compounds;* American Chemical Society: Washington, DC (in press).
27. Lipczynska-Kochany, E. Degradation of aqueous nitrophenols and nitrobenzene by means of the Fenton reaction. *Chemosphere* **1991**, *22*, 529–536.
28. Klecka, G.M.; Gonsior, S.J. Removal of 1,4-dioxane from wastewater. *J. Hazard. Mater.* **1986**, *13*, 161–168.

29. Watts, R.J.; Udell, M.D.; Rauch, P.A.; Leung, S.W. Treatment of pentachlorophenol-contaminated soils using Fenton's reagent. *Hazard. Waste Hazard. Mater.* **1990**, *7*, 335–345.

30. Lipczynska-Kochany, E. Novel method for a photocatalytic degradation of 4-nitrophenol in homogeneous aqueous solution. *Environ. Technol.* **1991**, *12 (Jan)*, 87–92.

31. Pignatello, J.J. Dark and photoassisted Fe^{3+}-catalyzed degradation of chlorophenoxy herbicides by hydrogen peroxide. *Environ. Sci. Technol.* **1992**, *26*, 944–951.

32. Hoigné, J.; Zuo, Y.; Nowell, L. Photochemical reactions in atmospheric waters; Role of dissolved iron species. In *Abstracts of the Division of Environmental Chemistry, San Francisco, April 5–10*; American Chemical Society: Washington, DC, 1992; pp. 147–149.

33. Faust, B.C.; Zepp, R.G.; Ceiler, D.L.; Garraty, J.M.; Hoigné, J. Photo-redox reactions of aqueous iron(III) complexes: Sources of oxidants in surface and atmospheric waters. In *Abstracts of the Division of Environmental Chemistry, San Francisco, April 5–10*; American Chemical Society: Washington, DC, 1992; pp. 436–439.

34. Emmett, G.C.; Michejda, C.J.; Sansone, E.B.; Keefer, L.K. Limitations of photodegradation in the decontamination and disposal of chemical carcinogens. In *Safe Handling of Chemical Carcinogens, Mutagens, Teratogens and Highly Toxic Substances*; Walters, D.B., Ed., Ann Arbor Science Publishers, Inc.: Ann Arbor, MI, 1980; pp. 535–553.

35. Weir, B.A.; Sundstrom, D.W.; Klei, H.E. Destruction of benzene by ultraviolet light-catalyzed oxidation with hydrogen peroxide. *Hazard. Waste Hazard. Mater.* **1987**, *4*, 165–176.

36. Borup, M.B.; Middlebrooks, E.J. Photocatalysed oxidation of toxic organics. *Water Sci. Tech.* **1987**, *19*, 381–390.

37. Peyton, G.R. Modeling advanced oxidation processes for water treatment. In *Emerging Technologies in Hazardous Waste Management*; Tedder, D.W., Pohland, F.G., Eds., American Chemical Society: Washington, DC, 1990; pp. 100–118.

38. Sundstrom, D.W.; Weir, B.A.; Redig, K.A. Destruction of mixtures of pollutants by UV-catalyzed oxidation with hydrogen peroxide. In *Emerging Technologies in Hazardous Waste Management*; Tedder, D.W., Pohland, F.G., Eds., American Chemical Society: Washington, DC, 1990; pp. 67–76.

39. Sundstrom, D.W.; Klei, H.E.; Nalette, T.A.; Reidy, D.J.; Weir, B.A. Destruction of halogenated aliphatics by ultraviolet catalyzed oxidation with hydrogen peroxide. *Hazard. Waste Hazard. Mater.* **1986**, *3*, 101–110.

40. Sundstrom, D.W.; Weir, B.A.; Klei, H.E. Destruction of aromatic pollutants by UV light catalyzed oxidation with hydrogen peroxide. *Environ. Prog.* **1989**, *8(1)(Feb)*, 6–11.

41. Mansour, M. Photolysis of aromatic compounds in water in the presence of hydrogen peroxide. *Bull. Environ. Contam. Toxicol.* **1985**, *34*, 89–95.

42. Ho, P.C. Photooxidation of 2,4-dinitrotoluene in aqueous solution in the presence of hydrogen peroxide. *Environ. Sci. Technol.* **1986**, *20*, 260–267.

43. Peyton, G.R.; Huang, F.Y.; Burleson, J.L.; Glaze, W.H. Destruction of pollutants in water with ozone in combination with ultraviolet radiation. 1. General principles and oxidation of tetrachloroethylene. *Environ. Sci. Technol.* **1982**, *16*, 448–453.

44. Prat, C.; Vicente, M.; Esplugas, S. Treatment of bleaching waters in the paper industry by hydrogen peroxide and ultraviolet radiation. *Water Res.* **1988**, *22*, 663–668.

45. Peyton, G.R.; Glaze, W.H. Destruction of pollutants in water with ozone in combination with ultraviolet radiation. 3. Photolysis of aqueous ozone. *Environ. Sci. Technol.* **1988**, *22*, 761–767.

46. Glaze, W.H.; Peyton, G.R.; Lin, S.; Huang, R.Y.; Burleson, J.L. Destruction of pollutants in water with ozone in combination with ultraviolet radiation. 2. Natural trihalomethane precursors. *Environ. Sci. Technol.* **1982**, *16*, 454–458.

47. McShea, L.J.; Miller, M.D.; Smith, J.R. Combining UV/ozone to oxidize toxics. *Pollut. Eng.* **1987**, *19(3)(Mar)*, 58–59.

48. Rapaport, D. UV/oxidation providers shed technical problems, fight cost perceptions. *Hazmat World* **1993**, *6(5)(May)*, 63–68.

49. Roy, K.A. UV-oxidation technology. *Hazmat World* **1990**, *3(6)(Jun)*, 35–50.

50. Anon. The Ultrox UV/oxidation process for treating contaminated ground water. *Hazard. Waste Consult.* **1990**, *8(4)(Jul–Aug)*,1/9–1/12.

51. Roy, K.A. Ultraviolet light. Researchers use UV light for VOC destruction. *Hazmat World* **1990**, *3(5)(May)*, 82–92.

52. Stevens, S. Rayox—A second generation enhanced oxidation process for the destruction of waterborne contaminants. In *Proceedings of the Thirty-sixth Ontario Waste Management Conference*; Ontario Ministry of the Environment: 1989; pp. 217–235.

53. Bolton, J.R.; Cater, S.R.; Safarzadeh-Amiri, A. The use of reduction reactions in the photodegradation of organic pollutants in waste streams. In *Abstracts of the Division of Environmental Chemistry, San Francisco, April 5–10*; American Chemical Society: Washington, DC, 1992; pp. 116–118.

54. Tseng, J.; Huang, C.P. Mechanistic aspects of the photocatalytic oxidation of phenol in aqueous solutions. In *Emerging Technologies in Hazardous Waste Management*; Tedder, D.W., Pohland, F.G., Eds., American Chemical Society: Washington, DC, 1990; pp. 12–39.

55. Pacheco, J.E.; Holmes, J.T. Falling-film and glass-tube solar photocatalytic reactors for treating contaminated water. In *Emerging Technologies in Hazardous Waste Management*; Tedder, D.W., Pohland, F.G., Eds., American Chemical Society: Washington, DC, 1990; pp. 40–51.

56. Heller, A.; Brock, J.R. Photoassisted oxidation of organic spills on water. *Preprints of Papers, Division of Environmental Chemistry, ACS National Meeting, San Francisco, CA* **1992**, *32*, 178–181.

57. Ollis, D.F.; Pelizzetti, E.; Serpone, N. Photocatalyzed destruction of water contaminants. *Environ. Sci. Technol.* **1991**, *25*, 1522–1529.

58. Izumi, I.; Dunn, W.W.; Wilbourn, K.O.; Fan, F.-R.F.; Bard, A.J. Heterogeneous photocatalytic oxidation of hydrocarbons on platinized TiO_2 powders. *J. Phys. Chem.* **1980**, *84*, 3207–3210.

59. Izumi, I.; Fan, F.-R.F.; Bard, A.J. Heterogeneous photocatalytic decomposition of benzoic acid and adipic acid on platinized TiO_2 powder. The photo-Kolbe decarboxylative route to the breakdown of the benzene ring and to the production of butane. *J. Phys. Chem.* **1981**, *85*, 218–223.

60. Fujihira, M.; Satoh, Y.; Osa, T. Heterogeneous photocatalytic oxidation of aromatic compounds on TiO_2. *Nature (London)* **1981**, *293*, 206–208.

61. Fujihira, M.; Satoh, Y.; Osa, T. Heterogeneous photocatalytic oxidation of aromatic compounds on semiconductor materials: The photo-Fenton type reaction. *Chem. Lett.* **1981**, 1053–1056.

62. Pichat, P.; Disdier, J.; Hermmann, J.-M.; Vaudano,P. Photocatalytic oxidation of liquid (or gaseous) 4-*tert*-butylbenzaldehyde by O_2 (or air) over TiO_2. *New J. Chem.* **1986**, *10*, 545–551.

63. Pruden, A.L.; Ollis, D.F. Degradation of chloroform by photoassisted heterogeneous catalysis in dilute aqueous suspensions of titanium dioxide. *Environ. Sci. Technol.* **1983**, *17*, 628–631.

64. Matthews, R.W. Photo-oxidation of organic material in aqueous suspensions of titanium dioxide. *Water Res.* **1986**, *20*, 569–578.

65. Barbeni, M.; Morello, M.; Pramauro, E.; Pelizzetti, E.; Vincenti, M.; Borgarello, E.; Serpone, N. Sunlight photodegradation of 2,4,5-trichlorophenoxy-acetic acid and 2,4,5-trichlorophenol on TiO_2. Identification of intermediates and degradation pathway. *Chemosphere* **1987**, *16*, 1165–1179.

66. Chemseddine, A.; Boehm, H.P. A study of the primary step in the photochemical degradation of acetic acid and chloroacetic acids on a TiO_2 photocatalyst. *J. Mol. Catal.* **1990**, *60*, 295–311.

67. Glaze, W.H.; Kenneke, J.F.; Ferry, J.L. Chlorinated byproducts from the titanium oxide-mediated photodegradation of trichloroethylene and tetrachloroethylene in water. *Environ. Sci. Technol.* **1993**, *27*, 177–184.

68. Heimbuch, J.A.; Wilhelmi, A.R. Wet air oxidation: a treatment means for aqueous hazardous waste streams. *J. Hazard. Mater.* **1985**, *12*, 187–200.

69. Freeman, H. *Innovative Thermal Hazardous Organic Waste Treatment Processes*; Noyes Publications: Park Ridge, NJ, 1985.

70. Chowdhury, A.K.; Copa, W.C. Wet air oxidation of toxic and hazardous organics in industrial wastewaters. *Indian Chem. Eng.* **1986**, *28(3)(Jul-Sep)*, 3–10.

71. Freeman, H.M.; Olexsey, R.A.; Oberacker, D.A.; Mournighan, R.E. Thermal destruction of hazardous waste: a state-of-the-art review. *J. Hazard. Mater.* **1987**, *14*, 103–117.

72. Katari, V.S.; Vatavuk, W.M.; Wehe, A. Incineration techniques for control of volatile organic compound emissions. Part I. Fundamentals and process design considerations. *J. Air Pollut. Control Assoc.* **1987**, *37*, 91–99.

73. Oppelt, E.T. Incineration of hazardous waste. A critical review. *J. Air Pollut. Control Assoc.* **1987**, *37*, 558–586.

74. Office of Technology Assessment. Cleanup technologies. In *Superfund Strategies*; Office of Technology Assessment: 1985; pp. 171–220.

75. Min, M.; Barbour, R.; Hwang, J. Treating land ban waste. *Pollut. Eng.* **1991**, *23(8)*, 64–70.

76. Dietrich, M.J.; Randall, T.L.; Canney, P.J. Wet air oxidation of hazardous organics in wastewater. *Environ. Prog.* **1985**, *4*, 171–177.

77. Shaw, R.W.; Brill, T.B.; Clifford, A.A.; Eckert, C.A.; Franck, E.U. Supercritical water. A medium for chemistry. *Chem. Eng. News*, **1991**, *23 Dec.*, 26–39.

78. Josephson, J. Supercritical fluids. *Environ. Sci. Technol.* **1982**, *16*, 548A.

79. Modell, M.; Gaudet, G.G.; Simson, M.; Hong, G.T.; Biemann, K. Supercritical water-testing reveals new process holds promise. *Solid Wastes Manage.* **1982**, *25(Aug)*, 26–76.

80. Thomason, T.B.; Modell, M. Supercritical water destruction of aqueous wastes. *Hazard. Waste* **1984**, *1*, 453–467.

81. Modell, M. Supercritical fluid technology in hazardous waste treatment. In *Management of Hazardous and Toxic Wastes in the Process Industries*; Kolaczkowski, S.T., Crittenden, B.D., Eds., Elsevier Applied Science: New York, 1987; pp. 66–93.

82. Hong, G.T.; Killilea, W.R.; Thomason, T.B. Method and Apparatus for Solids Separation in a Wet Oxidation Type Process, *World Patent WO 89/02874*, April 6, 1989.

83. Thomason, T.B.; Hong, G.T.; Swallow, K.C.; Killilea, W.R. The Modar supercritical water oxidation process. In *Innovative Hazard. Waste Treat. Technol. Ser. 1990, 1 (Therm. Processes)* 1990; pp. 31–42.

84. Modell, M. Processing methods for the oxidation of organics in supercritical water. *U. S. Patent 4,543,190*, September 4, 1985.

85. Modell, M. *Detoxification and Disposal of Hazardous Organic Chemicals by Processing in Supercritical Water*; US Army Medical Research and Development Command DAMD17-80-C-0078: NTIS AD-A179 005, 1987.

86. Timberlake, S.H.; Hong, G.T.; Simson, M.; Modell, M. Supercritical water oxidation for wastewater treatment: Preliminary study of urea destruction. *SAE Tech. Paper Ser. No. 820872* **1982.**

87. Staszak, C.N.; Malinowski, K.C.; Killilea, W.R. The pilot-scale demonstration of the MODAR oxidation process for the destruction of hazardous organic waste materials. *Environ. Prog.* **1987**, *6(1)*, 39–43.

88. Yang, H.H.; Eckert, C.C. Homogeneous catalysis in the oxidation of *p*-chlorophenol in supercritical water. *Ind. Eng. Chem. Res.* **1988**, *27*, 2009–2014.

89. Johnston, J.B.; Hannah, R.E.; Cunningham, V.L.; Daggy, B.P.; Sturm, F.J.; Kelly, R.M. Destruction of pharmaceutical and biopharmaceutical wastes by the Modar supercritical water oxidation process. *Bio/Technology* **1988**, *6*, 1423–1427.

90. Helling, R.K.; Tester, J.W. Oxidation of simple compounds and mixtures in supercritical water: Carbon monoxide, ammonia, and ethanol. *Environ. Sci. Technol.* **1988**, *22*, 1319–1324.

91. Swallow, K.C.; Killilea, W.R.; Malinowski, K.C.; Staszak, C.N. The Modar process for the destruction of hazardous organic wastes—Field test of a pilot-scale unit. *Waste Manage.* **1989**, *9*, 19–26.

92. Shanableh, A.; Gloyna, E.F. Supercritical water oxidation-wastewaters and sludges. *Water. Sci. Technol.* **1991**, *23*, 389–398.

93. Webley, P.A.; Tester, J.W.; Holgate, H.R. Oxidation kinetics of ammonia and ammonia-methanol mixtures in supercritical water in the temperature range 530-700°C at 246 bar. *Ind. Eng. Chem. Res.* **1991**, *30*, 1745–1754.

94. Lee, D.-S.; Gloyna, E.F. Hydrolysis and oxidation of acetamide in supercritical water. *Environ. Sci. Technol.* **1992**, *26*, 1587–1593.

95. Lee, D.-S.; Gloyna, E.F. *Supercritical Water Oxidation of Acetamide and Acetic Acid*; Center for Research in Water Resources, Bureau of Engineering Research, University of Texas: Austin, TX, 1990.

96. Barner, H.E.; Huang, C.Y.; Johnson, T.; Jacobs, G.; Martch, M.A.; Killilea, W.R. Supercritical water oxidation: An emerging technology. *J. Hazard. Mater.* **1992**, *31*, 1–17.

97. Webley, P.A.; Tester, J.W. Fundamental kinetics of methanol oxidation in supercritical water. In *Supercritical Fluid Science and Technology*; Johnson, K.P., Penninger, J.M.L., Eds., American Chemical Society: Washington, DC, 1989; pp. 259–275.

98. Huppert, G.L.; Wu, B.C.; Townsend, S.H.; Klein, M.T.; Paspek, S.C. Hydrolysis in supercritical water: Identification and implications of a polar transition state. *Ind. Eng. Chem. Res.* **1989**, *28*, 161–165.

99. Townsend, S.H.; Abraham, M.A.; Huppert, G.L.; Klein, M.T.; Paspek, S.C. Solvent effects during reactions in supercritical water. *Ind. Eng. Chem. Res.* **1988**, *27*, 143–149.

100. Helling, R.K.; Tester, J.W. Oxidation kinetics of carbon monoxide in supercritical water. *Energy Fuels* **1987**, *1*, 417–423.

101. Thornton, T.D.; Savage, P.E. Phenol oxidation in supercritical water. *J. Supercrit. Fluids* **1990**, *3*, 240–248.

102. Thornton, T.D.; LaDue, D.E., III; Savage, P.E. Phenol oxidation in supercritical water: Formation of dibenzofuran, dibenzo-*p*-dioxin, and related compounds. *Environ. Sci. Technol.* **1991**, *25*, 1507–1510.

103. Anon. A guide to innovative thermal hazardous waste treatment processes. *Hazard. Waste Consult* **1990**, *8(6)(Nov–Dec)*, 4/1–4/40.

MOLECULAR FORMULA INDEX

Compounds that are generally supplied as the hydrochloride or other salt forms are listed only as the free bases. Water of hydration is omitted. Inorganics are listed first. Some elements are listed although only the salts are discussed in the monographs.

CAS REGISTRY NUMBER INDEX

Some compounds may have more than one Registry Number depending on whether the free base, salt, D form, L form, and so on, is considered. We have tried to include as many of these Registry Numbers as possible but in each case we give only the name used in the monographs. Registry numbers for some elements are listed, although only the ions of those elements are discussed. For salts the registry numbers for the anhydrous and more common hydrated versions (including those used) are listed.

NAME INDEX

Although we have tried to include as many synonyms as possible, this list is not exhaustive. In particular we have not incorporated all possible minor variations, e.g., *p*-methyl as well as 4-methyl, and so the user should check all common variants. Note that we have used 2-chloroethyl instead of β-chloroethyl throughout and we have not included D, L, DL, or meso forms. Compounds that are generally supplied as the hydrochloride or other salt forms are listed only as the free bases. Water of hydration is ignored. The page numbers given are the first page of the monograph to emphasize the point that the entire chapter should be read before commencing any work. The compound in question, however, may be listed further down in the Related Compounds section or in the footnotes where synonyms are sometimes given.